Stability of Motion of Nonautonomous Systems
(Method of Limiting Equations)

Stability and Control: Theory, Methods and Applications
A Series of Books and Monographs on the theory of Stability and Control
Edited by A. A. Martynyuk, Institute of Mechanics, Kiev, Ukraine and V. Lakshmikantham,
Florida Institute of Technology, USA

This book is part of a series. The publisher will accept continuation orders which may be cancelled at any time and which provide for automatic billing and shipping of each title in the series upon publication. Please write for details.

Stability of Motion of Nonautonomous Systems (Method of Limiting Equations)

J. Kato
Mathematical Institute
Tohoku University, Japan

A. A. Martynyuk
Institute of Mechanics,
Kiev, Ukraine

A. A. Shestakov
Institute of Railway Transport,
Moscow, Russia

CRC Press
Taylor & Francis Group
Boca Raton London New York

CRC Press is an imprint of the
Taylor & Francis Group, an **informa** business

First published 1996 by Gordon and Breach Publishers

Published 2019 by CRC Press
Taylor & Francis Group
6000 Broken Sound Parkway NW, Suite 300
Boca Raton, FL 33487-2742

© 1996 by Taylor & Francis Group, LLC
CRC Press is an imprint of Taylor & Francis Group, an Informa business

First issued in paperback 2019

No claim to original U.S. Government works

ISBN 13: 978-0-367-45596-5 (pbk)
ISBN 13: 978-2-88449-035-1 (hbk)

Visit the Taylor & Francis Web site at
http://www.taylorandfrancis.com

and the CRC Press Web site at
http://www.crcpress.com

British Library Cataloguing in Publication Data

Kato, Junji
  Stability of Motion of Nonautonomous
  Systems: Method of Limiting Equations.
  (Stability & Control: Theory, Method &
  Applications Series, ISSN 1023–6155;
  Vol. 3)
  I. Title II Series
  515.352

Contents

Introduction to the Series

The problems of modern society are both complex and interdisciplinary. Despite the apparent diversity of problems, tools developed in one context are often adaptable to an entirely different situation. For example, consider the Lyapunov's well-known second method. This interesting and fruitful technique has gained increasing significance and has given a decisive impetus for modern development of the stability theory of differential equations. A manifest advantage of this method is that it does not demand the knowledge of solutions and therefore has great power in application. It is now well recognized that the concept of Lyapunov-like functions and the theory of differential and integral inequalities can be utilized to investigate qualitative and quantitative properties of nonlinear dynamic systems. Lyapunov-like functions serve as vehicles to transform the given complicated dynamic systems into a relatively simpler system and therefore it is sufficient to study the properties of this simpler dynamic system. It is also being realized that the same versatile tools can be adapted to discuss entirely different nonlinear systems, and that other tools, such as the variation of parameters and the method of upper and lower solutions, provide equally effective methods to deal with problems of a similar nature. Moreover, interesting new ideas have been introduced which would seem to hold great potential.

Control theory, on the other hand, is that branch of application-oriented mathematics that deals with the basic principles underlying the analysis and design of control systems. To control an object implies the influence of its behavior so as to accomplish a desired goal. In order to implement this influence, practitioners build devices that incorporate various mathematical techniques. The study of these devices and their interaction with the object being controlled is the subject to control theory. There have been, roughly speaking, two main lines of work in control theory which are complementary. One is based on the idea that a good model of the object to be controlled is available and that we wish to optimize its behavior, and the other is based on the constraints imposed by uncertainty about the model in which the object operates. The control tool in the latter is the use of feedback in order to correct for deviations from the desired behavior. Mathematically, stability theory, dynamic systems and functional analysis have had a strong influence on this approach.

Volume 1, *Theory of Integro-Differential Equations*, is a joint contribution by V. Lakshmikantham (USA) and M. Rama Mohana Rao (India).

Volume 2, *Stability Analysis: Nonlinear Mechanics Equations*, is by A. A. Martynyuk (Ukraine).

Volume 3, *Stability of Motion of Nonautonomous Systems: The Method of Limiting Equations*, is a collaborative work by J. Kato (Japan), A. A. Martynyuk (Ukraine) and A. A. Shestakov (Russia).

Due to the increased interdependency and cooperation among the mathematical sciences across the traditional boundaries, and the accomplishments thus far achieved in the areas of stability and control, there is every reason to believe that many breakthroughs await us, offering existing prospects for these versatile techniques to advance further. It is in this spirit that we see the importance of the "Stability and Control" series, and we are immensely thankful to Gordon and Breach Publishers for their interest and cooperation in publishing this series.

Foreword

The stability of systems involving small parameters has attracted the attention of investigators for a very long time. In particular, N. Krylov and N. N. Bogolyubov in their investigations of oscillations in systems close to the linear considered the stability problems of stationary amplitudes.

In the present book a number of new results are obtained using methods not previously employed in this area: the method of limiting equations constructed according to the Bebutov–Miller-Sell concept, the comparison method and the Lyapunov direct method based on scalar, vector and matrix functions. These methods allow various sufficient stability conditions to be derived for the solutions of systems with small parameters over both finite and infinite time intervals.

The aim of the authors to obtain results that can be applied to engineering problems is very much to be approved.

Yu. A. Mitropolsky

Preface to the English Edition

The theory of stability of motion, the fundamentals of which were laid down by A. Poincaré and A. M. Lyapunov, has recently been applied to new problems of science and technology. The most developed applications of stability theory have been to problems of the stability of orbiting satellites, rocket motion, vehicles (high-speed trains and cars), functioning of power grids, and nuclear reactors, for whose solution mainly autonomous systems of differential equations have been considered.

In modelling motions and processes by equations that contain the time variable explicitly and are nonperiodic with respect to it, difficulties arise in establishing sufficient stability conditions. However, it is necessary to apply such systems of equations in order to solve many important engineering problems. One of the key ideas in current investigations of these systems is the idea of limiting equations (systems).

The efforts of many investigators to apply limiting equations to qualitative problems resulted in the development of new techniques, which are described in this book.

This English edition differs considerably from the Russian one. Chapters 1–3 contain new results, while Chapters 4–6 have been rewritten. The book describes work done during 1984–93. Chapters 1, 2 and 5 were written by Kato and Martynyuk, Chapters 3, 4 and 6 by Martynyuk and Chapters 7–9 by Shestakov.

Acknowledgements

We should like to thank Professors M. Z. Litvin-Sedoy and V. M. Starzynsky for their careful reading and advice, which has proved to be very important for us. Our thanks go also to our colleagues in the Process Stability Department of the Institute of Mechanics, L. N. Chernetzkaya and S. N. Raschivalova, for their help in preparing this English edition. Finally, we wish to acknowledge the cooperation of Gordon and Breach Publishers, and to thank them for their patience and understanding.

J. Kato
A. A. Martynyuk
A. A. Shestakov
Sendai–Kiev–Moscow

Notation

R	the set of all real numbers
$R_+ = [0, \infty[\subset R$	the set of all nonnegative numbers
$R_- =]-\infty, 0]$	the set of all nonpositive numbers
R^n	the n-dimensional real vector space
$R \times R^n$	the Cartesian product of R and R^n
M	a metric space
H_0, H_1, H_2	Hilbert spaces
$A \setminus B$	the difference between sets A and B
$A \cup B, A \cap B$	the union, intersection of sets A and B
∂A	the boundary of the set A
$\overline{A}$	the closure of the set A
$\varnothing$	the empty set
$d(x, A) = \inf(\|x - y\| : y \in A)$	the distance from x to the set A
$\|\cdot\|$	the Euclidean norm
$f : R \times R^n \to R^n$	a vector function mapping $R \times R^n$ into R^n
$B_\Delta = \{x : \|x\| < \Delta\}$	the hyperball with centre at the origin and radius equal to Δ
$C^{(i,j)}(R \times R^n)$	the family of all functions i times differentiable on R and j times differentiable on R^n
$C(R_+ \times R^n)$	the family of all functions continuous on $R_+ \times R^n$
$x(t; t_0, x_0)$	a solution of a system at $t \in R$ iff $x(t_0) = x_0$, $x(t_0; t_0, x_0) = x_0$
$x(t)$	a state vector of a system at $t \in R$; $x = [x_1 \quad x_2 \quad \ldots \quad x_n]^T$

$$D^+ m(t_0) = \limsup \left\{ \frac{m(t) - m(t_0)}{t - t_0}, t \to t_0^+ \right\}$$

$$D_+ m(t_0) = \liminf \left\{ \frac{m(t) - m(t_0)}{t - t_0}, t \to t_0^+ \right\}$$

$T(f, \varepsilon) = \{\tau : |f(t+\tau) - f(t)| < \varepsilon \forall t\}$, namely the ε-translation set of f

$F_\theta(t, x) = f(t+\theta, x) \forall (t, x) \in R_+ \times D, \; D \subset R^n$, namely the translate f of the vector function f

B	the real linear vector space of functions mapping $]-\infty, 0[$ in R^n with seminorm $|\cdot|_B$
$H(F)$	the hull of the function $F(t, \phi)$
$t \in R$	a time variable
$t_0 \in R$	an initial time
$I_s = [s, \infty[$	the right semi-open unbounded time interval associated with s
$V \in C(R_+ \times R^n, R)$	a Lyapunov function
$V \in C(R \times M, R)$	an L functional of Dafermos type
$\mathscr{K}_{[0, \xi[}$	the class of comparison functions on $[0, \xi[$
$\lambda_{i(\cdot)}$	the ith eigenvalue of the matrix $(\cdot)$
$\Lambda_{M(\cdot)}$	the maximal eigenvalue of the matrix $(\cdot)$
$\lambda_{m(\cdot)}$	the minimal eigenvalue of the matrix $(\cdot)$

1 Stability Analysis of ODEs by the Method of Limiting Equations

1.0 Introduction

Systems with a finite number of degrees of freedom, including many of practical interest, are modelled well by ordinary differential equations—either linear or non-linear and autonomous or nonautonomous ones. However, the nonautonomous equations turn out to be the most difficult to study. Therefore many workers have concentrated their efforts on the development of methods allowing the investigation of such systems. One recent approach is based on the idea of constructing a limiting system for a given equation that incorporates the qualitative properties of its solutions. This chapter introduces results obtained in this direction.

1.0.1 *Nonautonomous equations*

Let $R_+ = [0, \infty[$ and let $\|\cdot\|$ denote a norm in the space R^n; let x and y be points in R^n and A a set in R^n. We shall use the distance $d(x, y) = \|x - y\|$, $d(x, A) = \inf\{d(x, y): y \in A\}$. We also introduce the notation

$$S(y, \sigma) = \{x \in R^n : d(x, y) < \sigma\}, \quad S_\sigma = \{y \in R^n : \|x\| < \sigma\},$$
$$A_\sigma = \{y \in R^n : d(y, A) < \sigma\}, \quad \sigma = \text{const} > 0.$$

For the set A, $\bar{A}$ is its closure, $\mathring{A}$ is its interior, A^c is its complement and ∂A is its boundary.

Let W be an open set in R^n, i.e. let each of its elements have an ε-neighbourhood.

For the vector function $f \in C(R_+ \times W, R^n)$ we consider the differential equation

$$\frac{dx}{dt} = f(t, x), \quad x(t_0) = x_0, \tag{1.0.1}$$

where $x(t)$ is a state vector of the system (1.0.1) at time $t \in R_+$. The following result is given in Yoshizawa [2], Theorem 5.1.

Theorem 1.0.1   Suppose that

(1) the vector function $f(t, x)$ is continuous on $R_+ \times W$ and bounded by a constant $M > 0$;
(2) any solution $x(t)$ of the system (1.0.1), starting at the point $(t_0, x_0) \in R_+ \times W$, is continuable from the right up to the value $t = t_1$ for a given $t_1 > t_0$;
(3) the set E of all points constituting the solution for $[t_0, t_1]$ starting at the point (t_0, x_0), is contained in the compact set $R_+ \times W$;

Then for every $\varepsilon > 0$ there exists a $\delta > 0$ such that if

$$d(P^*, E) \leqslant \delta \quad \text{and} \quad \int_{t_0}^{t_1} \| \psi(t) \| \, dt \leqslant \delta,$$

the estimate

$$\| x^*(t; t_0^*, x_0^*) - x(t; t_0, x_0) \| < \varepsilon$$

holds for all $t \in [t_0^*, t_1]$, where $\psi(t)$ is a continuous function and $x^*(t; t_0^*, x_0^*)$ is a solution, starting at the point P^* (t_0^*, x_0^*), $t_0^* \in [t_0, t_1]$, of the equation

$$\frac{dx}{dt} = f(t, x) + \psi(t), \tag{1.0.2}$$

on the interval $[t_0^*, t_1]$.
    Here

$$d(P^*, E) = \inf \{ d((t_0^*, x_0^*), (t, x)); (t, x) \in E \},$$
$$d(x^*, x) = \| x^* - x \|.$$

We shall need the following result to prove Theorem 1.0.1.

Lemma 1.0.1   Let the vector function $f(t, x)$ in the system (1.0.1) be continuous on $\mathscr{D}$: $t_0 \leqslant t \leqslant T$, $\| x \| < \infty$, and bounded by the constant $M > 0$ on $\mathscr{D}$. If E is the set of all points constituting the solution passing through the point (t_0, x_0), $x_0 \in R^n$, then E is a compact set.

Proof of Lemma 1.0.1   The set E is clearly bounded. We shall show that it is closed. Let the point (t', x') belong to E. Then there exists a sequence of points (t_k, x_k), $k = 1, 2, \ldots$, from the set E such that $(t_k, x_k) \to (t', x')$ as $k \to \infty$. By the definition of E, there exists a solution $x_k(t; t_0, x_0)$ of the system (1.0.1), passing through the point (t_k, x_k) and defined on $[t_0, T]$. Since

$$x_k(t; t_0, x_0) = x_0 + \int_{t_0}^{t} f(s, x_k(s; t_0, x_0)) \, ds$$

and $\| f(t, x) \| \leqslant M$, the subsequence converges uniformly by virtue of Ascoli's theorem. Let $x(t)$ be a limit of this convergent subsequence. Then we get for $k \to \infty$

$$x(t) = x_0 + \int_{t_0}^{t} f(s, x(s)) \, ds.$$

Therefore, $x(t)$ is a solution of the system (1.0.1) passing through the point (t_0, x_0). $x_k = x_k(t_k; t_0, x_0) = [x_k(t_k; t_0, x_0) - x(t_k)] + x(t_k)$ implies $x' = x(t')$. This shows that (t', x') is a point on the curve $x = x(t)$, being a solution of the system (1.0.1), i.e. $(t', x') \in E$. Hence E is closed. ∎

Proof of Theorem 1.0.1   By Lemma 1.0.1 the set E is closed. Suppose that for some $\varepsilon > 0$ there exists no δ such that the conditions of Theorem 1.0.1 are satisfied. One can suppose that $\bar{N}(\varepsilon, E) \subset R_+ \times W$, where $N(\varepsilon, E)$ is an ε-neighbourhood of the set E. Since E is a compact set, there exists a vector function $f^*(t, x)$ that is continuous and bounded for $|t| < \infty$, $x \in R^n$, and is equal to $f(t, x)$ on $\bar{N}(\varepsilon, E)$. A solution of the system (1.0.1) remaining in $\bar{N}(\varepsilon, E)$ is also a solution of the equation

$$\frac{dx}{dt} = f^*(t, x), \tag{1.0.3}$$

and the set of all points making up the integral curve of the system for $t \in [t_0, t_1]$ and passing through the point (t_0, x_0) coincides with E. Therefore one can suppose that for $\varepsilon > 0$ and for the equation

$$\frac{dx}{dt} = f^*(t, x) + \psi(t) \tag{1.0.4}$$

the assertion of Theorem 1.0.1 is not confirmed. It should be noted that any solution of the system (1.0.4) exists for all t. The assumptions made imply the existence of a sequence of points $\{P_k(t_k, x_k)\}$ and a sequence of functions $\{(t)\}$ such that

$$d(P_k, E) \to 0, \quad \int_J \| \psi_k(t) \| \, dt \to 0 \quad \text{as } k \to \infty$$

where $J = [t_0, t_1]$. Moreover, there exists a solution $\varphi_k(t)$ of the system

$$\frac{dx}{dt} = f^*(t, x) + \psi_k(t) \tag{1.0.5}$$

passing through the point P_k such that there exists no solution of the system (1.0.3) lying in E, the distance between the points of which and the points of the solution φ_k is smaller than ε.

Since,

$$\varphi_k(t) = \varphi_k(t_0) + \int_{t_0}^t f^*(s, \varphi_k(s)) \, ds + \int_{t_0}^t \psi_k(s) \, ds, \tag{1.0.6}$$

the sequence $\{\varphi_k(t)\}$ is uniformly bounded and equicontinuous on J. Therefore, by Ascoli's theorem, there exists a uniformly convergent subsequence of the sequence $\{\varphi_k(t)\}$. Let $\varphi(t)$ be a limit of the subsequence. Then (1.0.6) implies

$$\varphi(t) = \varphi(t_0) + \int_{t_0}^t f^*(s, \varphi(s)) \, ds,$$

and this shows, that the vector function $\varphi(t)$ is a solution of the system (1.0.3). If we denote by t' an accumulation point of points $\{t_k\}$, it becomes clear that $(t', \varphi(t')) \in E$, since E is closed. Hence there is a solution $\varphi^*(t)$ of (1.0.3) joining (t_0, x_0) and $(t', \varphi(t'))$. Hence $\varphi^*(t) \subset E$ and we may set $\varphi^*(t) = \varphi(t)$ for $t \geqslant t'$. If k is sufficiently large and t_k is sufficiently close to t', the distance between $\varphi_k(t)$ and $\varphi(t)$ is smaller than ε, becuase $\varphi_k(t)$ converges uniformly to $\varphi(t)$. This contradicts the previous assumption. Thus the theorem is proved.   ■

In the particular case $\psi(t) = 0$ the following corollary to this theorem holds.

Corollary to Theorem 1.0.1   Let the vector function $f(t, x)$ be continuous on the domain $a \leqslant t \leqslant b$, $\|x - x_0(t)\| \leqslant r$, $r > 0$, where $x_0(t)$ is a unique solution defined on $a \leqslant t \leqslant b$ and continuous from the right-hand side.
Then for any $\varepsilon_1 > 0$ there exists a $\delta_1 > 0$ such that if $a \leqslant t_0 \leqslant b$ and $\|x_0 - x_0(t_0)\| < \delta_1$ then $\|x(t; t_0, x_0) - x_0(t)\| < \varepsilon_1$ for $t \in [t_0, b]$.

1.1   Limiting Equations

We consider a system of ordinary differential equations

$$\frac{dx}{dt} = f(t, x), \tag{1.1.1}$$

where the function $f: R_+ \times R^n \to R^n$ is continuous. $x(t; t_0, x_0; f)$ denotes a solution of (1.1.1) with $(t_0, x_0) \in R_+ \times R^n$, and the initial conditions

$$x(t_0) = x_0. \tag{1.1.2}$$

Definition 1.1.1   The vector function f in the system (1.1.1) is *admissible* if it is continuous on $R_+ \times W$ and for any pair $(t_0, x_0) \in R_+ \times W$ there exists a unique solution $x_f(t; t_0, x_0)$ satisfying the initial condition $x_f(t_0; t_0, x_0) = x_0$.

For any $\tau \in R_+$ we shall consider the translations (shifts) f_τ of an admissible function f, which are defined by
$$f_\tau(t, x) = f(t + \tau, x) \quad \text{for } (t, x) \in R_+ \times W. \tag{1.1.3}$$

It is clear that any translation f_τ is admissible, and therefore the set

$$T(f) = \{f_\tau : \tau \in R_+\}$$

is a subset of the space $C(R_+ \times W, R^n)$. If the space $C(R_+ \times W, R^n)$ is endowed with the compact open topology, the following result holds.

Theorem 1.1.1 (see Sell [3])   The set $T(f)$ is relatively compact in the compact open topology iff the function f is bounded and uniformly continuous for any set $R_+ \times D$, where D is a compact set of W.

Definition 1.1.2   The function $g: R_+ \times W \to R^n$ will be referred to as *limiting* for the function $f(t, x)$ if there exists a sequence of numbers $\tau_1, \tau_2, \ldots, \tau_k > 0$, $\tau_k \to \infty$ as $k \to \infty$, such that the sequence of functions $f_{\tau_1}, f_{\tau_2}, \ldots$ tends to g on $R_+ \times W$ in the compact open topology.

The set of all limiting functions for the vector function $f(t, x)$ will be denoted by

$$\Omega^*(f) = \{g \in C(R_+ \times W, R^n): \exists \{\tau_k\} \subset R_+, \tau_k \to +\infty,$$

$$f_{\tau_k} \to g \text{ in the compact open topology}\}. \tag{1.1.4}$$

It is known that the set Ω^* is non-empty if the set $T(f)$ is relatively compact in the compact open topology.

Definition 1.1.3   The system of differential equations

$$\frac{dx}{dt} = g(t, x), \tag{1.1.5}$$

where $x \in R^n$ and $g \in \Omega^*(f)$, is called a *limiting system of* the system (1.1.1).

Definition 1.1.4   The vector function f in the system (1.1.1) is referred to as *regular*, if any vector function $g \in \Omega^*(f)$ is admissible.

The following result contains sufficient conditions for regularity of the function f.

Theorem 1.1.2 (see Sell [3])   Let the vector function f in the system (1.1.1) be locally Lipschitzian in x with Lipschitz constant independent of t.
Then any function $g \in \Omega^*(f)$ is locally Lipschitzian in x with the same constant.

Definition 1.1.5   Equation (1.1.1) is referred to as *positive precompact* if for the sequence $\{\tau_k\}, \tau_k \to +\infty, k \to \infty$, there exists a subsequence $\{\sigma_k\}$ such that the sequence of functions f_{σ_k} converges.

We shall present sufficient conditions for precompactness and regularity of the system (1.1.1).

Theorem 1.1.3   Suppose that

(1) for any fixed compact $K \subset R^n$ the vector function $f(t, x)$ is bounded and satisfies the condition

$$\| f(t, x) - f(t, y) \| \leqslant v(\| x - y \|, t) \quad \forall x, y \in K,$$

where $v(r, t)$ is a nondecreasing function of r, continuous in t, $v(0, t) = 0$;
(2) there is a function $v(r)$ such that

$$\int_t^{t+1} v(r, \sigma) \, d\sigma \leqslant N(r), \quad N(r) \to 0 \quad \text{as } r \to 0 \quad \text{for all } t \in R_+.$$

Then the system (1.1.1) is positive precompact.

Remark 1.1.1   The case where the function $f(t, x)$ is continuous in x uniformly in t is a special case of Theorem 1.1.3.

Theorem 1.1.4   Suppose that for any compact $K \subset R^n$ the vector function $f(t, x)$ is bounded and satisfies the condition

$$\| f(t, x) - f(t, y) \| \leq m(t) \| x - y \| \quad \forall (x, y) \in K,$$

where the function $m(t)$ is locally integrable and such that for all s

$$\int_s^{s+1} m(\sigma) \, d\sigma \leq M,$$

and $M < \infty$ is fixed.

Then the system (1.1.1) is positive precompact and regular.

Remark 1.1.2   If the conditions of Theorem 1.1.4 are satisfied, the convergence to the limiting equation is convergence in the sense of the metric.

In order to define the limiting set $\Omega^*(f)$ it is possible to consider various topology. The following is due to Artstein [6]: We shall suppose that f is continuous in x and measurable in t and satisfies the local Carathéodory condition, i.e. for all x in a bounded domain the condition

$$\| f(t, x) \| \leq h(t)$$

is satisfied, where $h(t)$ is a locally integrable function. Moreover, we shall also suppose that the following assumption on the global behaviour of the function $f(t, x)$ is satisfied.

Assumption 1.1.1   For any compact set $K \subset R^n$ there exists a nondecreasing function $\mu: R_+ \to R_+$ that is continuous at 0, with $\mu(0) = 0$, and such that for any continuous function $u: [a, b] \to K$ the integral $\int_a^b f(s, u(s)) \, ds$ is defined and

$$\left\| \int_a^b f(s, u(s)) \, ds \right\| \leq \mu(a - b). \tag{1.1.6}$$

Let G be the set of all functions $g_k(t, x)$ that are continuous in x, measurable in t and satisfy Assumption 1.1.1; Also, assume that

$$\int_a^b g_k(s, u_k(s)) \, ds \to \int_a^b g_0(s, u_0(s)) \, ds, \tag{1.1.7}$$

if the sequence $g_k \in G$ converges to g_0 and u_k be a sequence of continuous functions on $[a, b]$ uniformly convergent to u_0. The condition (1.1.3) for the functions $g_k \in G$ means that the convergence in (1.1.4) is uniform on compact sets.

We shall say that a sequence of functions $g_1, g_2, \ldots$ tends to the function g_0 on $R_+ \times W$ as $k \to \infty$ in the compact open topology, if this sequence converges uniformly to the function g_0 on every compact $C \subset R_+ \times W$.

Example 1.1.1 (Bondi, Moauro, Visentin [2]). Consider the second-order differential equation

$$a\ddot{x} + b\dot{x} + cx = x\sin\sqrt{t},\qquad(1.1.8)$$

where $x\in R^1$, $t\in R_+$, $a>0$, $b>0$, $0<c<1$. By Theorem 7 of Sell [3], we have that for every $\mu\in[-1,1]$ the equation

$$a\ddot{x} + b\dot{x} + cx = \mu x,\qquad(1.1.9)$$

is a limiting equation (1.1.10).

Example 1.1.2 (Artstein [6]). Consider the damped oscillator equation $\ddot{x} + h(t)\dot{x} + x = 0$, or rather its equivalent system

$$\frac{dx}{dt} = y,$$

$$\frac{dy}{dt} = -h(t)y - x.\qquad(1.1.10)$$

Assume that $h(t)\geqslant 0$ and that the indefinite integral of h is uniformly continuous (i.e. $\int_a^b h(s)\,ds \leqslant \mu(a-b)$, where μ is continuous at 0 and $\mu(0)=0$). Then (1.1.8) satisfies Assumption 1.1.1, and we can deduce that all the limiting equations of (1.1.8) have the same form, namely

$$\frac{dx}{dt} = y,$$

$$\frac{dy}{dt} = -g(t)y - x,\qquad(1.1.11)$$

where g satisfies

$$\int_0^t g(s)\,ds = \lim \int_0^t h(t_k + s)\,ds$$

for a certain sequence $t_k \to \infty$.

1.2 Some Properties of Solutions of Limiting Equations

We introduce some notation. Let $M = R_+ \times R^n$, or let M be an interval in R_+. We denote by $C(M, R^n)$ the space of continuous functions defined on M with values in R^n, with the compact open topology.

A sequence $\{f_k\}$ in the space $C(M, R^n)$ is called *convergent to g, k-uniformly on M*, if f_k converges to g uniformly for any compact set in M as $k \to \infty$, which is equivalent with the convergence on the compact open topology.

A function $f(t, x)\in C(R_+ \times R^n, R^n)$ is referred to as *positively precompact*, if for any sequence $\{t_k\}$ in R_+ such that $t_k \to \infty$ as $k \to \infty$, the sequence $\{f(t + t_k, x)\}$ contains a k-uniformly convergent subsequence.

We denote by $\Omega(f)$ the set of all limiting functions g such that the sequence $\{f(t + t_k, x)\}$ converges k-uniformly to g for some sequence $\{t_k\}$ such that $t_k \to \infty$ as $k \to \infty$ and $H(f)$ denotes the closure of $T(f) = \{f(t + \tau, x); \tau \in R_+\}$.

We consider the system

$$\frac{dx}{dt} = f(t, x),$$ (1.2.1)

where $f(t, x) \in G(R_+ \times R^n, R^n)$ and the function $f(t, x)$ is assumed to be positive precompact.

Definition 1.2.1   The system

$$\frac{dx}{dt} = g(t, x)$$ (1.2.2)

is referred to as a *limiting* equation for the system (1.2.1) if $g \in \Omega(f)$.

Definition 1.2.2   The system (1.2.1) is called *regular* if the solutions of any limiting equation for it are unique for the initial problem.

Remark 1.2.1 $\Omega(f)$ and the regularity are equivalent with $\Omega^*(f)$ and the regularity in Definition 1.1.4. However, note that the admissibility is not assumed here for $f(t, x)$. Without supposing $f \in \Omega(f)$, the uniqueness property in this definition is not a necessary assumption on the solutions of the system (1.2.1).

Definition 1.2.3   Let D be an open set in R^n. The continuous vector function $f(t, x)$ is *almost-periodic in t uniformly in* $x \in D$ iff for any sequence $\{\tau_k\}$ there exist a subsequence $\{\tau_{kj}\}$ relative to $\{\tau_k\}$ and a continuous function $g(t, x)$ such that $f(t_1 + \tau_{kj}, x) \to g(t, x)$ uniformly on $R \times S$, where S is any compact set in D for $j \to \infty$.

Lemma 1.2.1   Let the sets $H(f)$, $T(f)$ and $\Omega(f)$ be defined as above. Then the following assertions hold.

(a) $H(f) = T(f) \cup \Omega(f)$.
(b) For any function $g \in H(f)$ the inclusion $H(g) \subset H(f)$ is satisfied. Moreover, $H(g) \subset \Omega(f)$ for any function $g \in \Omega(f)$ and $\Omega(f) = H(f)$ iff $f \in \Omega(f)$.
(c) If the vector function $f(t, x)$ is almost-periodic in t uniformly in $x \in R^n$ then $f \in \Omega(g)$ for any vector function $g \in H(f)$.

Proof   (a) and (c) are obvious. We shall prove (b). By (a), the vector function $g \in T(f)$ or $g \in \Omega(f)$. If $g \in T(f)$ then $T(g) \subset T(f)$ while $T(g) \subset \Omega(f)$ if $g \in \Omega(f)$. Therefore the first part of the assertion follows immediately. To prove the second part, it is sufficient to show that the set $\Omega(f)$ is closed. Let $\{g_m(t, x)\}$ be a sequence in $\Omega(f)$ that converges to the vector function $g \in G(R_+ \times R^n, R^n)$, k-uniformly on $R_+ \times R^n$. We choose a sequence of bounded open sets D_j such that $\bar{D}_j \subset D_{j+1}$, $\bigcup_{j=1}^{\infty} = R_+ \times R^n$. On the subsequence we can suppose that if $(t, x) \in \bar{D}_m$ then the inequality $\|g_m(t, x) - g(t, x)\| < 1/m$. Since $g_m \in \Omega(f)$, there exists a sequence $\{t_{mk}\}$, $t_{mk} \to \infty$ as $k \to \infty$ such that $f(t + t_{mk}, x) \to g_m(t, x)$ k-uniformly on $R_+ \times R^n$.

Here one can assume that $t_{mm} \geqslant m$ and the estimate

$$\| f(t + t_{mm}, x) - g(t, x) \| < 2/m$$

holds on the set $\bar{D}_j$ if $m \geqslant j$. At the same time, for any compact set K in $R_+ \times R^n$ there exists a set $\bar{D}_j$ such that $K \subset \bar{D}_j$. Therefore we can see that $g \in \Omega(f)$. The third part of the assertion follows from the second part. ∎

Lemma 1.2.2 Let $f(t + t_k, x)$ converge to a function $g(t, x)$, k-uniformly on $R_+ \times R^n$ for the sequence $\{t_k\} \to \infty$ as $k \to \infty$, and $x_k(t)$ be a solution of the system (1.2.1) such that $x_k(t_k)$ converges to the value $\xi \in R^n$.

Then the sequence $\{x_k(t + t_k)\}$ contains some subsequence that converges to the solution $x(t)$ of the system (1.2.2), passing through the point $\xi \in R^n$, at $t = 0$, k-uniformly in the domain of definition of the solution $x(t)$. Furthermore, if $x(t)$ is the unique solution of the system (1.2.2) passing through the point ξ for $t = 0$ then the sequence $x_k(t + t_k)$ must also converge to $x(t)$.

Proof See Yoshizawa [1]. It should be noted that the function $x(t)$ may not be continuable up to $t = \infty$ even if every solution $x_k(t)$ is continuable up to $t = \infty$. ∎

Lemma 1.2.3 Let α and σ be positive constants. Suppose that for any $g \in H(f)$ every solution $x(t)$ of the system (1.2.2) is continuable up to $t = \sigma$ whenever $\| x(0) \| \leqslant \alpha$.

Then there exists a constant $\beta > 0$ such that the solution $x(t)$ of the system (1.2.1) satisfies the estimate $\| x(t) \| \leqslant \beta$ on the interval $[\tau, \tau + \sigma]$ whenever $\| x(\tau) \| \leqslant \alpha$ for any $\tau \geqslant 0$.

Proof Suppose that the assertion is false. Then there exists sequences $\{\tau_k\}$ and $\{t_k\}$ and a sequence of solutions $\{x_k(t)\}$ of the system (1.2.1) such that for $\tau_k \geqslant 0$, $\tau_k \leqslant t_k \leqslant \tau_k + \sigma$, the estimates $\| x_k(\tau_k) \| \leqslant \alpha$ and $\| x_k(t_k) \| \geqslant k$ are satisfied. We suppose further that $t_k - \tau_k$ converges to some $\tau \in [0, \sigma]$ and $x_k(\tau_k)$ converges to some $\xi \in R^n$, $\| \xi \| \leqslant \alpha$ and $f(t + \tau_k, x) \to g(t, x) \in H(f)$, k-uniformly on $R_+ \times R^n$. Applying Lemma 1.2.1, we see that an appropriate subsequence $\{x_k(t + \tau_k)\}$ converges to a solution $x(t)$ of (1.2.2) passing through the point ξ for $t = 0$ uniformly on $[0, \sigma]$ because the solution $x(t)$ is defined on $[0, \sigma]$. Since $x_k(t_k) = x_k(t_k - \tau_k + \tau_k)$, $x(\tau)$ must be a finite point, and this contradicts the assumption that $\| x_k(t_k) \| \geqslant k$. Thus the lemma is proved. ∎

1.3 Uniform Asymptotic Stability

The following result shows the relationship between the properties of zero solutions of the systems (1.1.1) and (1.1.7).

Proposition 1.3.1 (see Sell [2]) Let

(1) the vector function f be regular and $f(t, 0) = 0 \ \forall t \in R_+$;
(2) the set $T(f)$ be relatively compact in the compact open topology;

(3) there exist a positive number σ and two continuous positive and monotone functions $\alpha: [0, \sigma] \to R_+$ and $\beta: R_+ \to R$, with $\alpha(r) \to 0$ as $r \to 0$ and $\beta(t) \to 0$ as $t \to +\infty$, such that the norm of the solution of the limiting system (1.1.7) satisfies the estimate

$$\| x(t; 0, x_0; g) \| \leqslant \alpha(\| x_0 \|) \beta(t), \qquad (1.3.1)$$

for every $x_0 \in \bar{S}_\sigma = \{ x \in R^n : \| x \| \leqslant \sigma \}$, $t \in R_+$ and $g \in \Omega^*(f)$.

Then the zero solution of the system (1.1.1) is uniformly asymptotically stable.

This proposition is proved by the following result.

Theorem 1.3.1 Let

(1) the vector function f be locally Lipschitzian in x with Lipschitz constant independent of t, $f(t, 0) = 0 \, \forall t \in R_+$;
(2) the set $T(f)$ be relatively compact in the compact open topology;
(3) there exists a positive number σ, and for any $v > 0$ there exists a $T(v) \geqslant 0$ whenever $\| x_0 \| < \sigma$ and $g \in \Omega^*(f)$, every solution $x_g(t; 0, x_0)$ of the limiting equation (1.1.7) satisfies the conditions
  (i) the solution $x(t; 0, x_0; g)$ exists for all $t \geqslant 0$;
  (ii) the inequality $\| x(t; 0, x_0; g) \| < v$ holds for $t \geqslant T(v)$.

Then the zero solution of (1.1.1) is uniformly asymptotically stable.

Proof By condition (2), we find that for any $\eta > 0$ and every compact subset $K \subset R_+ \times W$ there exists a value $\omega(\eta, K) \geqslant 0$ such that for any $\tau \geqslant \omega(\eta, K)$ there exists a $g \in \Omega^*(f)$ such that

$$\| f_\tau(t, x) - g(t, x) \| < \eta \qquad (1.3.2)$$

for all $(t, x) \in K$. We shall now prove that for any $\varepsilon > 0$ there exists a $\delta(\varepsilon) > 0$ such that $\| x_0 \| < \delta(\varepsilon)$ and $g \in \Omega^*(f)$ imply

$$\| x(t; 0, x_0; g) \| < \varepsilon \quad \forall t \geqslant 0, \qquad (1.3.3)$$

i.e. the zero solution of any limiting equation is uniformly stable and the values of $\delta(\varepsilon)$ can be taken independently of $g \in \Omega^*(f)$. By Theorem 1.1.2, any function $g \in \Omega^*(f)$ is locally Lipschitzian in x with the same constant as f. We set $\varepsilon \in]0, \sigma[$ and take $\delta \in]0, \varepsilon \exp(-LT)[$. Here L is a Lipschitz constant of the function f relative to the compact set P_ε, and $T = T(\varepsilon)$ is the positive number mentioned in condition (3) of the theorem. Applying the Gronwall–Bellman lemma to the inequality

$$\| x(t; 0, x_0; g) \| \leqslant \| x_0 \| + \int_0^t L \| x(t, 0, x_0; g) \| \, dt,$$

we get

$$\| x(t; 0, x_0; g) \| \leqslant \delta \exp(LT) < \varepsilon \quad \forall t \in [0, T],$$

whenever $\|x_0\| \leqslant \delta$ and $g \in \Omega^*(f)$. This proves that condition (3) ensures uniform stability of the zero solution of the limiting system.

Further, we shall prove that the zero solution of the system (1.1.1) is uniformly stable. We take $\varepsilon \in]0, \sigma[$ and $\delta = \delta(\frac{1}{2}\varepsilon)$, $T = T(\frac{1}{2}\delta)$, and L as a Lipschitz constant of the function f relative to the compact set S_ε. By virtue of the inequality (1.3.2) for a fixed $\eta \in]0, \delta \exp(-LT)/2T[$, there exists a constant $\omega = \omega(\eta) \geqslant 0$ such that for any $\tau \geqslant \omega$ there exists a $g \in \Omega^*(f)$ and

$$\|f_\tau(t, x) - g(t, x)\| < \eta \quad \forall (t, x) \in [0, T] \times S_\varepsilon. \tag{1.3.4}$$

Let $x_0 \in S_\delta$ and $t_0 = \omega$. We shall prove that $\|x(t; t_0, x_0; f)\| < \varepsilon \,\forall t \geqslant t_0$. If this is false, there exist values t_1 and t_2, with $t_0 \leqslant t_1 < t_2$, such that

$$\|x(t_1; t_0, x_0; f)\| = \delta, \|x(t_2; t_0, x_0; f)\| = \varepsilon$$
$$\delta \leqslant \|x(t; t_0, x_0; f)\| \leqslant \varepsilon \quad \forall t \in [t_1, t_2]. \tag{1.3.5}$$

We let $x_1 = x(t_1; t_0, x_0; f)$ and note that $\forall t \geqslant 0$

$$x(t_1; 0, x_1; f_{t_1}) = x(t + t_1; t_1, x_1; f) = x(t + t_1; t_0, x_0; f). \tag{1.3.6}$$

Since for $t_1 \geqslant t_0 \geqslant \omega$ the inequality (1.3.4) is valid, there exists a $g \in \Omega^*(f)$ such that

$$\|f_{t_1}(t, x) - g(t, x)\| < \eta \quad \forall (t, x) \in [0, T] \times S_\varepsilon.$$

Again applying the Gronwall–Bellman lemma, we get

$$\|x(T; 0, x; f_{t_1}) - x(T; 0, x_1; g)\| \leqslant T\eta \exp(LT),$$

and therefore

$$\|x(T; 0, x; f_{t_1})\| \leqslant \|x(T; 0, x_1; g)\| + T\eta \exp(LT) < \delta. \tag{1.3.7}$$

In view of (1.3.6) and (1.3.7), we obtain $\|x(t_1 + T; t_0, x_0; f)\| < \delta$, and, since (1.3.5) holds, $0 < t_2 - t_1 < T$.

Applying the Gronwall–Bellman lemma once again, we arrive at the estimate

$$\|x(t_2 - t_1; 0, x_1; f_{t_1})\| - \|x(t_2 - t_1; 0, x_1; g)\|$$
$$\leqslant (t_2 - t_1)\eta \exp[L(t_2 - t_1)] \leqslant T\eta \exp(LT).$$

Hence

$$\varepsilon = \|x(t_2; t_0, x_0; f)\| = \|x(t_2 - t_1; 0, x_1; f_{t_1})\| < \tfrac{1}{2}\varepsilon + \tfrac{1}{2}\delta < \varepsilon,$$

which is a contradiction. To complete the proof, it is sufficient to show the existence of a $\delta' = \delta'(\varepsilon) > 0$ such that for $\|x_0\| < \delta'$, $t_0 \in [0, \omega[$ the inequality

$$\|x(t_1; t_0, x_0; f)\| \leqslant \delta' \quad \forall t \in [t_0, \omega]$$

holds. This will be true if $\delta' = \delta \exp(-L\omega)$.

It remains to show that the zero solution of the system (1.1.1) is uniformly attractive. Let $v \in (0, \sigma)$ and let $\delta'(\frac{1}{2}\delta)$ and $\delta'(v)$ be positive numbers related to the numbers $\frac{1}{2}\delta$ and v in the definition of uniform stability of the zero solution of the system (1.1.1). If we take $T = T(\frac{1}{2}\delta(v))$, then by condition (3) of the theorem, in the inequality (1.3.2) a non-negative number $\omega = \omega(\eta) \geqslant 0$ will be found such that for all $\tau \geqslant \omega$ for a $g \in \Omega^*(f)$

the estimate

$$\| f_\tau(t, x) - g(t, x) \| < \eta \qquad (1.3.8)$$

is satisfied for all $(t, x) \in [0, \tau] \times S_\sigma$, where $\eta \in]0, \delta'(v) \exp(-LT)/2T[$ (the value L is the Lipschitz constant for the function f relative to the compact set S_σ). To prove uniform attraction of the zero solution of the system (1.1.1), it is sufficient to show that for $\| x_0 \| < \delta'(\frac{1}{2}\sigma)$ and $t_0 \in R_+$ the estimate

$$\| x(t_0 + T + \omega; t_0, x_0; f) \| < \delta'(v) \qquad (1.3.9)$$

is satisfied.

Let $x_1 = x(t_0 + \omega; t_0, x_0; f)$. It is clear that

$$x(t; 0, x_0; f_{t_0 + \omega}) = x(t + t_0 + \omega; t_0, x_0; f) \quad \forall t \geqslant 0. \qquad (1.3.10)$$

By the condition (1.3.8), there exists a $g \in \Omega^*(f)$ such that

$$\| f_{t_0 + \omega}(t, x) - g(t, x) \| < \eta \quad \forall (t, x) \in [0, T] \times S_\sigma.$$

We can easily find the estimate

$$\| x(T; 0, x_1; f_{t_0 + \omega}) x(T; 0, x_1; g) \| \leqslant \eta T \exp(LT),$$

which implies

$$\| x(T; 0, x_1; f_{t_0 + \omega}) \| \leqslant \| x(T; 0, x_1; g) \| + \eta T \exp(LT) < \delta'(v). \qquad (1.3.11)$$

The estimate (1.3.11) together with (1.3.10) proves (1.3.9) and completes the proof of Theorem 1.3.1. ∎

1.4 Eventual Properties of Solutions

In this section we investigate the eventual (possible) properties of solutions of the system (1.1.1) under more general (to some extent) assumptions on its right-hand side. Namely, it is assumed that the vector function f is not defined for $x = 0$, while any vector function $g \in \Omega(f)$ is continuous for $x = 0$ and $t \in R_+$.

Definition 1.4.1 The point $0 \in R^n$ is referred to as *eventually stable* if for any $\varepsilon > 0$, and $t_0 \in R_+$ there exist a $\mu(\varepsilon) \geqslant 0$ and a $\delta(t_0, \varepsilon) > 0$ such that for any $t_0 \geqslant \mu(\varepsilon)$, $x_0 \in S_\delta \setminus \{0\}$ the norm of the solution $x(t; t_0, x_0; f)$ satisfies the estimate $\| x(t; t_0, x_0; f) \| < \varepsilon$ for all $t \geqslant t_0$.

Definition 1.4.2 The point $0 \in R^n$ is referred to as *eventually asymptotically stable with domain of attraction independent of t_0*, if being eventually stable it is also eventually attractive with domain of attraction independent of t_0, i.e. there exist constants $\lambda > 0$ and $\mu > 0$, and for any $v > 0$, $t_0 \geqslant \mu$ and any $x_0 \in S_\lambda \setminus \{0\}$ a $T(v, t_0, x_0)$ will be found such that $\| x(t; t_0, x_0; f) \| < v$ for $t \geqslant t_0 + T$, where $x(t; t_0, x_0; f)$ is a solution of the system (1.1.1) with initial conditions (t_0, x_0).

We have the following result.

Theorem 1.4.1   Let

(1) the set W be open in R^n and contain the point $\{0\}$;
(2) the vector function $f \in C(R_+ \times (W \setminus \{0\}), R^n)$;
(3) the set $T(f)$ be relatively compact in the compact open topology;
(4) the vector function $g \in C(R_+ \times W, R^n)$, $g(t, 0) = 0$ for all $g \in \Omega^*(f)$ and $t \in R_+$;
(5) there exist a number $\sigma > 0$, continuous functions $\psi : [0, \sigma] \to R_+$ and $\gamma : R_+ \to R_+$, with $\gamma \to 0$ as $t \to +\infty$, such that
   (i) for all $g \in \Omega^*(f)$

$$\| g(t, x) - g(t, y) \| \leqslant \psi(\| x - y \|) \quad \forall (x, y) \in \bar{S}_\sigma \times \bar{S}_\sigma, \ t \in R_+; \qquad (1.4.1)$$

   (ii) the scalar equation

$$\frac{dr}{dt} = \psi(r), \quad r(0) = 0 \qquad (1.4.2)$$

   admits only one solution $r(t) \equiv 0$ for $t \geqslant 0$;
   (iii) for all $g \in \Omega^*(f)$ the solution of the limiting equation (1.1.7) with the initial condition $x_0 \in S_\sigma$ satisfies the estimate

$$\| x(t; 0, x_0; g) \| \leqslant \gamma(t) \quad \forall t \in R_+. \qquad (1.4.3)$$

Then the point $0 \in R^n$ for the equation (1.1.1) is

(a) eventually uniformly stable;
(b) eventually asymptotically stable with domain of attraction independent of t_0.

Proof   First, it should be noted that under conditions (5i, ii), the unique solution of limiting system (1.1.7) passes through the point $(t_0, x_0) \in R_+ \times W$. Therefore $g \in \Omega^*(f)$ is admissible and the vector f is regular. We begin with the proof of assertion (a). For any $\varepsilon \in (0, \sigma)$ there exists a $\delta \in (0, \varepsilon)$ such that for any $\| x_0 \| \leqslant \delta$ and any $g \in \Omega^*(f)$

$$\| x(t; 0, x_0; g) \| \leqslant \varepsilon \quad \forall t > 0. \qquad (1.4.4)$$

In fact, we find from condition (5iii) that for any $\varepsilon \in (0, \sigma)$ there exists a $T(\varepsilon) > 0$ such that for any $\| x_0 \|$ and $\forall g \in \Omega^*(f)$

$$\| x(t; 0, x_0; g) \| \leqslant \varepsilon \quad \forall t \geqslant T(\varepsilon). \qquad (1.4.5)$$

We note that

$$D^+ \| x(t; 0, x_0; g) \| = \lim_{h \to 0+} \sup \frac{\| x(t + h; 0, x_0; g) \| - \| x(t; 0, x_0; g) \|}{h}$$

$$\leqslant \lim_{h \to 0+} \sup \frac{1}{h} \int_t^{t+h} \| g(\tau, x(\tau; 0, x_0; g)) \| \, d\tau$$

$$= \| g(t; x(t; 0, x_0; g)) \|.$$

This relation and the condition (1.4.1) yield

$$D^+ \| x(t; g) \| \leqslant \psi(\| x(t; g) \|),$$

for all $g \in \Omega^*(f)$ and $t \in [0, T]$.

Further, we consider the problems

$$\frac{dr}{dt} = \psi(r),$$

$$r(0) = r_0 = \|x_0\|, \quad t \in [0, T]; \tag{1.4.6}$$

$$\frac{dr}{dt} = \psi(r),$$

$$r(0) = 0, \quad t \in [0, T]. \tag{1.4.7}$$

Let $r^+(t, r_0)$ be a maximal solution of the problem (1.4.6). Besides, for any $\varepsilon > 0$ there exists a $\delta \in (0, \varepsilon)$ such that for $|r_0| \le \delta$ every solution $r^+(t, r_0)$ of the problem (1.4.6) exists on the interval $[0, T]$ and, moreover, $|r^+(t, r_0)| < \varepsilon$.

The comparison principle (see Lakshmikantham and Leela [1]) implies

$$\|x(t; 0, x_0; g)\| \le r^+(t, r_0) < \varepsilon$$

for all $t \in [0, T]$, provided $\|x_0\| < \delta(\varepsilon)$. This estimate together with the above arguments proves the inequality (1.4.3).

Further, we shall prove that the point $0 \in R^n$ is eventually uniformly stable for the system (1.1.1), i.e. for any $\varepsilon \in (0, \sigma)$ there exists a $\delta(\varepsilon) \in (0, \varepsilon)$ and a constant $\omega(\varepsilon) > 0$ such that for all $x_0 \in S_\delta \backslash \{0\}$

$$\|x(t; t_0, x_0; f)\| < \varepsilon \quad \forall t \ge \omega, \ t_0 \ge 0. \tag{1.4.8}$$

We take $\delta = \delta(\frac{1}{2}\varepsilon)$ and $T = T(\frac{1}{2}\delta)$, where T is the value in the inequality (1.4.5). Consider the two problems

$$\frac{dx}{dt} = f_\tau(t, x),$$

$$x(0) = x_1; \tag{1.4.9}$$

$$\frac{dx}{dt} = g(t, x),$$

$$x(0) = x_1. \tag{1.4.10}$$

Let the solutions be $x(t; f_\tau) = x(t; 0, x_1; f_\tau)$ and $x(t; g) = x(t; 0, x_1; g)$ respectively. The estimate

$$D^+ \|x(t; f_\tau) - x(t)\| \le \|f_\tau(t, x(t; g)) - g(t, x(t; g))\|$$

$$\le \|f_\tau(t, x(t; f_\tau)) - g(t, x(t; f_\tau))\| + \|g(t, x; (t; f_\tau)) - g(t, x(t; g))\|$$

is easily obtained. Hence, in view of condition (5i), we get

$$D^+ \|x(t; f_\tau) - x(t; g)\| \le \|f_\tau(t, x(t; f_\tau)) - g(t, x(t; f_\tau))\|$$

$$+ \psi(\|x(t; f_\tau) - x_g(\tau; g)\|).$$

Now for t from some interval $[0, c]$, for which $x(t; f_\tau)$ and $x(t; g)$ remain in the set $S_\sigma \backslash \{0\}$, consider the problem

$$\frac{dr}{dt} = \psi(r) + \eta, \quad \eta = \text{const} > 0, \quad t \in [0, T], \quad r(0) = 0. \tag{1.4.11}$$

Since by condition (5iii), and we can choose a value $\beta(\alpha)$ for any $\alpha > 0$ such that, provided $\int_0^T \eta \, dt \leq \beta$, the maximal solution of the problem (1.4.11) will satisfy the condition $|r(t)| < \alpha$, $t \in [0, T]$. Therefore, for $\alpha = \frac{1}{2}\delta(\frac{1}{2}\varepsilon)$ there exists a $\beta(\frac{1}{2}\delta(\frac{1}{2}\varepsilon)) > 0$ such that for any $\eta \in \,]0, \beta/T[$ the maximal solution $r(t)$ of the problem (1.4.11) satisfies the estimate $|r(t)| < \frac{1}{2}\delta$. We take $\eta \in \,]0, \beta/T[$. By condition (3), there exists a number $\omega(\eta, [0, T] \times (\bar{S}_\sigma \backslash S_\delta)) = \omega(\varepsilon)$ such that for any $\tau \geq \omega$ a $g \in \Omega^*(f)$ can be found such that

$$\| f_\tau(t, x) - g(t, x) \| < \eta \quad \forall (t, x) \in [0, T] \times (\bar{S}_\sigma \backslash S_\delta). \tag{1.4.12}$$

We now take $x_0 \in S_\delta$, $t_0 \geq \omega$, and prove that $\| x(t; t_0, x_0; f) \| < \varepsilon \, \forall t \in [t_0, \infty[$. Indeed, if this inequality does not hold, there exist values t_1 and t_2, $t_0 < t_1 < t_2$, such that

$$\| x(t_1; t_0, x_0; f) \| = \delta, \| x(t_2; t_0, x_0; f) \| = \varepsilon, \tag{1.4.13}$$

$$\delta \leq \| x(t; t_0, x_0; f) \| \leq \varepsilon \quad \forall t \in [t_1, t_2]. \tag{1.4.14}$$

Setting $x_1 = x(t_1; t_0, x_0; f)$, we have $x(t + t_1; t_0, x_0; f) = x(t + t_1; t_1, x_1; f) = x(t; 0, x_1; f_{t_1})$. Since $t_1 > t_0 \geq \omega(\varepsilon)$, by the estimate (1.4.12), there exists a function $g \in \Omega^*(f)$ such that

$$\| f_\tau(t, x) - g(t, x) \| < \eta \quad \forall (t, y) \in [0, T] \times (\bar{S}_\sigma \backslash S_\delta).$$

We consider two problems of the type (1.4.9) and (1.4.10) again:

$$\frac{dx}{dt} = f_{t_1}(t, x),$$

$$x(0) = x_1; \tag{1.4.15}$$

$$\frac{dx}{dt} = g(t, x),$$

$$x(0) = x_1. \tag{1.4.16}$$

In the same way, we find that $\| x(t; f_{t_1}) - x(t; g) \|$ satisfies the conditions

$$D^+ \| x(t; f_{t_1}) - x(t; g) \| \leq \psi(\| x(t; f_{t_1}) - x(t; g) \|) + \eta,$$

$$\| x(0; f_{t_1}) - x(0; g) \| = 0.$$

Further, it should be verified that neither of the following holds:

(j) $T \leq t_2 - t_1$;
(jj) $T > t_2 - t_1$.

If (j) holds then, in the same way as above, we get $\| x(T; f_{t_1}) - x(T; g) \| < \frac{1}{2}\delta$. At the same time, $\| x(T; 0, x_1; g) \| < \frac{1}{2}\delta$ whenever $\| x_1 \| = \delta$ and $g \in \Omega(f)$. Therefore

$$\| x(T; 0, x_1; f_{t_1}) \| \leq \| x(T; 0, x_1; f_{t_1}) - x(T; 0, x_1; g) \| + \| x(T; 0, x_1; g) \| < \delta,$$

but this contradicts (1.4.14), since $t_1 + T \in [t_1, t_2]$.

Now we shall verify (jj). Similarly to the above, we get the inequality

$$\| x(t_2 - t_1; f_{t_1}) - x(t_2 - t_1; g) \| < \frac{1}{2}\delta$$

and the estimate $\|x(t_2 - t_1; 0, x_1; f_{t_1})\| \leqslant \|x(t_2 - t_1; 0, x_1; f_{t_1}) - x(t_2 - t_1; 0, x_1; g)\| + \|x(t_2 - t_1; 0, x_1; g)\| < \frac{1}{2}\delta + \frac{1}{2}\varepsilon < \varepsilon$, which contradicts the conditions (1.4.13) and (1.4.14). This proves the eventual uniform stability of the point $0 \in R^n$.

We now prove assertion (b). To this end, we must show that the set $\{0\}$ is eventually attractive with domain of attraction independent of t_0, i.e. there exist $\lambda \in]0, \sigma[$ and $\mu > 0$ such that for any $v \in]0, \frac{1}{2}\sigma[$, $t_0 \geqslant \mu$ and $\forall x_0 \in S_\lambda \backslash \{0\}$ there exists a $\tilde{T}(v, t_0, x_0) > 0$ such that $\|x(t; t_0, x_0; f)\| < v$ for $t \geqslant t_0 + \tilde{T}$, $t \in [t_0, c[$.

Consider the following cases

(h) $c < \infty$;
(hh) $c = \infty$.

In case (h) the set

$$\{x(t; t_0, x_0; f) : t \in [t_0, c[, \text{ with } t_0 \geqslant \omega(\tfrac{1}{2}\sigma), x_0 \in S_{\delta(\sigma/2)} \backslash \{0\}\}$$

is bounded owing to eventual stability. The positive limit set

$$\Lambda^+(x(\cdot; f)) = \{y \in R^n : \exists \{t_n\} \subset I(x(\cdot; f)), t_n \to c, x(t_n; t_0, x_0; f) \to y \text{ as } n \to +\infty\}$$

of the solutions $x(\cdot; f)$ is non-empty, connected and compact, and, moreover, $x(t; t_0, x_0; f) \to \Lambda^+(x(\cdot; f))$ as $t \to c$ which is a contradiction, since $\Lambda^+(x(\cdot; f)) = \{0\}$ for $c < +\infty$.

Consider now the case (hh). We denote by $(\delta(\tfrac{1}{2}\sigma), \omega(\tfrac{1}{2}\sigma), \delta(v)$ and $\omega(v)$ the values of δ and ω corresponding to $\tfrac{1}{2}\sigma$ and v introduced in the condition (1.4.8). It is sufficient to show that for $\|x_0\| < \lambda = \delta(\tfrac{1}{2}\sigma), t \geqslant \mu = \omega(\tfrac{1}{2}\sigma)$ and, correspondingly, for $\delta(v)$ there exists a $\tilde{T}(v) = \omega(v)$ such that for some $\tilde{t} \in [\omega, \tilde{T}]$

$$\|x(t_0 + \tilde{t}; t_0, x_0; f)\| < \delta(v), \tag{1.4.17}$$

since at every moment after the initial $t_0 + \tilde{t}$ the estimate $\|x(t; t_0, x_0; f)\| < v$ can be obtained for $t \geqslant t_0 + \tilde{T}$. This finally proves Theorem 1.4.1.

We shall prove the inequality (1.4.17).

By the estimate (1.4.5), under the assumption of Theorem 1.4.1, given $\frac{1}{2}\delta(v)$ there exists a $T_1 = T(\frac{1}{2}\delta(v))$ such that for any $x_0 \in S_\sigma$ and some function $g \in \Omega(f)$

$$\|x(t; 0, x_0; g)\| < \tfrac{1}{2}\delta(v) \quad \forall t \geqslant T_1. \tag{1.4.18}$$

At the same time, the eventual stability and the choice of the value $\frac{1}{2}\delta(v)$ imply that if $x_0 \in S_{\delta(\sigma/2)} \backslash \{0\}$ and $t_0 \geqslant \omega(\frac{1}{2}\sigma)$ then

$$\|x(t; t_0, x_0; f)\| < \tfrac{1}{2}\sigma \quad \forall t \geqslant t_0. \tag{1.4.19}$$

Consider the problem (1.4.11) again and use the notation v and T_1 instead of $\frac{1}{2}\varepsilon$ and T. In addition, given $\frac{1}{2}\delta(v) > 0$, there exists a $\beta(\frac{1}{2}\delta(v)) > 0$ such that if $\eta \in]0, \beta/T_1[$ then

$$|r(t)| < \tfrac{1}{2}\delta \quad \forall t \in [0, T_1]. \tag{1.4.20}$$

Choosing η appropriately, we define $\omega(\eta, [0, T_1] \times (\bar{S}_\sigma \backslash S_{\delta(v)})) = \omega'(v)$ such that for any $\tau \geqslant \omega'$ for a some function $g \in \Omega(f)$

$$\|f_\tau(t, x) - g(t, x)\| < \eta \quad \forall (t, x) \in [0, T_1] \times (\bar{S}_\sigma \backslash S_{\delta(v)}). \tag{1.4.21}$$

Denoting $\Theta = \max(\omega(v), \omega'(v))$ and $x_1 = x(t_0 + \Theta; t_0, x_0; f)$, we get $x(t + t_0 + \Theta; t_0, x_0; f) = x(t; 0, x_1; f_{t_0 + \theta})$ and, as a consequence of (1.4.19), $\|x_1\| < \frac{1}{2}\sigma$.

Similary to the above, we set $\tau = t_0 + \Theta$ and obtain

$$D^+ \|x(t; 0, x_1; f_{t_0 + \theta}) - x(t; 0, x_1; g)\|$$
$$\leqslant \psi(\|x(t; 0, x_1; f_{t_0 + \theta}) - x(t; 0, x_1; g)\|) + \eta$$

with the initial condition

$$\|x(0; 0, x_1; f_{t_0 + \theta}) - x(0; 0, x_1; g)\| = 0.$$

According to the comparison principle (see Lakshmikantham and Leela [1]), we find

$$\|x(t; 0, x_1; f_{t_0 + \theta}) - x(t; 0, x_1; g)\| < \frac{1}{2}\delta(v) \quad \forall t \in [0, T_1]. \tag{1.4.22}$$

In view of the estimates (1.4.18)–(1.4.21), we have the inequality

$$\|x(t_0 + \tilde{T}; t_0, x_0; f)\| \leqslant \|x(T_1; 0, x_1; f_{t_0 + \theta}) - x(T_1; 0, x_1; g)\| + \|x(T_1; 0, x_1; g)\| < \delta(v),$$

which proves the inequality (1.4.17).   ∎

1.5   Converse Theorems

In contrast to Sections 1.3 and 1.4, where some properties of the solutions of the system (1.1.1) were investigated in terms of the limiting system (1.1.7), in this section we consider the conditions under which the properties of the solutions of the system (1.1.1) become a solution of a limiting equation. In this sence the theorems presented in this section are called *converse*, while the theorems containing the conditions that ensure some dynamical properties of the initial system in terms of limiting equations are understood as the *direct* ones.

1.5.1   Boundedness

We introduce the following notions.

Definition 1.5.1   The solutions of the system (1.2.1) are

(a) *bounded* if there exists a $\beta > 0$ such that $\|x(t)\| < \beta$ for all $t \geqslant t_0$, where β may depend on each solution where and after (t_0, x_0) denotes the initial condition for $x(t)$;

(b) *equibounded* if for any $\alpha > 0$ and $t_0 \in R_+$ there exists a $\beta(t_0, \alpha) > 0$ such that if $x_0 \in S_\alpha$ then $\|x(t)\| < \beta(t_0, \alpha)$ for all $t \geqslant t_0$;

(c) *uniformly bounded* if β in (b) is independent of t_0;

(d) *ultimately bounded* if there exists a constant $B > 0$ such that every solution $x(t)$ of the system (1.2.1) satisfies the condition $\overline{\lim}_{t \to \infty} \|x(t)\| < B$;

(e) *equiultimately bounded* if there exists a constant $B > 0$ and for some $\tau \geqslant 0$ and $\alpha > 0$ there exists a $\sigma(\tau, \alpha) \geqslant 0$ such that every solution $x(t)$ of the system (1.2.1) satisfies the estimate $\|x(t)\| \leqslant B$ for all $t \geqslant \tau + \sigma(\tau, \alpha)$, provided $\|x(\tau)\| \leqslant \alpha$;

(f) *uniformly ultimately bounded* if the value $\sigma(\tau, \alpha)$ in (e) can be taken independent of τ.

The constant B is called a *bound* of the ultimately boundedness. The solutions are obviously continuous up to $t = \infty$, provided they are bounded.

Defnition 1.5.2   The zero solution of the system (1.2.1) is *globally uniformly asymptotically stable* if it is uniformly stable and the solutions of the system (1.2.1) are uniformly ultimately bounded with arbitrarily small bound.

Theorem 1.5.1   Suppose that the solutions of any limiting equation (1.2.2) are continuable up to $t = \infty$. Then the solutions of the system (1.2.1) are uniformly bounded if they are uniformly ultimately bounded.

Proof   Since for any vector function $g \in H(f)$ the solutions of (1.2.2) are continuable up to $t \to \infty$, a $\beta(\alpha) > 0$ can be found by Lemma 1.2.1 such that the solution $x(t)$ of the system (1.2.1) satisfies the estimate $\| x(t) \| \leqslant \beta(\alpha)$ on $[\tau, \tau + \sigma(\alpha)]$, provided $\| x(\tau) \| \leqslant \alpha$ for $\tau \geqslant 0$, where $\sigma(\alpha)$ is mentioned in the definition of uniformly ultimate boundedness. Moreover, $\| x(t) \| \leqslant B$ for all $t \geqslant \tau + \sigma(\alpha)$ for some constant $B > 0$. Therefore if $\| x(t) \| < \alpha$ for $\tau \geqslant 0$, one has the estimate $\| x(t) \| \leqslant \max(\beta(\alpha), B)$ for $t \geqslant \tau$. This shows that the solutions of the system (1.2.1) are uniformly bounded. ∎

Corollary to Theorem 1.5.1   If the solutions of the system (1.2.2) are uniformly ultimately bounded for any vector function $g \in H(f)$, they are uniformly bounded.

Proof   In view of Lemma 1.2.1, we have $H(g) \subset H(f)$ for any $g \in H(f)$, and therefore $\Omega(g) \subset H(f)$. Hence the corollary is obtained through Theorem 1.5.1. ∎

Remark 1.5.1   In this corollary $g \in H(f)$ can be replaced by $g \in \Omega(f)$. However, in this case the solutions of the system (1.2.1) are not necessarily uniformly bounded, and the conclusion about eventual uniform boundedness is the best one possible.

The following definition is well known (see Nemytsky, Stepanov [1]).

Definition 1.5.3   A set A is referred to as *invariant* (with respect to the dynamical system $f(p, t)$) if it satisfies the condition

$$f(A, t) = A \quad (-\infty < t < +\infty).$$

Definition 1.5.4   A set A is referred to as *minimal* if it is nonempty, closed and invariant and does not contain a proper subset possessing these three properties.

Theorem 1.5.2   Let

(1) the inclusion $f \in \Omega(g)$ hold for any $g \in \Omega(f)$, or an equivalent condition $H(f) = \Omega(f)$ is a minimal set;
(2) the solutions of any limiting equation (1.2.2) be uniformly bounded;
(3) the solutions of the system (1.2.1) be ultimately bounded with bound B.

Then for any $g \in H(f)$ the solutions of the system (1.2.2) are uniformly ultimately bounded.

Proof According to Lemma 1.2.1, if $g \in \Omega(f)$ then $\Omega(g) \subset H(g) \subset \Omega(f)$. On the other hand, $f \in \Omega(g)$ by assumption (1). Therefore we have $f \in \Omega(f)$ and $\Omega(f) = H(f)$ by Lemma 1.2.1. Thus the system (1.2.1) is a limiting equation by itself. Then, by our assumption, the solutions of the system (1.2.1) are uniformly bounded. Besides, for any vector function $g \in H(f)$ the solutions of the system (1.2.2) are uniformly bounded.

Suppose there exists a $g \in H(f)$ for which the solutions of the system (1.2.2) are nonuniformly ultimately bounded. Then there exists a constant $\alpha > 0$, sequences $\{\tau_k\}$ and $\{t_k\}$ and a sequence of solutions $\{x_k(t)\}$ of the system (1.2.2) such that $\tau_k \geqslant 0$, $t_k - \tau_k \geqslant k$, $\|x_k(\tau_k)\| \leqslant \alpha$ and $\|x_k(t)\| \geqslant B$ on $[\tau_k, t_k]$. Otherwise, $\beta_g(B + 1)$ must be a bound for uniform ultimate boundedness, where $\beta_g(\cdot)$ denotes the number associated with the uniform boundedness of the solutions of the system (1.2.2). Here we note that $\|x_k(t)\| \leqslant \beta_g(\alpha)$ for all $t \geqslant \tau_k$. Applying Lemma 1.2.2 we assume that $g(t_k + \tau_k, x) \rightarrow h(t, x)$, k-uniformly on $R_+ \times R^n$, and $x_k(t + \tau_k) \rightarrow x(t)$, k-uniformly on R_+, where $x(t)$ is a solution of the equation

$$\frac{dx}{dt} = h(t, x)$$

such that $B + 1 \leqslant \|x(t)\| < \beta_g(\alpha)$ on $[0, \infty[$. Since $h \in H(g)$ and $h(g) \subset H(f)$, according to Lemma 1.2.1 and the fact that $H(f) = \Omega(f)$, we have, by condition (1), $h \in \Omega(f)$ and $f \in \Omega(h)$. Therefore there exists a sequence $\{s_k\}$, $s_k \rightarrow \infty$ as $k \rightarrow \infty$, for which $h(t + s_k, x) \rightarrow f(t, x)$ in k-uniform on $R_+ \times R^n$. Applying Lemma 1.2.2 once again, we can assume that the sequence $\{x(t, s_k)\}$ converges to the solution $y(t)$ of the system (1.2.1). It is clear that $B + 1 \leqslant \|y(t)\|$ for all $t \geqslant 0$. But this leads to a contradiction, thus proving the theorem. ∎

Remark 1.5.2   For the general system, if the solutions are uniformly bounded and ultimately bounded, they are equiultimately bounded.

Theorem 1.5.2 is false without assumption (1). This can easily be shown from the following example.

Example 1.5.1   We set

$$g(x) = \begin{cases} -2\sqrt{(x - k)(k + 1 - x)} & \text{for } k \leqslant x \leqslant k + 1, \quad k = 0, 1, 2, \ldots, \\ -g(-x) & \text{for } x < 0, \end{cases}$$

and

$$f(t, x) = g(x) - \frac{x}{1 + t^2}.$$

Clearly $f(t, x) \in C(R_+ \times R^n, R^n)$ and $\Omega(f) = g$. It is not difficult to see that the solutions of the limiting equation are uniformly bounded and the solutions of (1.2.2) with $g(t, x) = g(x)$ are not ultimately bounded. On the other hand, since

$$t_{k+1} - t_k = \int_{k+1}^{k} \frac{dx}{-2\sqrt{(x - k)(k + 1 - x)}} = \frac{1}{2}\pi,$$

every minimal solution $x(t)$ of (1.2.2) is such that $|x(t)| \leqslant 1$ for all $t \geqslant \frac{1}{2}\pi\alpha$ as soon as $|x(0)| \leqslant \alpha$. Thus we can see that the solutions of (1.2.1) are uniformly ultimately bounded. Obviously, the system (1.2.1) given in this example is not regular.

If the right-hand side of the system (1.2.1) is regular, the following result holds, when assumption (1) in Theorem 1.5.2 is omitted.

Theorem 1.5.3   Let

(1) the system (1.2.1) be regular and for any $g \in \Omega(f)$ the solutions of the system (1.2.2) be continuable u, to $t = \infty$;
(2) the solutions of the system (1.2.1) be uniformly ultimately bounded, with bound B.

Then for any vector function $g \in H(f)$ the solutions of the system (1.2.2) are uniformly ultimately bounded, with the same bound B.

Proof   It is clear that for a $g \in T(f)$ the solutions of the system (1.2.2) are uniformly ultimately bounded, with bound B. Let $\sigma(\cdot)$ be the number associated with the uniform ultimate boundedness of the solutions of the system (1.2.1). Suppose now that the conclusion is wrong for some $g \in \Omega(f)$. Then there exists a solution $x(t)$ of the system (1.2.1) such that $\|x(\tau)\| \leqslant \alpha$ for some $\tau \geqslant 0$, but $\|x(s)\| > B$ for $s > \tau + \sigma(\alpha)$. On the other hand, there exists a sequence $\{t_k\}$, $t_k \to \infty$ as $k \to \infty$, for which $f(t + t_k, x) \to g(t, x)$, k-uniformly on $R_+ \times R^n$, because $g \in \Omega(f)$. Let $x_k(t)$ be a solution of the system (1.2.1) such that $x_k(\tau + t_k) = x(\tau)$. Since $x(t)$ is the only solution of the system (1.2.2) through the point $x(\tau)$ for $t = \tau$ and $x(t)$ exists for all $t \geqslant \tau$, according to our assumptions the sequence $\{x_k(t + t_k)\}$ is to converge to $x(t)$, k-uniformly on $[\tau, \infty[$, owing to Lemma 1.2.2.

Since $s + t_k > \tau + t_k + \sigma(\alpha)$, we have the estimate $\|x_k(s + t_k)\| \leqslant B$, and therefore $\|x(s)\| \leqslant B$, which contradicts the inequality $\|x(s)\| > B$. Thus the solutions of the system (1.2.2) are uniformly ultimately bounded, with constant bound B.   ∎

The following results are proved in the same manner.

Theorem 1.5.4   Let

(1) the system (1.2.1) be regular;
(2) the solutions of the system (1.2.1) be uniformly bounded.
Then the solutions of any limiting equations associated with the system (1.2.1) are uniformly bounded, with the same pair of constants $(\alpha, \beta(\alpha))$.

Theorem 1.5.5   Let

(1) the system (1.2.1) be regular and $f(t, 0) = 0$;
(2) the zero solution of the system (1.2.1) be uniformly stable.

Then the zero solution of any limiting equation associated with the system (1.2.1) is uniformly bounded.

Remark 1.5.3   Theorem 1.5.5 is well known (see Sell [1]), under the assumption of uniqueness for the solutions of the system (1.2.1).

The next result follows immediately from Theorems 1.5.3 and 1.5.4.

Theorem 1.5.6   Let

(1) the system (1.2.1) be regular;
(2) the solutions of the system (1.2.1) be uniformly bounded and uniformly ultimately bounded.

Then for any $g \in H(f)$ the solutions of the system (1.2.2) are uniformly ultimately bounded, with the same bound as the system (1.2.1).

Remark 1.5.4   In view of the definition of global uniform asymptotic stability, Theorems 1.5.3 and 1.5.6 can be reformulated for this case.

We shall present one more result of this kind.

Theorem 1.5.7   Let

(1) the system (1.2.1) be regular and $f(t,0) = 0$;
(2) the zero solution of the system (1.2.1) be globally uniformly asymptotically stable and uniformly bounded.

Then the zero solution of any limiting equation associated with the system (1.2.1) is globally asymptotically stable.

Remark 1.5.5   The corresponding result in case of uniform asymptotic stability was obtained by Sell [1].

Remark 1.5.6   In Theorem 1.5.7 neither regularity nor uniform boundedness can be omitted, even for almost-periodic systems. The following counterexample shows this.

Example 1.5.2   Define a function $h(t, x)$ which is 4-periodic in t as follows:

$$h(t, x) = \begin{cases} -x & \text{for } 0 \leqslant x \leqslant 1, \ 0 \leqslant t \leqslant \infty, \\ -x^2 & \text{for } 1 \leqslant x, \quad 0 \leqslant t \leqslant 1, \\ -x^2(2-t) + (x^2-2)(t-1) & \text{for } 1 \leqslant x, \quad 1 \leqslant t \leqslant 2, \\ x^2 - 2 & \text{for } 1 \leqslant x, \quad 2 \leqslant t \leqslant 3, \\ (x^2-2)(4-t) - x^2(t-3) & \text{for } 1 \leqslant x, \quad 3 \leqslant t \leqslant 4. \end{cases}$$

and $h(t, x) = -h(t, -x)$ for $x \leqslant 0$ and $0 \leqslant t < \infty$. Let $f(t, x) = h(t, x) - a(t)x^3$, where $a(t)$ is an almost-periodic function (due to Sibuya [1]) with the following properties

(a) $a(t) > 0$ for all $t \geqslant 0$;
(b) there is a $b \in \Omega(a)$ such that $b(t) = 0$ on $[0, 5]$.

Then $f(t, x) \in C(R_+ \times R^n, R)$ and $f(t, x)$ is almost-periodic in t uniformly for $x \in R$.
    Consider equation (1.2.1) with $f(t, x)$ as defined above. Clearly, equation (1.2.1) is regular, since every $g \in \Omega(f)$ has the form $h(t + m, x) - b(t)x^3$ with a constant $m \in [0, 4]$ and a $b \in \Omega(a)$. Since $a(t) > 0$ for all $t \geqslant 0$, for any compact interval $J \subset I$ we can find an $r > 0$ such that $xf(t, x) < 0$ for all $t \in J$ and $|x| \geqslant r$. This quarantees that every solution of (1.2.1) is continuable up to $t = \infty$. Moreover, the region $I \times \{x : |x| < 1\}$ can be shown to be positively invariant, because $xf(t, x) < 0$ if $|x| = 1$. On the other hand, we have

$xx' \leqslant -x^2|x|$ on $[4k, 4k+1] \times R$ for any integer k, and hence every solution $x(t)$ of (1.2.1) starting at $t = 4k$ must satisfy $|x(4k+1)| < 1$. Thus we can see that every solution $x(t)$ of (1.2.1) starting at $t = \tau$ reaches the line $t = 4k$, satisfies $|x(4k+1)| < 1$, remains in $|x| < 1$ for all $t \geqslant 4k+1$ and satisfies the inequality $x(t)x'(t) \leqslant -x^2(t)$ there, where k is the smallest integer such that $4k \geqslant \tau$. Therefore it is not difficult to show that the zero solution of (1.2.1) is globally uniformly asymptotically stable. However, by property (b) of $a(t)$, we can find a $g \in \Omega(f)$ such that

$$g(t, x) = (x^2 - 2) \operatorname{sgn} x \quad \text{for} \ 0 \leqslant t \leqslant 1 \ \text{and} \ |x| \geqslant 1.$$

Therefore every solution $x(t)$ of (1.2.1) satisfying $|x(0)| > \sqrt{2}[\exp(2\sqrt{2})+1]/[\exp(2\sqrt{2})-1]$ cannot be continuable up to $t = 1$.

Remark 1.5.7   If the right-hand side of the system (1.2.1) is periodic in t, regular, and its solutions are ultimately bounded, then, according to Yoshizawa [1], the solutions of the system (1.2.1) are uniformly bounded and uniformly ultimately bounded. However, this assertion is not necessarily true for almost-periodic systems (see Conley, Miller [1]).

1.5.2   Stability and attraction

We shall present a converse theorem on the property of uniform stability.

Theorem 1.5.8   Let

(1) the set W contain a point $0 \in R^n$ and the vector function $f \in C(R_+ \times W, R^n)$ satisfy $f(t, 0) = 0 \ \forall t \in R_+$;
(2) the set $T(f)$ be relatively compact in the compact open topology;
(3) the zero solution of the system (1.2.1) be uniformly stable or uniformly attractive.

Then the zero solution of any limiting equation associated with the system (1.2.1) is also uniformly stable or uniformly attractive.

Proof   Let $\varepsilon \in \,]0, \infty[$. We take $\delta(\varepsilon)$ corresponding to the property of uniform stability of the zero solution of equation (1.2.1). Suppose that $g \in \Omega^*(f)$. Then by condition (2) $g = \lim_{n \to \infty} f_{\tau_n}$ for some sequence $\{\tau_n\} \subset R_+$, with $\tau_n \to +\infty$. Let $t_0 \in R_+$; then $g_{t_0} = \lim_{n \to \infty} f_{t_0 + \tau_n}$. Therefore $\lim_{n \to \infty} x(t; 0, x_0; f_{t_0} + \tau_n) = x(t; 0, x_0; g_{t_0})$ uniformly in t on any compact subset of R_+. By an appropriate choice of subsequence $\{\tau_n\}$, one may assume for every $n \in \mathbb{N}$, the set of positive integers

$$\| x(t; 0, x_0; g_{t_0}) - x(t; 0, x_0; f_{t_0 + \tau_n}) \| < 1/n \quad \forall t \in [0, n]. \tag{1.5.1}$$

Let $\| x_0 \| < \delta$; then $\| x(t; 0, x_0; f_{t_0 + \tau_n}) \| < \varepsilon$ for all $t \in R_+$ and $n \in \mathbb{N}$. Hence for $\| x_0 \| < \delta$

$$\| x(t; 0, x_0; g_{t_0}) \| < \frac{1}{n} + \varepsilon$$

for any $n \in \mathbb{N}$ and $t \in [0, n]$. Setting $n \to +\infty$, we obtain $\| x(t; 0, x_0; g_{t_0}) \| \leqslant \varepsilon$ for all $t \in R_+$. This proves the uniform stability of the zero solution of (1.1.7).

Let U be a domain of uniform attraction of the zero solution of the system (1.2.2) and $g \in \Omega^*(f)$. We shall show that U is also a domain of uniform attraction of the system (1.2.2) with the same estimate of the value T as for (1.2.1). Let $\varepsilon > 0$ and $K \subset U$ be compact. We take $T = T(\varepsilon, K)$ corresponding to uniform attraction of the zero solution of the system (1.2.1). Suppose $(t_0, x_0) \in R_+ \times K$. We state the estimate (1.5.1) as before. Since the function $x(t; t_0 + \tau_n, x_0; f_{t_0 + \tau_n})$ can be considered as a solution of the system (1.1.1) through the point $(0, x_0)$, our assumption yields $\| x(t; t_0 + \tau_n, x_0; f_{t_0 + \tau_n}) \| < \varepsilon$ for any $n \in R$ and $t \geqslant T$. Therefore $\| x(t; t_0, x_0; g_{t_0}) \| < 1/n + \varepsilon$ for any $n \in \mathbb{N}$ and $t \in [T, n]$. Letting $n \to +\infty$, we obtain the desired result.   ∎

Theorem 1.5.9   Let

(1) conditions (1) and (2) of Theorem 1.5.8 be satisfied;
(2) the zero solution of the system (1.2.1) be uniformly stable;
(3) the zero solution of the limiting system (1.2.2) have a domain of attraction U.

Then U is a domain of attraction of the zero solution of the initial system (1.2.1).

Proof   Let U be a domain of attraction of the zero solution of any limiting equation (1.2.2). Suppose that U is not a domain of attraction of the zero solution of the system (1.2.1). Then, there must exist an $\varepsilon > 0$, a compact set $K \subset U$, a sequence $\{x_n\} \subset K$ ($\{x_n\}$ can converge to $x_0 \in K$), and sequences $\{\tau_n\}$, $\{T_n\} \subset R_+$, as $\tau_n \to +\infty$, $T_n \to +\infty$, such that

$$\| x(T_n; 0, x_n; f_{\tau_n}) \| \geqslant \varepsilon.$$

We take $\delta = \delta(\tfrac{1}{2}\varepsilon)$ as mentioned in the definition of uniform stability of the zero solution of the system (1.2.1). Then we get

$$\| x(t; 0, x_n; f_{\tau_n}) \| \geqslant \delta \quad \text{for } t \in [0, T_n].$$

However, if $\| x(t; 0, x_n; f_{\tau_n}) \| < \delta$ for some $t \in [0, T_n]$, setting $y_n = x(0; 0, x_n; f_{\tau_n})$, uniform stability of the zero solution of the system (1.2.1) yields

$$\| x(t; 0, x_n; f_{\tau_n}) \| = \| x(t + \tau_n, \tau_n, y_n; f) \| < \tfrac{1}{2}\varepsilon \quad \text{for } t \geqslant \tau.$$

Hence for $t = T_n$ we get an estimate that contradicts the assumption we have made.

Since $\Omega^*(f)$ is compact we may set $\lim_{n \to \infty} f_{\tau_n} = g \in \Omega^*(f)$. Since the convergence $\lim_{n \to \infty} x(t; x_n; f_{\tau_n}) = x(t, x_0; g)$ is uniform in t on compact subsets of R_+ and $\| x(t; x_n; f_{\tau_n}) \| \geqslant \delta$ for any $t \in [0, T_n]$, we conclude that $\| x(t, x_0; g) \| \geqslant \delta$ for all $t \in R_+$. But this contradicts the assumption that U is a domain of attraction of the zero solution of the equation (1.1.7).   ∎

1.6   Stability under Perturbations

Along with investigation of the stability of systems of the type (1.1.1) in terms of the limiting system constructed in some topology examination of stability properties of the system (1.1.1) under persistent perturbations (PP) in the sense of Duboshin [1] is of interest.

1.6.1   Stability in the sense of Duboshin

Definition 1.6.1. For a sequence $\{f_k\} \subset C(R \times W, R^n)$ (or $\subset C(R_+ \times W, R^n)$) we shall say that f_k converges to a g in the Bohr topology if $f_k(t,x)$ converges to $f(t,x)$ uniformly on $R \times K$ (or on $R_+ \times K$) for any compact set $K \subset W$.

Together with the set defined by (1.1.6), we shall consider the set

$$\Lambda^*(f) = \{g \in C(R_+ \times W, R^n) : \exists \{\tau_n\} \subset R_+,$$
$$\tau_n \to +\infty, \{f_{\tau_n}\} \to g \text{ in the Bohr topology}\}. \qquad (1.6.1)$$

We recall that $\Lambda^*(f) \subset \Omega^*(f)$, and shall consider the limiting system defined by (1.1.7).

Definition 1.6.2 The equilibrium state $x = 0$ of the system (1.1.7) is *uniformly stable under PP* if for any $\varepsilon > 0$ there exist constants $\delta_1 > 0$ and $\delta_2 > 0$ such that for any $t_0 \in R_+$ and $x_0 \in S_{\delta_1}$, under any perturbations $r \in C(R_+ \times W, R^n)$ such that $\| r(t,x) - g(t,x) \| < \delta_2$ for all $(t,x) \in R_+ \times S_\varepsilon$, the solution $x(t; t_0, x_0; r)$ of the perturbed system the inclusion $x(t; t_0, x_0; r) \in S_\varepsilon$ holds when all $t \geq t_0$.

The following result holds.

Theorem 1.6.1   Let

(1) the vector function f be locally Lipschitzian in x uniformly in t, $f(t,0) = 0$;
(2) the set $\Lambda^*(f)$ be nonempty;
(3) there exist a vector function $g \in \Lambda^*(f)$, $g(t,0) = 0$ such that the zero solution of equation (1.1.7) is uniformly stable under PP.

Then the zero solution of the system (1.1.1) is uniformly stable under PP.

Proof   Let $f_{\tau_n} \to g$ in the Bohr topology. Then for any $\varepsilon > 0$ and $\eta = \frac{1}{2}\delta_2$ there exists a $v > 0$ such that, for $n \geq v$, $\| f_{\tau_n}(t,x) - g(t,x) \| < \eta$ at all $(t,x) \in R_+ \times \bar{S}_\varepsilon$, where δ_1 and δ_2 are those in Definition 1.6.2. Taking this into account, it can easily be shown that if $\| r(t,x) - f_{\tau_n}(t,x) \| < \frac{1}{2}\delta_2$ for a $\tau_n \geq t_v$, $t_0 \geq \tau_v$ and $x_0 \in S_{\delta_1}$, then $x(t; t_0, x_0; r) \in S_\varepsilon$. Further, by choosing appropriate $\delta_1' < \delta_1$ and $\delta_2' < \frac{1}{2}\delta_2$, one can show that if $x_0 \in S_\delta$, $t_0 \in [0, \tau_v]$, and the vector function $r \in C(R_+ \times W, R^n)$, with $\| r(t,x) - f(t,x) \| < \delta$ for all $(t,x) \in R_+ \times S_\varepsilon$, then the inclusion $x(t; t_0, x_0, r) \in S_\delta$ is satisfied for all $t \in [t_0, \tau_v]$.   ∎

Theorem 1.6.1 has a converse.

Theorem 1.6.2   Let

(1) conditions (1) and (2) of Theorem 1.6.1 be satisfied;
(2) the zero solution of the system (1.1.1) be uniformly stable under PP.

Then the zero solution of any equation (1.1.7) with the vector function $g \in \Lambda^*(f)$ is uniformly stable under PP, and, moreover, the values of δ_1 and δ_2 can be chosen independently of $g \in \Lambda^*(f)$.

1.6.2   Stability under PP in almost-periodic systems

Consider the system

$$\frac{dx}{dt} = f(t, x) \tag{1.6.2}$$

with $f \in C(R \times W, R^n)$. We denote by $H(f)$ the closure of the set $T(f)$ (see Section 1.1) in the Bohr topology.

Along with the vector function $g \in H(f)$, we shall consider a limiting system

$$\frac{dx}{dt} = g(t, x), \tag{1.6.3}$$

whose solution $\psi: R_+ \to R^n$ as well as the solution $\varphi: R_+ \to R^n$ of the system (1.6.1) is bounded. We shall write $(\psi, g) \in \Lambda^*(\varphi, f)$ if there exists a sequence $\{\tau_n\} \to \infty$ as $n \to \infty$ such that $\{f_{\tau_n}\} \to g \in \Lambda^*(f)$ in the Bohr topology and $\{\varphi_{\tau_n}\} \to \psi$ uniformly on any compact set from R_+. Let μ be a positive number and K be a compact subset in R^n such that $K \subset W, d(\partial K, \partial W) > \mu$, and $\varphi \in K(R_+, R^n)$ is a solution of the system (1.6.2) such that $\varphi(t) \in K$ for any $t \in R_+$.

Together with the systems (1.6.2) and (1.6.3), we shall consider a perturbed system of equations

$$\frac{dx}{dt} = r(t, x), \tag{1.6.4}$$

where $r \in C(R_+ \times W, R^n)$.

Definition 1.6.3   The solution $\varphi(t)$ of the system (1.6.2) is

(a) *uniformly stable under PP* if for any $\varepsilon \in]0, \mu[$ there exists a $\delta \in]0, \varepsilon[$ such that if $(s, y) \in R_+ \times B(\varphi(s), \delta), r \in C(R_+ \times W, R^n)$ and

$$\| r(t, x) - f(t, x) \| < \delta \quad \text{for} \quad t \in [s, \infty[, \| x - \varphi(t) \| \leqslant \varepsilon,$$

then for any solution $x(t; s, y; r)$ of the system (1.6.4)

$$\| x(t; s, y; r) - \varphi(t) \| < \varepsilon \quad \forall t \geqslant s;$$

(b) *uniformly asymptotically stable under PP* if it is uniformly stable under PP and there exist two numbers σ_0 and ε_0 $(0 < \sigma_0 < \varepsilon_0 < \mu)$ and for each $v > 0$ there exist $T(v) > 0$ and $\eta(v) > 0$ such that if $(s, y) \in R_+ \times B(\varphi(s), \sigma_0), r \in C(R_+ \times W, R^n)$ and $\| r(t, x) - f(t, x) \| < \eta$ for $t \in [s, \infty[, \| x - \varphi(t) \| \leqslant \varepsilon_0$, then for every solution $x(t; s, y; r)$ of the system (1.6.4)

$$\| x(t; s, y; r) - \varphi(t) \| < v \quad \forall t \geqslant s + T.$$

Remark 1.6.1   When $\varphi(t) = 0$, Definition 1.6.3(a) is equivalent with Definition 1.6.2. A special case of the system (1.6.4) is the system

$$\frac{dx}{dt} = f(t, x) + q(t), \tag{1.6.5}$$

where $q \in C(R_+, R^n)$.

The following result holds.

Lemma 1.6.1   Let the set $H(f)$ be compact and the vector function f regular, i.e. the solution of (1.6.3) is unique for the initial condition and for all $g \in H(f)$, and let $x(t)$ be a solution of (1.6.2) and K a compact set such that $K \subset W$, $d(\partial K, \partial W) > \chi$ for a $\chi > 0$.

Then for each $T > 0$ and $\varepsilon \in \,]0, \chi[$ there exists a $\delta \in \,]0, \varepsilon[$ such that for any $s \in R_+$, $\| x(t; s, y; f + q) - x(t) \| < \varepsilon$ for all $t \in [s, s + T]$ when $x(t) \in K$ for $t \in [s, s + T]$ and if $\| y - x(s) \| < \delta$ and $q \in C(R_+, R^n)$ such that $\| q(t) \| < \delta$ on $[s, s + T]$.

With regard to the systems (1.6.3) and (1.6.5), this result is equivalent to the following.

Lemma 1.6.2   Suppose that the set $H(f)$ is compact and the vector function f regular. Let $x(t)$ be a solution of (1.6.2) and K a compact set such that $K \subset W$, $d(\partial K, \partial W) > \chi$ for a $\chi > 0$.

Then for each T and $\varepsilon \in \,]0, \chi[$ there exists a $\delta \in \,]0, \varepsilon[$ such that for any $s \in R_+$, $\| x(t; s, y; r) - x(t) \| < \varepsilon$ for all $t \in [s, s + T]$ when the solution $x(t)$ of the system (1.6.2) satisfies the condition $x(t) \in K$ for $t \in [s, s + T]$ and if the vector function $r \in C(R_+ \times W, R^n)$ satifies $\| r(t, x) - f(t, x) \| < \delta$ for $t \in [s, s + T]$, $\| x - x(t) \| \leqslant \varepsilon$, and

$$\| y - x(s) \| < \delta,$$

where $x(t; s, y, r)$ is a solution of (1.6.4).

Further, we shall need the following results.

Theorem 1.6.3 (continuous dependence of solutions). Let

(1) the vector function $f \in C(R_+ \times W, R^n)$;
(2) the solution $x(t)$ of the system (1.6.2) through the point $(t_0, x(t_0)) \in R_+ \times W$ is defined on $[t_0, t_1]$, unique and, moreover, $x(t) \in K$ on the same interval, where K is a compact set, $K \subset W$, $d(\partial K, \partial W) > \chi$ for a $\chi > 0$.

Then for each $\varepsilon \in \,]0, \chi[$ there exists a $\delta \in \,]0, \varepsilon[$ such that for any $s \in [t_0, t_1]$ a solution $x(t; s, y; r)$ of the system (1.6.4) exists on $[s, t_1]$ and satisfies the estimate

$$\| x(t; s, y, r) - x(t) \| < \varepsilon$$

whenever $\| y - x(s) \| < \delta$,   the   vector   function $r \in C(R_+ \times W, R^n)$ satisfies $\| r(t, x) - f(t, x) \| < \delta$ for $t \in [s, t_1]$ and $\| x - x(t) \| < \varepsilon.$

Theorem 1.6.4   Let

(1) in the system (1.6.2) the vector function $f \in C(R_+ \times W, R^n)$ be regular and admissible;
(2) the set $H(f)$ be compact;
(3) the solution $x(t) \in K$ for a compact set $K \subset W$ of the system (1.6.2) be uniformly asymptotically stable.

Then the solution $x(t)$ is stable under PP.

Now assume that the function f in (1.6.2) is almost periodic on $R \times W$ in the sense of Bohr, that is, $H(f) \subset C(R \times W, R^n)$ is compact in the Bohr topology, and regular.

Theorem 1.6.5   Let

(1) the vector function f in the system (1.6.2) be almost-periodic in t in the sense of Bohr;

(2) the solution $\varphi(t)$ of the system (1.6.2) satisfy $\varphi(t) \in K$ for a compact set $K \subset W$ such that $\mathrm{dist}\,(\partial K, \partial W) \geqslant \mu > 0$ for a $\mu > 0$.

(3) there exist a such that $\psi(t)$ is pair $(\psi, g) \in \Lambda^*(\varphi, f)$ uniformly stable under PP for (1.6.3).

Then the solution $\varphi(t)$ is uniformly stable under PP.

Proof   Since $\varphi(t) \in K$, $\psi(t) \in K$. Then uniform stability under PP of the solution $\psi(t)$ implies that for $\varepsilon \in]0, \tfrac{1}{2}\mu[$ there exists a $\delta'(\tfrac{1}{2}\varepsilon) \in]0, \tfrac{1}{2}\varepsilon[$ such that when

$$\| y - \psi(s) \| < \delta'(\tfrac{1}{2}\varepsilon) \quad \text{for } s \in R_+$$

and $r \in C(R_+ \times W, R^n)$ is such that

$$\| r(t, x) - g(t, x) \| < \delta'(\tfrac{1}{2}\varepsilon) \quad \text{for } t \geqslant s \text{ and } \| x - \psi(t) \| \leqslant \tfrac{1}{2}\varepsilon,$$

we have

$$\| x(t; s, y; r) - \psi(t) \| \leqslant \tfrac{1}{2}\varepsilon. \tag{1.6.6}$$

Since $(\psi, g) \in \Lambda^*(\varphi, f)$, there exists a sequence $\{\tau_i\}$, $\tau_i \to \infty$ as $i \to \infty$, for which, a number $M'(\varepsilon, \mu)$ can be found so that for $(\tau, x) \in R_+ \times \bar{B}(K, \mu)$ and $h \geqslant M'$

$$\| f_{\tau_h}(t, x) - g(t, x) \| < \delta'(\tfrac{1}{2}\delta'(\tfrac{1}{2}\varepsilon)). \tag{1.6.7}$$

Moreover, the sequence $\{\varphi(\tau_i)\} \to \psi(0)$ as $i \to \infty$, and therefore for $\delta'(\tfrac{1}{2}\delta'(\tfrac{1}{2}\varepsilon))$ a number $M(\varepsilon, \mu) \geqslant M'(\varepsilon, \mu)$ can be found so that

$$\| \varphi(\tau_h) - \psi(0) \| < \delta'(\tfrac{1}{2}\delta'(\tfrac{1}{2}\varepsilon)) \quad \text{for } h \geqslant M(\varepsilon, \mu).$$

In view of this, we find from the conditions (1.6.6) and (1.6.7)

$$\| \varphi(t + \tau_h) - \psi(t) \| < \tfrac{1}{2}\delta'(\tfrac{1}{2}\varepsilon) \quad \text{for } h \geqslant M(\varepsilon, \mu), \; t \geqslant 0. \tag{1.6.8}$$

Consider now a vector function $r \in C(R_+ \times W, R^n)$ such that

$$\| r(t, x) - f(t, x) \| < \tfrac{1}{2}\delta'(\tfrac{1}{2}\varepsilon) \quad \text{for } t \geqslant s \text{ and } \| x - \varphi(t) \| \leqslant \varepsilon.$$

Hence we have

$$\| r_{\tau_m}(t, x) - f_{\tau_m}(t, x) \| < \tfrac{1}{2}\delta'(\tfrac{1}{2}\varepsilon) \quad \text{for } t \geqslant s \text{ and } \| x - \varphi(t + \tau_m) \| \leqslant \varepsilon \tag{1.6.9}$$

Now, in terms of the inequalities (1.6.7) and (1.6.8), we get

$$\| r_{\tau_m}(t, x) - g(t, x) \| \leqslant \| r_{\tau_m}(t, x) - f_{\tau_m}(t, x) \| + \| f_{\tau_m}(t, x) - g(t, x) \| < \delta'(\tfrac{1}{2}\varepsilon)$$
$$\text{for } t \geqslant s \quad \text{and} \quad \| x - \varphi(t + \tau_m) \| \leqslant \varepsilon \tag{1.6.10}$$

In addition, we find from the inequality (1.6.8) that the estimate $\| x - \psi(t) \| \leqslant \tfrac{1}{2}\varepsilon$ implies

$$\| x - \varphi(t + \tau_m) \| \leqslant \| x - \psi(t) \| + \| \psi(t) - \varphi(t + \tau_m) \| < \tfrac{1}{2}\varepsilon + \tfrac{1}{2}\delta' < \varepsilon.$$

Then, in view of the estimates (1.6.6) and (1.6.10), we see that the conditions

$$\| y - \psi(s) \| < \delta'(\tfrac{1}{2}\varepsilon) \quad \text{for } s \in R_+,$$

$$\| r_{\tau_m}(t, x) - g(t, x) \| < \tfrac{1}{2}\delta'(\varepsilon) \quad \text{for } r \geqslant s \text{ and } \| x - \psi(t) \| \leqslant \tfrac{1}{2}\varepsilon$$

imply the estimate

$$\| x(t; s, y, r_{\tau_m}) - \psi(t) \| < \tfrac{1}{2}\varepsilon. \tag{1.6.11}$$

Note that the estimate (1.6.8) yields the inequality

$$\| y - \varphi(s + \tau_m) \| < \tfrac{1}{2}\delta'(\tfrac{1}{2}\varepsilon),$$

from which we obtain the estimate $\| y - \psi(s) \| < \delta'(\tfrac{1}{2}\varepsilon)$.

Taking into account the fact that the inequalities (1.6.7) and (1.6.9) imply the estimate (1.6.10), and using (1.6.8) and (1.6.9), we see that the conditions

$$\| y - \varphi(s + \tau_m) \| < \tfrac{1}{2}\delta'(\tfrac{1}{2}\varepsilon) \quad \text{for } s \in R_+,$$

$$\| r_{\tau_m}(t, x) - f_{\tau_m}(t, x) \| < \tfrac{1}{2}\delta'(\tfrac{1}{2}\varepsilon) \quad \text{for } t \geqslant s \text{ and } \| x - \psi(t + \tau_m) \|$$

leads to the estimate

$$\| x(t; s, y; r_{\tau_m}) - \varphi(t + \tau_m) \| \leqslant \| x(t; s, y; r_{\tau_m}) - \psi(t) \|$$
$$+ \| \psi(t) - \varphi(t + \tau_m) \| < \tfrac{1}{2}\varepsilon + \tfrac{1}{2}\delta'(\tfrac{1}{2}\varepsilon) < \varepsilon.$$

Replacing in these estimates $t + \tau_m$ and $s + \tau_m$ by t and s respectively, we find that, provided

$$\| y - \varphi(s) \| < \tfrac{1}{2}\delta'(\tfrac{1}{2}\varepsilon) \quad \text{for } s \geqslant \tau_m,$$

$$\| r(t, x) - f(t, x) \| < \tfrac{1}{2}\delta'(\tfrac{1}{2}\varepsilon) \quad \text{for } t \geqslant s \text{ and } \| x - \varphi(t) \| \leqslant \varepsilon \tag{1.6.12}$$

the estimate $\| x(t; s, y; r) - \varphi(t) \| < \varepsilon$ is satisfied.

On the other hand, by Theorem 1.6.5, there exists a $\delta \in]0, \tfrac{1}{2}\delta'(\tfrac{1}{2}\varepsilon)[$ such that the conditions

$$\| y - \varphi(s) \| < \delta, \quad \text{for } s \in [0, \tau_m[,$$

$$\| r(t, x) - f(t, x) \| < \delta \quad \text{for } t \in [s, \tau_m] \text{ and } \| x - \varphi(t) \| \leqslant \varepsilon \tag{1.6.13}$$

ensure the inequality $\| x(t; s, y; r) - \varphi(t) \| < \tfrac{1}{2}\delta'(\tfrac{1}{2}\varepsilon)$.

Finally, using the estimates (1.6.12) and (1.6.13), we find from the conditions

$$\| y - \varphi(s) \| < \delta, \quad \text{for } s \in R_+,$$

$$\| r(t, x) - f(t, x) \| < \delta, \quad \text{for } t > s \text{ and } \| x - \varphi(t) \| \leqslant \varepsilon$$

that the inequality $\| x(t; s, y; r) - \varphi(t) \| \leqslant \varepsilon$ is valid, which corresponds to Definition 1.6.3(a). Thus Theorem 1.6.5 is proved. ∎

Theorem 1.6.6 Let
(1) the vector function f in the system (1.6.2) be almost-periodic in t in the sense of Bohr and regular;
(2) the solution of the system (1.6.2) $\varphi(t) \in K$ for a compact set $K \subset W$ such that dist $(\partial K, \partial W) \geqslant \mu > 0$;

(3) there exist a such that $\psi(t)$ pair $(\psi, g) \in \Lambda^*(\varphi, f)$ is uniformly asymptotically stable for (1.6.3).

Then the solution $\varphi(t)$ is uniformly asymptotically stable under PP.

Proof Suppose that the solution $\psi(t)$ of the system (1.6.3) is uniformly asymptotically stable, and, in addition, the condition $\varphi(t) \in K$ implies $\psi(t) \in K$. In this regard, the conditions of Theorem 1.6.4 are satisfied, and therefore the solution $\psi(t)$ is uniformly stable under PP. Applying Theorem 1.6.5, we can see the same property for the solution $\varphi(t)$. Let $\varepsilon_0 \in]0, \mu[$. For each $v \in]0, \frac{1}{4}\varepsilon_0[$ a number $\delta(v) > 0$ is chosen associated with v in the definition of uniformly stability under PP of the solutions $\varphi(t)$ and $\psi(t)$. In the presence of uniform attraction for the solution $\psi(t)$, there exists a $\varepsilon_0 \in]0, \mu[$ and for the corresponding $\frac{1}{3}\delta(v)$ there exists a number $T = T(\frac{1}{3}\delta(v)) > 0$ such that the solution of the limiting equation (1.6.3) under the condition

$$\| z - \psi(s) \| < \varepsilon_0 \quad \text{for } s \in R_+, \quad t \geqslant s + T,$$

satisfies the estimate

$$\| x(t; s, z; g) - \psi(t) \| < \tfrac{1}{3}\delta(v). \tag{1.6.14}$$

By the uniform stability of the solution $\varphi(t)$, the conditions $s \in R_+$, $\| z - \psi(t) \| < \delta(\frac{1}{4}\varepsilon_0)$, $t \geqslant s$, imply the estimate

$$\| x(t; s, z; g) - \psi(t) \| < \tfrac{1}{4}\varepsilon_0, \tag{1.6.15}$$

and $x(t; s, z; g) \in \bar{B}(K, \frac{1}{2}\varepsilon_0) \equiv K$ for all $t \geqslant s$. This allows Lemma 1.6.2 to be applied in the investigation of the solution $x(t; s, z; g)$. Moreover, we obtain from the inequality (1.6.15) the inclusion

$$\{ x : \| x - \psi(t) \| \leqslant \tfrac{1}{2}\varepsilon_0, t \in [s, s + T] \}$$
$$\supset \{ x : \| x - x(t; s, z; g) \| \leqslant \tfrac{1}{3}\delta(v), t \in [s, s + T] \}.$$

Therefore, to the value $\frac{1}{3}\delta(v)$ there corresponds a $\xi(v) \in]0, \delta(\frac{1}{3}\delta(v))[$ such that for $s \in R_+$, $r \in C(R_+ \times W, R^n)$ such that

$$\| r(t, x) - g(t, x) \| < \xi(v) \quad \text{for } t \in [s, s + T] \text{ and } \| x - \psi(t) \| \leqslant \tfrac{1}{2}\varepsilon_0$$

imply

$$\| x(t; s, z; r) - x(t; s, z; g) \| < \tfrac{1}{3}\delta(v). \tag{1.6.16}$$

Since $(\psi, g) \in \Lambda^*(\varphi, f)$, there exists a sequence $\{\tau_i\}$, $\tau_i \to \infty$ as $i \to \infty$, for which, given $\xi(v)$, there exists an $m_1(v) > 0$ such that

$$\| \varphi(\tau_i) - \psi(0) \| < \xi(v) \quad \text{for } i > m_1(v). \tag{1.6.17}$$

Besides, for the corresponding $\xi(v)$ and a compact set K, an $m(v, K) \geqslant m_1(v)$ can be found such that

$$\| f_{\tau_h}(t, x) - g(t, x) \| < \tfrac{1}{2}\xi(v) \tag{1.6.18}$$

for $(t, x) \in R_+ \times K$, $h \geqslant m$.

Since the solution $\psi(t)$ is stable under PP, we find in terms of the estimates (1.6.17) and (1.6.18)

$$\| \varphi(t + \tau_h) - \psi(t) \| < \tfrac{1}{3}\delta(v) \quad \text{for } h \geqslant m(v, K), \ t \geqslant 0. \tag{1.6.19}$$

Considering a vector function $r \in C(R_+ \times W, R^n)$ and the time $s \in R_+$ such that

$$\| r(t, x) - f(t, x) \| < \eta(v) = \tfrac{1}{2}\xi(v) \quad \text{for } t \geqslant s, \quad \| x - \varphi(t) \| \leqslant \varepsilon_0, \tag{1.6.20}$$

we note that (1.6.20) yields the estimate

$$\| r(t + \tau_m, x) - f(t + \tau_m, x) \| < \eta \quad \text{for } t \geqslant s \text{ and } \| x - \varphi(t + \tau_m) \| \leqslant \varepsilon_0.$$

From this and (1.6.18) and (1.6.19) we deduce the estimate

$$\| r_{\tau_m}(t, x) - g(t, x) \| \leqslant \| r(t + \tau_m, x) - f(t + \tau_m, x) \|$$
$$+ \| f(t + \tau_m, x) - g(t, x) \| < \xi(v) \tag{1.6.2.1}$$

for $t \geqslant s$ and $\| x - \psi(t) \| \leqslant \tfrac{1}{2}\varepsilon_0$.

Now taking $\sigma_0 \in \,]0, \delta(\tfrac{1}{2}\delta(\tfrac{1}{4}\varepsilon_0))[$, we see that to complete the proof of the theorem it is sufficient to show that for every $s \in R_+$ the following result holds: the condition $\| y - \varphi(s) \| < \sigma_0$ implies the existence of a $t' \in [s + \tau_m, s + \tau_m + T]$ such that

$$\| x(t'; s, y; r) - \varphi(t') \| < \delta(v). \tag{1.6.22}$$

Then by t, we get the estimates from the definition of the stability of the solution $\varphi(t)$ under PP in the form

$$\| x(t; s, y; r) - \varphi(t) \| < v \quad \text{for } t \geqslant s + \tilde{T}(v),$$

where $\tilde{T}(v) = \tau_{m(v)} + T(\tfrac{1}{3}\delta(v))$.

Let the inequality (1.6.22) be false under the above conditions. Then there exists a time $\hat{s} \in R_+$ and a point $\hat{y} \in B(\varphi(\hat{s}), \sigma_0)$ such that

$$\| x(t; \hat{s}, \hat{y}; r) - \varphi(t) \| \geqslant \delta(v) \quad \text{for } t \in [\hat{s} + \tau_m, \hat{s} + \tau_m + T]. \tag{1.6.23}$$

Consider the point $\hat{x} = x(\hat{s} + \tau_m; s, \hat{y}; r)$. Since by the conditions of the theorem the solution $\varphi(t)$ is uniformly stable under PP, we have $\| x(t; \hat{s}, \hat{y}; r) - \varphi(t) \| < \tfrac{1}{2}\delta(\tfrac{1}{4}\varepsilon_0)$ provided $\hat{s} \in R_+$, $\| \hat{y} - \varphi(\hat{s}) \| < \sigma_0 \leqslant \delta(\tfrac{1}{2}\delta(\tfrac{1}{4}\varepsilon_0))$, $\| r(t, x) - f(t, x) \| < \eta < \delta(\tfrac{1}{3}\delta(v))$ for $t \geqslant s$ and $\| x - \varphi(t) \| < \tfrac{1}{2}\delta(\tfrac{1}{4}\varepsilon_0) < \varepsilon$, $t \geqslant \hat{s}$. From here, we find that

$$\| \hat{x} - \varphi(\hat{s} + \tau_m) \| < \tfrac{1}{2}\delta(\tfrac{1}{4}\varepsilon_0). \tag{1.6.24}$$

From the inequality (1.6.19) and the fact $\hat{y} \in B(\varphi(\hat{s}), \sigma_0)$ we get

$$\| \hat{x} - \psi(\hat{s}) \| \leqslant \| \hat{x} - \varphi(\hat{s} + \tau_m) \| + \| \varphi(\hat{s} + \tau_m) - \psi(\hat{s}) \|$$
$$< \tfrac{1}{2}\delta(\tfrac{1}{4}\varepsilon_0) + \tfrac{1}{3}\delta(v) < \delta(\tfrac{1}{4}\varepsilon_0). \tag{1.6.25}$$

In view of the estimates (1.6.14) and (1.6.25), we obtain

$$\| x(\hat{s} + T; \hat{s}, \hat{x}, g) - \psi(\hat{s} + T) \| < \tfrac{1}{3}\delta(v). \tag{1.6.26}$$

On the other hand, the estimates (1.6.15) and (1.6.25) yield

$$\| x(t; \hat{s}, \hat{x}, g) - \psi(t) \| < \tfrac{1}{4}\varepsilon_0 \quad \text{for } t \geqslant \hat{s},$$

which implies $x(t; \hat{s}, \hat{x}, g) \in K$, and therefore the condition (1.6.16) is valid. But, when r_{τ_m} satisfies the estimate (1.6.21),

$$\| x(\hat{s} + T; \hat{s}, \hat{x}, r_{\tau_m}) - x(\hat{s} + T; \hat{s}, \hat{x}, g) \| < \tfrac{1}{3}\delta(v). \tag{1.6.27}$$

Taking into account that

$$x(\hat{s} + t; \hat{s}, \hat{x}, r_{\tau_m}) = x(\hat{s} + \tau_m + t; \hat{s} + \tau_m, \hat{x}, r)$$

$$= x(\hat{s} + \tau_m + t; \hat{s}, \hat{y}, r) \quad \text{for } t \geqslant 0,$$

and applying the inequalities (1.6.19), (1.6.26) and (1.6.27), we arrive at

$$\| x(\hat{s} + \tau_m + T; \hat{s}, \hat{y}, r) - \varphi(\hat{s} + \tau_m + T) \|$$
$$\leqslant \| x(\hat{s} + \tau_m + T; \hat{s}, \hat{y}, r) - x(\hat{s} + T, \hat{s}, \hat{x}; g) \|$$
$$+ \| x(\hat{s} + T; \hat{s}, \hat{x}; g) - \psi(\hat{s} + T) \|$$
$$+ \| \psi(\hat{s} + T) - \varphi(\hat{s} + \tau_m + T) \| < \delta(v).$$

This inequality contradicts (1.6.23). Thus Theorem 1.6.6 is proved. ∎

1.7 Lipschitz Stability

We shall introduce for system of equations (1.0.1) some assumptions under which the Lipschitz stability of its zero solution will be investigated.

Assumption 1.7.1 For the system (1.0.1)

(1) the vector function f is uniformly continuous and bounded in the set $R_+ \times C$, where C is an arbitrary compact set of W;
(2) for all $t \geqslant 0$ we have $f(t, 0) = 0$ and $0 \in W$;
(3) the function f is locally Lipschitz continuous with respect to x in W, with Lipschitz constant independent of t.

Definition 1.7.1 The zero solution of the system (1.0.1) is

(a) *uniformly Lipschitz stable* if there exist $M > 0$ and $\delta > 0$ such that for all $(t_0, x_0) \in R_+ \times (B_\delta \cap W)$, $\| x(t; \tau_0, x_0; f) \| \leqslant M \| x_0 \|$ for all $t \geqslant \tau_0$;
(b) *strongly Lipschitz stable* if for all $\tau_0 \geqslant 0$ there exist $M = M(t; \tau_0) > 0$, $M \in C([\tau_0, \infty[, R_+), \sigma = \sigma(\tau_0) > 0, \rho = \rho(\tau_0), 0 < \rho < 1$, with $M(t, \tau_0) \leqslant \rho$ for all $t \geqslant \tau_0 + \sigma$ and $\delta = \delta(\tau_0) > 0$ such that for all $x_0 \in B_\delta \cap W$

$$\| x(t; \tau_0, x_0; f) \| \leqslant M(t, \tau_0) \| x_0 \|$$

when all $t \geqslant \tau_0$;
(c) *strongly uniformly by Lipschitz stable* if there exist $M = M(t) > 0, M \in C([0, \infty[, R_+)$ a $\sigma > 0, 0 < \rho < 1$, with $M(t) < \rho$ for all $t \geqslant t_0 + \sigma$ and a $\delta > 0$ such that for all

$$(t_0, x_0) \in R_+ \times (B_\delta \cap W)$$

$$\| x(t; t_0, x_0; f) \| < M(t) \| x_0 \|$$

for all $t \geq t_0$.

Consider a family of continuous functions

$$F = \{ f_\alpha : R_+ \times W \to R^n, \ f_\alpha(t, 0) = 0, \quad \forall t \geq 0, \alpha \in A \},$$

where A is a finite set.

Definition 1.7.2 The zero solution of the system

$$\frac{dx}{dt} = f_\alpha(t, x), \quad f_\alpha \in F$$

is *strongly* (respectively *uniformly*) *Lipschitz equistable* on the right-hand side if for every $f_\alpha \in F$ it is strongly (respectively uniformly) Lipschitz stable and the function M and the constants σ, ρ and δ from Definition 1.7.1(a) (respectively Definition 1.7.1(c)) are independent of the function f_α.

Together with the system (1.0.1) we shall consider a system of limiting equations (1.1.7).

Theorem 1.7.1 (Refer to Th. 1.1.1) The set $F = \{ f_\theta : \theta \geq 0 \}$ is relatively compact in the compact open topology iff Assumption 1.7.1(1) holds.

Corollary 1.7.1 Let Assumption 1.7.1(1) hold. Then $\Omega^*(f) \neq \emptyset$.

Theorem 1.7.2 Let Assumptions 1.7.1(2) and (3) hold. Then each of the functions $f^* \in \Omega^*(f)$ satisfies the same conditions. Moreover, the Lipschitz constant of the function f is common for each of the functions $f^* \in \Omega^*(f)$.

Theorem 1.7.3 Let

(1) Assumption 1.7.1 hold;
(2) the zero solution of the system (1.1.7) be strongly Lipschitz equistable $g \in \Omega^*(f)$.

Then the zero solution of the system (1.0.1) is stable.

Proof Suppose that this is not true; that is,

$$(\exists \varepsilon_0 > 0)(\exists \tau_0 \geq 0)(\forall \delta > 0)(\exists x_0 \in B_\delta \cap W)(\exists t_{x_0} > \tau_0)$$

such that

$$\| x(t_{x_0}; \tau_0, x_0; f) \| \geq \varepsilon_0.$$

Let $M^*(t) = M(t; \tau_0)$, $\sigma^* = \sigma(\tau_0)$, $\rho^* = \rho(\tau_0)$ and $\delta^* = \delta(\tau_0)$ be the function and the constants in Definition 1.7.2. Let $L = L(m, \delta^*) > 0$ be the Lipschitz constant for the function f for $(t, x) \in R_+ \times (B_{m\delta^*} \cap W)$, where $m = \max \{ M^*(t) : t \geq \tau_0 \}$. It follows from

Theorem 1.7.2 that L is also a Lipschitz constant for any of the functions $f^* \in \Omega^*(f)$ on the set $R_+ \times (B_{m\delta^*} \cap W)$. Let $T > \sigma^*$ be an arbitrary constant. Without loss of generality, we assume that $\delta^* \leqslant \varepsilon_0 \exp(-LT)$. Let the constant $\omega > T$ be sufficiently large that for $\theta > \omega$ there exists a function $f^* \in \Omega^*(f)$ such that

$$\| f_\theta(t, x) - f^*(t, x) \| < \frac{1}{T}(1 - \rho^*)\delta^{**} \exp(-LT), \tag{1.7.1}$$

where

$$(t, x) \in [\tau_0, \tau_0 + T] \times (B_{m\delta^*} \cap W), \quad 0 < \rho^* < 1, \quad 0 < \delta^{**} < \delta^*.$$

For $\tau_0 \leqslant t \leqslant \tau_0 + \omega$ we have

$$\| x(t; \tau_0, x_0; f) \| = \| x(t; \tau_0, x_0; f) - x(t; \tau_0, 0: f) \|$$

$$\leqslant \| x_0 \| + \int_{\tau_0}^t L \| x(\tau; \tau_0, x_0; f) - x(\tau; \tau_0, 0; f) \| \, d\tau$$

$$= \| x_0 \| + L \int_{\tau_0}^t \| x(\tau; \tau_0, x_0; f) \| \, d\tau$$

whence, by means of the Gronwall–Bellman inequality, we deduce the estimate

$$\| x(t; \tau_0, x_0; f) \| \leqslant \| x_0 \| \exp(L(t - \tau_0)) \leqslant \| x_0 \| \exp(L\omega). \tag{1.7.2}$$

Let $0 < \delta < \delta^{**} \exp(-L\omega)$. Then it follows from (1.7.2) that for a point $x_0 \in B_\delta \cap W$ satisfying the assumption the following inequality holds:

$$\| x(t; \tau_0, x_0; f) \| < \delta^{**} < \varepsilon_0, \quad \tau_0 \leqslant t \leqslant \tau_0 + \omega. \tag{1.7.3}$$

It is seen from (1.7.3) that $t_{x_0} > \tau_0 + \omega$. Hence there exists a point $t_2 > \tau_0 + \omega$ such that

$$\| x(t_2; \tau_0, x_0; f) \| = \varepsilon_0.$$

Moreover, we assume that t_2 is the smallest number satisfying this equality. We conclude from (1.7.3) that there exists a point $t_1, \tau_0 + \omega < t_1 < t_2$, such that

$$\| x(t_1; \tau_0, x_0; f) \| = \delta^{**}, \quad \| x(t; \tau_0, x_0; f) \| \geqslant \delta^{**}, \quad t_1 \leqslant t \leqslant t_2. \tag{1.7.4}$$

Since $\theta = t_1 - \tau_0 > \omega$, it follows that there exists a function $f^* \in \Omega^*(f)$ such that the inequality (1.7.1) holds. We introduce the notation $x_1 = x_f(t_1; \tau_0, x_0)$. We find from (1.7.4) that for $\tau_0 \leqslant t \leqslant \tau_0 + T$, using (1.7.1) we find that

$$\| x(t; \tau_0, x_1; f_\theta) - x(t; \tau_0, x_1; f^*) \| < \frac{1}{T}(1 - \rho^*)\delta^{**}(t - \tau_0) \exp(-LT)$$

$$+ \int_{\tau_0}^t L \| x(\tau; \tau_0, x_1; f_\theta) - x(\tau; \tau_0, x_1; f^*) \| \, d\tau,$$

whence we derive the estimate

$$\| x(t; \tau_0, x_1; f_\theta) - x(t; \tau_0, x_1; f^*) \| < \frac{1}{T}(1 - \rho^*)\delta^{**}(t - \tau_0) \exp(L(t - \tau_0 - T)).$$

For $t = \tau_0 + T$ it follows from the above inequality that

$$\| x(t_0 + T; \tau_0, x_1; f_\theta) \| < \| x(\tau_0 + T; \tau_0, x_1; f^*) \| + (1 - \rho^*)\delta^{**}$$
$$\leqslant M^*(t)\| x_1 \| + (1 - \rho^*)\delta^{**} \leqslant \delta^{**}.$$

Then, in view of the equality

$$x(\tau_0 + T; \tau_0, x_1; f_\theta) = x(\tau_0 + \theta + T; \tau_0, x_0; f),$$

we conclude that $t_2 < \tau_0 + T$; that is, $t_2 - t_1 < T$. It is seen from this inequality that

$$\| x(t_2; \tau_0, x_0; f) \| = \| x(t_2; t_1, x_1; f) \|$$
$$\leqslant \| x_1 \| \exp(L(t_2 - t_1)) < \delta^* \exp(LT) \leqslant \varepsilon_0$$

This contradicts the way in which the point t_2 was chosen, and thus completes the proof of Theorem 1.7.3. ∎

Theorem 1.7.4 Let

(1) Assumption 1.7.1 hold;
(2) the zero solution of the system (1.1.7) be strongly uniform Lipschitz equistable on the right-hand side $g = f_* \in \Omega^*(f)$.

Then the zero solution of the system (1.0.1) is uniformly stable.

Proof This is analogous to that of Theorem 1.7.3. ∎

Theorem 1.7.5 Let

(1) Assumption 1.7.1 hold;
(2) the zero solution of the system (1.0.1) be uniformly Lipschitz stable.

Then for any function $f_* \in \Omega^*(f)$ the zero solution of the system (1.1.7) with $g = f_*$ is uniformly Lipschitz stable.

Proof It follows from the definition of a limiting function that

$$(\forall f_* \in \Omega_f^*)(\forall \eta > 0) \, (\forall T > 0)(\forall C \text{ a compact subset of } W) \, (\exists \theta = \theta(f_*, \eta, T, C) > 0)$$

such that

$$\| f_\theta(t, x) - f_*(t, x) \| < \eta \quad \forall(t, x) \in [0, T] \times C.$$

Let f_* be an arbitrary function of $\Omega^*(f)$. Assume that the zero solution of the system (1.1.7) is not uniformly Lipschitz stable. Then

$$(\forall M^* > 0)(\forall \delta^* > 0)$$
$$(\exists(\tau_0^*, x_0^*) \in R_+ \times (B_{\delta^*} \cap W)), \quad \tau_0^* = \tau_0^*(M^*, \delta^*), \quad x_0^* = x_0^*(M^*, \delta^*)$$
$$(\exists t^* > \tau_0^*)$$

such that

$$\| x(t^*; \tau_0^*, x_0^*; f^*) \| \geqslant M^* \| x_0^* \|.$$

Let M and δ be the constants from Definition 1.7.1(a) let $\bar{B}_{M\delta} \subset W$, and let $L = L(M, \delta) > 0$ be the Lipschitz constant for the functions f and f_* for $(t, x) \in R_+ \times B_{M\delta}$, $M^* > 2M$, $\tau_0^* = \tau_0^*(M^*, \delta)$ and $x_0^* = x_0^*(M^*, \delta)$. There exists a point $t^* > \tau_0^*$ such that

$$\| x(t^*; \tau_0^*, x_0^*; f^*) \| \geqslant M^* \| x_0^* \|.$$

Let $\theta = \theta(f_*, \eta, t^*, \bar{B}_{M\delta})$, where

$$\eta = \frac{M^* \| x_0^* \| \exp(-L(t^* - \tau_0^*))}{2(t^* - \tau_0^*)}.$$

Then for $(t, x) \in [0, t^*] \times \bar{B}_{M\delta}$ we have

$$\| f_\theta(t, x) - f_*(t, x) \| < \frac{M^* \| x_0^* \| \exp[-L(t^* - \tau_0^*)]}{2(t^* - \tau_0^*)}.$$

For any $t \geqslant \tau_0^*$ the following inequality holds:

$$\| x(t; \tau_0^*, x_0^*; f_\theta) \| = \| x(t + \theta; \tau_0^* + \theta, x_0^*; f) \| \leqslant M \| x_0^* \| < \tfrac{1}{2} M^* \| x_0^* \|. \quad (1.7.5)$$

On the other hand, for $\tau_0^* \leqslant t \leqslant t^*$ we have

$$\| x(t; \tau_0^*, x_0^*; f_\theta) - x(t; \tau_0^*, x_0^*; f^*) \| \leqslant \eta(t - \tau_0^*)$$

$$+ \int_{\tau_0^*}^{t} L \| x(\tau; \tau_0^*, x_0^*; f_\theta) - x(\tau; \tau_0^*, x_0^*; f^*) \| \, d\tau,$$

whence for $t = t^*$ we deduce the estimate

$$\| x(t^*; \tau_0^*, x_0^*; f_\theta) - x(t^*; \tau_0^*, x_0^*; f^*) \| \leqslant \eta(t^* - \tau_0^*) \exp(L(t^* - \tau_0^*))$$

$$= \tfrac{1}{2} M^* \| x_0^* \|. \quad (1.7.6)$$

Finally, from (1.7.5) and (1.7.6), we obtain

This last inequality contradicts the assumption. Thus Theorem 1.7.5 is proved. ■

1.8   An Application of Lyapunov's Direct Method in Conjunction with the Method of Limiting Equations

1.8.1   *Invariant properties and the principle of invariance*

We shall continue the investigation of the nonautonomous equation (1.1.1) and its associated limiting system (1.1.7). Here the convergence defined in Section 1.1 is assumed.

Let Ω be a limiting set of functions $x = x(t)$ defined as the set of limits $\lim x(\tau_k)$ for sequences $\tau_k \to \infty$. We shall denote it by $\Omega(x)$.

Theorem 1.8.1   Let $x(t)$ be a solution of equation (1.1.1). Suppose that $\Omega(f)$ is positive precompact.

Then for any $y_0 \in \Omega(x)$ there exists a limiting equation

$$\frac{dx}{dt} = g(t, x)$$

associated with the system (1.1.1), and there exists a solution $y(t)$ of the problem

$$\frac{dx}{dt} = g(t, x), \quad x(0) = y_0,$$

that remains in the domain $\Omega(x)$ for all t for which it is defined.

In the case of Bohr topology, this is expressed by $(y, g) \in \Delta^*(x, f)$, see p. 25.

Proof   Choosing a sequence $x(\tau_k) \to y_0$, we can easily show that there exists a convergent subsequence f_{τ_k} with limit g. Further, applying Lemma 1.2.2, we get the assertion of Theorem 1.8.1.

Remark 1.8.1.   This property of solutions is called the *locally semi-quasi-invariant property*. The qualifiers "semi-" and "quasi-" comply with the fact that not every limiting equation is admissible, since there may be other solutions through the point y_0 for $t = 0$ that do not remain in the set $\Omega(x)$. The prefix "quasi-" can be omitted if the solution of the initial-value problem for the limiting equation is unique, and, "semi-" can be omitted if the limiting equation is unique (for example, together with positive precompactness, this means that system (1.1.1) is asymptotically autonomous).

One more property of "semi-quasi-invariance" is contained in the following result (Artstein [3]).

Theorem 1.8.2   Let $x(t)$ be a solution of equation (1.1.1). Suppose that the set $\Omega(x)$ is nonempty and compact. Then for every limiting equation $dx/dt = g(t, x)$ associated with (1.1.1) there exists a vector $y_0 \in \Omega(x)$ such that the solution $y = y(t)$ of the problem

$$\frac{dx}{dt} = g(t, x), \quad x(0) = y_0,$$

exists for $y(t) \in \Omega(x)$ for all $t \in R$.

Proof   This is similar to that of Theorem 1.8.1.

Corollary to Theorem 1.8.2   If the solution $x(t)$ of equation (1.1.1) converges to some point $y_0 \in R^n$ as $t \to \infty$ then the point is an equilibrium point for any limiting equation associated with (1.1.1).

1.8.2   *Localization of compact sets*

The property of invariance of an ω-limiting set in conjunction with Lyapunov's direct method allows investigation of the stability of solution properties and estimation of the corresponding domains. We shall present some results obtained in this area.

Definition 1.8.1   A set $Q \subset R^n$ is *locally semi-quasi-invariant relative to* $\Omega(f)$ if for any $y_0 \in Q$ there exists a $g(t, x) \in \Omega(f)$, and the maximally defined solution $y = y(s)$ through the point y_0 for $t = 0$ of the problem

$$\frac{dx}{dt} = g(t, x), \quad x(0) = y_0,$$

is such that $y(t) \in Q$ for any t for which the solution is defined.

Theorem 1.8.3   Suppose that the solution $x(t)$ of the system (1.1.1) is bounded and, moreover, convergent to a set $E \subset R^n$ as $t \to \infty$. If $\Omega(f)$ is positive precompact then $x(t)$ converges to the largest set M in E that is semi-quasi-invariant relative to $\Omega(f)$.

Proof   This follows immediately from Definition 1.8.1 and Theorem 1.8.2.   ∎

The qualitative properties of solutions of the system (1.1.1) will be connected with au63xiliary functions called Lyapunov functions.

Definition 1.8.2   A function $V: R_+ \times R^n \to R$ is referred to as *positive-definite* if there exists a connected time-invariant neighborhood N of the point $x = 0$ such that

(a) V is continuous in $(t, x) \in R_+ \times N$, $V \in C(R_+ \times N, R^n)$;
(b) V is nonnegative on N: $V(t, x) \geqslant 0 \, \forall (t, x) \in R_+ \times N$;
(c) V vanishes for $x = 0$;
(d) there exists a positive-definite function w on $N, w: N \to R_+$, satisfying the inequality

$$w(x) \leqslant V(t, x) \quad \forall (t, x) \in R_+ \times N.$$

We shall define the *right-hand upper Dini derivative* of the function $V(t, x)$ along the solutions of the system (1.2.1) as

$$D^+ V(t, x) = \lim \sup \{ [V(t + \theta, x(t + \theta)) - V(t, x)] \theta^{-1} : \theta \to 0+ \},$$

where the (lim)sup also ranges over the solution $x(t + \theta)$ of (1.2.1) satisfies $x(t) = x$.

Definition 1.8.3   The function $V: R_+ \times R^n \to R$ is called a *Lyapunov function* if

(a) it is positive-definite on N;
(b) the function $D^+ V(t, x)$ is nonpositive on N and $D^+ V(t, 0) = 0 \, \forall t \in R_+$.

Theorem 1.8.4   Let the system (1.1.1) be such that

(1) the conditions of Assumption 1.1.1 are satisfied;
(2) there exists a Lyapunov function for (1.1.1) satisfying for $(t, x) \in R_+ \times N$

$$D^+ V(t, x) \leqslant w(x) \leqslant 0$$

where w is a continuous function.
    Then any bounded solution remaining in N of (1.1.1) converges to the set $E = \{x: w(x) = 0\}$ as $t \to +\infty$.

Proof This theorem is a generalization of the result of Yoshizawa [2] due to LaSalle [1]. We shall only note here that LaSalle assumed that the function $f(t, x)$ is bounded in t for bounded x, but his proof can also be extended for this result as well. ∎

We shall refer one more theorem on localization, which is due to Haddock [1].

Theorem 1.8.5 Suppose that there exists a Lyapunov function for (1.1.1) on $R_+ \times R^n$. Let $H \subseteq R^n$ be a closed set. Assume that for any $\varepsilon > 0$ and every compact $K \subseteq R^n$ there exist a $\delta = \delta(\varepsilon, K) > 0$ and a $\pi = \pi(\varepsilon, K) \geqslant 0$ such that

$$D^+ V(t, x) \leqslant -\delta \| f(t, x) \| + e(t) \qquad (1.8.1)$$

for $t \geqslant \pi$ and $x \in K \cap S^c(H, \varepsilon)$, where $\int_0^\infty |e(s)| \, ds < \infty$ and $S^c(H, \varepsilon)$ is the complement of the ε-neighbourhood of the set H. Then the solution $x(t)$ of the system (1.1.1) satisfies one of the two properties:

(a) $x(t) \to p$ as $t \to +\infty$, where $p \bar{\in} H$;
(b) $x(t) \to H_\infty = H \cup \{\infty\}$ as $t \to \bar{\omega}, t_0 < \omega \leqslant \infty$.

Example 1.8.1 (Haddock [2])   Consider the system

$$\frac{dx}{dt} = q(t)y,$$

$$\frac{dy}{dt} = -q(t)x - p(t)y, \qquad (1.8.2)$$

where for $t \geqslant 0$ the functions p and q are continuous, $p(t) \geqslant 0$ and $|q(t)| \leqslant \alpha p(t)$ for some $\alpha > 0$. The Lyapunov function is taken in the form $2V = x^2 + y^2$. Then $D^+ V = -p(t)y^2$. Let H be a set of points on the x axis. For every $\varepsilon > 0$ and each compact set K in R^2 we have $D^+ V(t, x) \leqslant -p(t)\varepsilon^2$ and $\| f(t, x, y) \| \leqslant p(t)[(\alpha + 1)|y| + \alpha|x|] \leqslant p(t)\beta$ for all $t \geqslant 0$ and all $(x, y) \in K \cap S^c(H, \varepsilon)$. Here $\beta > 0$ depends on K. It follows from this that $D^+ V(t, x) \leqslant -\delta \| f(t, x, y) \|$ for $t \geqslant 0$ and $(x, y) \in K \cap S^c(H, \varepsilon)$, where $\delta = \varepsilon^2/\beta$. Therefore the condition (1.8.1) of Theorem 1.8.5 is satisfied, and the solution of the system (1.8.2) has one of the properties (a) or (b). In particular, if $\int_0^\infty p(t) \, dt = \infty$ then it can be shown that property (a) does not hold, but property (b) does.

1.8.3   Uniform asymptotic stability

We shall start with two theorems involving the properties of the set $E = \{x : D^+ V = 0\}$.

Theorem 1.8.6   Suppose that $h(t, x)$ in the system

$$\frac{dx}{dt} = g(x) + h(t, x) \qquad (1.8.3)$$

tends to zero as $t \to \infty$. Let $V(x)$ be a Lyapunov function for the system

$$\frac{dx}{dt} = g(x). \qquad (1.8.4)$$

If the set E is compact and asymptotically stable relative to the system (1.8.4) then any bounded solution $x(t)$ of the system (1.8.3) converges to E.

Proof The ω-limiting set $\Omega(x)$ is for the system (1.8.4) semi-invariant. Let $y_0 \in \Omega(x)$ and let $y(t)$ be a solution of (1.8.4) that remains in $\Omega(x)$. In addition, the inclusions $\Omega(y) \subset \Omega(x)$ and $\Omega(y) \subset E$ hold. Therefore $x(t)$ approaches "closely" to E, an infinite number of times. Now it can easily be shown that if the solution $x(t)$ is frequently found in the neighborhood of E, it is "trapped" in this neighborhood. This follows from the uniform stability of the set E and the fact that the systems (1.8.3) and (1.8.4) are "close" for sufficiently large t, since $h(t, x) \to 0$ as $t \to +\infty$. ∎

Remark 1.8.2 Theorem 1.8.7 can be proved by a perturbation method. In our presentation only the fact that the right-hand sides of (1.8.3) and (1.8.4) are close for large t is used. This approach can be extended to systems that are not asymptotically autonomous and to which perturbation theory is not applicable.

Theorem 1.8.7 Let

(1) the vector function f in the system (1.1.1) be given in the form $f(t, x) = p(t, x) + h(t, x)$, where $p(t, x)$ is a periodic function in t with period T and $h(t, x) \to 0$ as $t \to \infty$;

$$\frac{dx}{dt} = p(t, x) \qquad (1.8.5)$$

(2) there exist a periodic function $V(t, x)$ for the system (1.8.5) such that

$$D^+ V(t, x) \leqslant w(x) \leqslant 0,$$

where $w(x)$ is a continuous function;

(3) the set $E = \{x : w(x) = 0\}$ be compact and asymptotically stable relative to equation (1.8.5).

Then any bounded solution of the system (1.1.1) converges to the set E.

Proof This is mostly the same as that of Theorem 1.8.6. The difference is that the closeness of the translations f_τ and p for large τ is used. ∎

In terms of the result of Morgan and Narenda [1], Artstein [2] has obtained sufficient conditions for uniform asymptotic stability of the state $x = 0$ of the system (1.1.1).

Theorem 1.8.8 Let $\Omega(f)$ be compact in the sense of Artstein [6] and

(1) all conditions of Assumption 1.1.1 be satisfied;
(2) there exist a continuous function $V(x)$ and a time-invariant neighbourhood N of the state $x = 0$ such that

$$V(0) = 0, \qquad V(x) > 0 \quad \forall x \in N \setminus \{0\}$$

(3) for each $x \in N$ there exists a convex compact set $K(x) \subset R^n$ such that
   (i) $f(t, x) \in K(x)$ $\forall t \geqslant t_0 \geqslant 0$;
   (ii) the inclusion $y \in K(x)$ implies either $(\partial/\partial x)V(x)^T y < 0$ or $y = 0$;
(4) for any $\delta > 0$ there exist numbers a and $b > 0$ such that

$$\int_{t_0}^t \| f(s, x) \| \, ds \geqslant a(t - t_0) + b$$

for all $x \in N$ when $\| x \| \geqslant \delta$.

Then the equilibrium state $x = 0$ of the system (1.1.1) is uniformly asymptotically stable.

Proof   Conditions (2) and (3) imply that $V(x)$ is a Lyapunov function for the system (1.1.1), because $(\partial/\partial x)V(x)^T f(t, x) \leqslant 0$ in the neighbourhood of N, and therefore the state $x = 0$ is uniformly stable. To complete the proof, we must show that N is a domain of attraction of the state $x = 0$ for any limiting equation associated with the system (1.1.1) and apply Theorem 1.5.9. For a $g \in \Omega(f)$ consider the limiting equation associated with the system (1.1.1)

$$\frac{dx}{dt} = g(t, x), \tag{1.8.6}$$

and, moreover, $g(t, x) \in K(x)$ for any $t \geqslant t_0$. In fact, for s from a bounded interval, $g(s, x)$ is a weak L_1 limit of a sequence of functions with values in $K(x)$, and therefore $g(t, x)$ is a limit in the L_1 norm of a sequence of convex combinations of elements from the initial sequence. Since the set $K(x)$ is convex $g(t, x) \in K(x)$. If $\varphi(t)$ is a solution of equation (1.8.6) near the state $x = 0$ then $V(\varphi(t))$ does not increase, owing to the condition $(\partial/\partial x)V(x)^T g(\varphi, t) \leqslant 0$. Therefore the function $V(x)$ is constant on the ω-limiting set $\Omega(\varphi)$. According to the principle of invariance for (1.8.6) in the L_1 norm, there exists a limiting equation

$$\frac{dx}{dt} = h(t, x) \tag{1.8.7}$$

whose solution $\psi(t)$ takes values in $\Omega(\varphi)$. Since the growth of $(\partial/\partial x)V(x)^T h(\psi, t)$ on the set N is measured by the degree of variation of the function $V(\psi(t))$, which is equal to 0, and since $h(t, x) \in K(x)$, we have $h(t(t), \psi) = 0$ for almost all t. This implies the invariability of the function $\psi(t)$, i.e. $\psi(t) = c$. Let $c \neq 0$. Since (1.8.7) is a limiting equation for (1.1.1) for any $r > 0$ the norm

$$\left\| \int_{t_0}^{t_0+r} f(s, c) \, ds \right\| \tag{1.8.8}$$

can be made arbitrarily small by an the appropriate choice of t_0. Hence we find that if the norm (1.8.8) is small then so is the expression

$$\int_{t_0}^{t_0+r} \| f(s, c) \| \, ds.$$

This contradicts condition (4). Thus the proof of Theorem 1.8.8 is complete.   ■

1.9   Comments and References

1.0   Together with the other results Theorem 1.0.1 and its proof can be found in Yoshizawa [2].

1.1   The Bebutov–Miller–Sell concept of constructing limiting equations is presented according to Bebutov [1, 2], Artstein [1–6] and Miller and Sell [1, 2].

1.2   The global properties of solutions to limiting equations are given according to Kato and Yoshizawa [3]. See also Artstein [2] and Marcus [1].

1.3   Proposition 1.3.1 is known as Sell's hypothesis (Sell [2, 3]). Theorem 1.3.1 was established by Bondi, Moaro and Visentin [1].

1.4   Eventual properties of solutions of nonautonomous system are from D'Anna, Maio and Monte [1]. See also D'Anna [1], D'Anna, Maio and Moauro [1], D'Anna and Monte [1]

1.5   The basic results of this section are due to Kato and Yoshizawa [3].

1.6   The stability conditions for solutions under PP in this section correspond to the results of Visentin [2], Duboshin [1], Kato and Yoshizawa [1, 2], D'Anna [2] and Bondi, Moauro and Visentin [1]. Other results on stability under PP see in Strauss and Yorke [1–2], Zubov [3], Hahn [1], Rouche and Mawhin [1].

1.7   The results is this section are according to Dishliev and Bainov [1] on the basis of ideas and results of Dannan and Elaydi [1].

1.8   The direct Lyapunov method (Lyapunov [1], Chetaev [1], Zubov [1–3]) is effectively applied in the qualitative theory of SODEs in terms of limiting equations. The results covered in this section are taken from LaSalle [4], Artstein [2], Shestakov [6], Haddock [2] and Saperstone [1].

2 Limiting Equations and Stability of Infinite Delay Systems

2.0 Introduction

It is well known that many processes allow natural modelling by equations with deviating argument. Numerous examples of such processes can be found, for example, in Krasovski [1], Rezvan [1] and Hale [2]. Systems with infinite delay (after-effect) (SID) have recently been investigated. In contrast to finite delay systems, spaces with particular properties must be considered for systems with infinite delay. Coleman–Meizel [1], Kappel and Schappacher [1], Schumacher [1], Hale and Kato [1] etc. introduced an axiomatic definition of phase space for SID, and at the same time initiated a unified approach to the analysis of a such systems. Lyapunov's direct method based on Lyapunov–Razumikhin functions has also been used in the stability analysis of SID. In this case the corresponding definitions are modified to take account of the concept of stability developed for R^n in a B space defined by specific hypotheses. These definitions are given below.

2.1 Phase Space of a System with Infinite Delay

2.1.1 B space

Let R^n be an n-dimensional Euclidean space and $\|x\|$ a Euclidean norm in it. For a function $x(t)$ defined on $]-\infty, a[$ and for each $t \in]-\infty, a[$ the function

$$x_t(s) = x(t + s), \quad s \leqslant 0$$

is defined. Let B be a real linear vector space of functions mapping $]-\infty, 0]$ into R^n with a seminorm $|\cdot|_B$.

Assumption 2.1.1 The space B satisfies the following conditions.

(1) If $x(t)$ is defined on $]-\infty, \tau + \alpha[$ and continuous on $[\tau, \tau + \alpha[, \alpha > 0$, and if $x_\tau \in B$ then for every $t \in [\tau, \tau + \alpha[$

(i) $x_t \in B$ and x_t is continuous in t with respect to $|\cdot|_B$;
(ii) there exist a constant $K > 0$ and a continuous non-negative function $M(\beta)$; $M(\beta) \to 0$ for $\beta \to \infty$, such that

$$|x_t|_B \leqslant K \sup_{\tau \leqslant s \leqslant t} \|x(s)\| + M(t - \tau)|x_\tau|_B;$$

(iii) there exists a constant $N > 0$ such that $\|x(t)\| \leqslant N|x_t|_B$.
(2) The quotient space $B/|\cdot|_B$ is a separable Banach space, i.e. it has a countable base.

2.1.2   Examples of B spaces

The following are typical examples of B spaces:
(a) the space $C_\gamma, \gamma > 0$, of continuous functions φ such that $e^{\gamma s}\varphi(s)$ is bounded and uniformly continuous on $]-\infty, 0]$ with the norm

$$|\varphi|_{C_\gamma} = \sup\{e^{\gamma s}\|\varphi(s)\|, s \leqslant 0\};$$

here the constant K from condition (1) (ii) of Assumption 2.1.1 is equal to 1;
(b) the space $C([-h, 0])$, $h \geqslant 0$, of functions φ continuous on $[-h, 0]$ with seminorm

$$|\varphi|_{C([-h,0])} = \sup_{-h \leqslant s \leqslant 0} \|\varphi(s)\|;$$

(c) the space M_γ of measurable functions φ with finite norm

$$|\varphi|_{M_\gamma} = \|\varphi(0)\| + \int_{-\infty}^{0} e^{\gamma s}\|\varphi(s)\| ds, \quad \gamma > 0.$$

Fundamental theorems on existence and uniqueness can be proved.

2.2   Definitions of Stability in B Space

We consider the system of functional differential equations

$$\frac{dx}{dt} = F(t, x_t), \tag{2.2.1}$$

where dx/dt is a right-hand derivative with a function $F(t, \varphi)$ defined on $R \times B$. Let $x(t, F)$ be a solution of (2.2.1). In particular, $x(t, s, \varphi, F)$ denotes a solution of (2.2.1) passing through the point (s, φ). Suppose that (2.2.1) has a bounded solution $u(t)$ defined on I, and assume

$$L = \sup\{|F(t, \varphi)|: t \geqslant 0 \text{ and } |\varphi|_B \leqslant 2H\} < \infty,$$

where H is a positive constant such that $|u_t|_B \leqslant H$ for $t > 0$.

Let $u(t): R_+ \to R_+$ be the solution of (2.2.1) whose stability is under investigation. We denote $I = [0, +\infty[$ and $I_s = [s, +\infty[, s \geqslant 0$. In view of results of Hale and Kato [1] and Hino [2] and similarly to the concept of stability given by Grujič, Martynyuk and Ribbens-Pavella [1], we make the following definitions.

Definition 2.2.1   A solution $u(t)$ of (2.2.1) is called

(a) *stable in B (SB)* iff for any $s \in I$ and any $\varepsilon \in]0, + \infty[$ there exists a $\delta(s, \varepsilon) > 0$ such that $|x_t(s, \varphi, F) - u_t|_B < \varepsilon$ for all $t \in I_s$ as long as $|\varphi - u_s|_B < \delta(s, \varepsilon)$;

(b) *uniformly stable in B (USB)* if it is stable in B and for any $\varepsilon \in]0, + \infty[$ the number δ in (a) can be chosen such that

$$\inf\{\delta(s, \varepsilon) : s \in I\} > 0;$$

(c) *globally stable in B (GSB)* iff (a) is satisfied and $\delta(s, \varepsilon) \to + \infty$ as $\varepsilon \to + \infty$ for any $s \in I$;

(d) *uniformly globally stable in B (UGSB)* iff

$$\inf\{\delta(s, \varepsilon) : s \in I\} \to \infty \quad \text{as } \varepsilon \to \infty;$$

(e) *unstable in B* iff there exist $s \in I$ and $\varepsilon > 0$ such that for any $\delta \in]0, + \infty[$ there exists $\varphi : |\varphi - u_s|_B < \delta$ such that $|x_t(s, \varphi, F) - u_t|_B \geqslant \varepsilon$ for a $t > s$.

Definition 2.2.2 A solution $u(t)$ of (2.2.1) is called

(a) *attractive in B (AB)* iff there exist $\Delta(s) > 0$ for any $s \in I$ and $\tau(s; \varphi, \rho) \in]0, + \infty[$ for any $\rho > 0$ and any φ, $|\varphi - u_s| < \Delta(s)$, such that the inequality $|x_t(s, \varphi, F) - u_t|_B < \rho$ holds for any $t \in [s + \tau(s, \varphi, \rho), + \infty[$;

(b) *φ-uniformly attractive in B (φ-UAB)* iff in (a) there exists $\tau_u[s, \Delta(s), \rho] \in [0, + \infty[$ for any $\rho \in]0, + \infty[$ such that $\sup[\tau(s, \varphi, \rho) : \varphi \in Q_{\Delta(s)}] = \tau_u[s, \Delta(s), \rho]$, where $Q_\Delta = \{\varphi : |\varphi|_B < \Delta\}$;

(c) *s-uniformly attractive in B (s-UAB)* if and only if (a) is satisfied and $\Delta > 0$ exists, and for any $(\varphi, \rho) \in Q_\Delta \times]0, + \infty[$ there exists a $\tau_u(I, \varphi, \rho) \in [0, + \infty[$ such that

$$\sup[\tau(s; \varphi, \rho) : s \in I] = \tau_u(I, \varphi, \rho);$$

(d) *uniformly attractive in B (UAB)* iff (a) holds and $\Delta > 0$ exists, and for any $\rho \in]0, + \infty[$ there exists a $\tau_u(I, \Delta, \rho) \in [0, + \infty[$ such that

$$\sup[\tau(s, \varphi, \rho) : (s, \varphi) \in I \times Q_\Delta] = \tau_u(I, \Delta, \rho);$$

(e) *globally (a), (b), (c) or (d) in B* iff the respective condition holds for any $\Delta(s) \in]0, + \infty[$ and any $s \in I$.

Definition 2.2.3 A solution $u(t)$ of (2.2.1) is called

(a) *asymptotically stable in B (ASB)* iff it is SB and AB;

(b) *equiasymptotically stable in B(EASB)* iff it is SB and φ-UAB;

(c) *quasiuniformly asymptotically stable in B (QUASB)* iff it is USB and s-UAB;

(d) *uniformly asymptotically stable in B (UASB)* iff it is USB and UAB;

(e) *globally (a), (b), (c) or (d) in B* iff the corresponding condition holds globally;

(f) *exponentially stable in B (ESB)* iff there exist $\Delta > 0$ and real numbers $\alpha \geqslant 1$ and $\beta > 0$ such that $|\varphi - u_s|_B < \Delta$ implies

$$|x_t(s, \varphi, F) - u_t|_B \leqslant \alpha |\varphi - u_s|_B \exp[-\beta(t - s)] \quad \forall t \in I_s;$$

(g) *globally exponentially stable in B (EGSB)* iff (f) holds for $\Delta = + \infty$.

2.3   Limiting Systems

Along with (2.2.1), we shall consider a system constructed by a special algorithm.

Let $F(t, \varphi)$ be an R^n-valued continuous function defined on $[0, \infty[\times B$, where B is a Banach space, and let $T(F)$ be the set of continuous functions $F(t + s, \varphi)$ for all $s \geqslant 0$, which is a subset of $C([0, \infty[\times B, R^n)$. Here $C(\Omega, R^n)$ denotes the set of all continuous R^n-valued functions defined on Ω, and is a topological space with the compact open topology. Let $H(F)$ be the closure of $T(F)$. As will be shown below, if $F(t, \varphi)$, defined on $]-\infty, +\infty[\times B$, is almost-periodic in t uniformly for $\varphi \in B$ and if B is separable then $H(F)$ consists of the restriction to $[0, \infty[$ of the elements of the hull in the usual sense. For this reason we may call $H(F)$ the *hull* of F.

Definition 2.3.1   A function $F(t, \varphi)$ defined on $R \times B$ is said to have a *compact hull* if for any compact set $W \subset R \times B$ and any sequence $\{t'_n\}$, $t'_n \geqslant 0$, there exists a subsequence $\{t_n\}$ such that the sequence $\{F(t + t_n, \varphi)\}$ converges uniformly on W.

Definition 2.3.2   The *hull* $H(F)$ (or $H^+(F)$ respectively) is a set of pairs (G, Ω), $\Omega \subset R \times B$, such that there exists a sequence $\{t_n\}$, $t_n \geqslant 0$ (or $t_n \to \infty$ when $n \to \infty$ respectively) for which $\{F(t + t_n, \varphi)\}$ converges to $G(t, \varphi)$ uniformly on Ω.

The following assertion is well known when B is a Euclidean space. However, it is not trivial in the general case, where we should note that the translation numbers can be determined only on each compact subset of B and that the argument used in the case where B is a Euclidean space in no longer valid, since there may not exist compact subsets K_m satisfying $\bigcup_m K_m = B$.

Lemma 2.3.1   Let $F(t, \varphi)$ be almost-periodic in t uniformly for $\varphi \in B$, where B is a separable Banach space.

Then for a given real sequence $\{h_m\}$ there exist a subsequence $\{\tau_m\}$ of $\{h_m\}$ and a function $G(t, \varphi)$ such that the sequence $\{F(t + \tau_m, \varphi)\}$ converges to $G(t, \varphi)$ uniformly on $]-\infty, \infty[\times S$ for any compact subset S of B, and, furthermore, for this G there is a sequence $\{\sigma_m\}$, $\sigma_m \to \infty$ as $m \to \infty$, such that $\{F(t + \sigma_m, \varphi)\}$ converges to $G(t, \varphi)$ uniformly on $]-\infty, \infty[\times S$ for any compact subset S of B.

Proof   Since B is separable, there exists a countable set of points $K_\infty = \{\varphi^1, \varphi^2, \ldots\}$ that is dense in B. Clearly the s-point set $K_s = \{\varphi^1, \ldots, \varphi^s\}$ is compact for any s. Hence, from the almost-periodicity, we can find a subsequence $\{\tau_m\}$ of $\{h_m\}$ such that

$$\{F(t + \tau_m, \varphi)\} \text{ converges uniformly on }]-\infty, \infty[\times K_s \text{ for any } s. \qquad (2.3.1)$$

First, we shall show that the sequence $\{F(t + \tau_m, \varphi)\}$ converges uniformly on $]-\infty, \infty[\times S$ for any compact subset S of B.

Let S be an arbitrary compact subset of B, and let $\{U^1_m, \ldots, U^m_{P_m}\}$ be a $(1/m)$-net of S. Since the set K_∞ is dense in B, there is an integer $k(m, l)$ such that $\varphi^{k(m,l)} \in U^m_l$ for any m, l. Let S^* be the union of S and the set $\{\varphi^{k(m,l)}; l = 1, \ldots, p_m, m = 1, 2, \ldots\}$. Then we can prove that S^* is compact, because S^* has no accumulation point outside the compact set S. As is well known, $F(t, \varphi)$ is uniformly continuous on $]-\infty, \infty[\times S^*$, and therefore for any $\varepsilon > 0$ there exists a number $\delta(\varepsilon) > 0$ such that if $\|\varphi - \psi\| < \delta(\varepsilon)$ then

$$|F(t, \varphi) - F(t, \psi)| < \varepsilon \qquad (2.3.2)$$

for all $(t, \varphi), (t, \psi) \in \,]-\infty, \infty[\times S^*$. Put

$$N(\varepsilon) = \max\{k(m(\varepsilon), l); \, l = 1, 2, \ldots, p_{m(\varepsilon)}\},$$

where $m(\varepsilon)$ is an integer greater than $1/\delta(\frac{1}{3}\varepsilon)$. By (2.3.1), there is an integer $N_0(\varepsilon)$ such that if $\mu, \nu \geq N_0(\varepsilon)$ then

$$|F(t + \tau_\mu, \varphi) - F(t + \tau_\nu, \varphi)| < \tfrac{1}{3}\varepsilon \quad \forall (t, \varphi) \in \,]-\infty, \infty[\times K_{N(\varepsilon)}.$$

For a given $(t, \varphi) \in \,]-\infty, \infty[\times S$ we can find a $\varphi^k \in S^* \cap K_{N(\varepsilon)}$ such that $\|\varphi - \varphi^k\| < \delta(\frac{1}{3}\varepsilon)$. Hence we have

$$|F(t + \tau_\mu, \varphi) - F(t + \tau_\mu, \varphi^k)| < \tfrac{1}{3}\varepsilon \quad \forall \mu.$$

Therefore we can see that if $\mu, \nu \geq N_0(\varepsilon)$ then

$$\begin{aligned}
|F(t + \tau_\mu, \varphi) - F(t + \tau_\nu, \varphi)| &\leq |F(t + \tau_\mu, \varphi) - F(t + \tau_\mu, \varphi^k)| \\
&\quad + |F(t + \tau_\mu, \varphi^k) - F(t + \tau_\nu, \varphi^k)| \\
&\quad + |F(t + \tau_\nu, \varphi^k) - F(t + \tau_\nu, \varphi)| < \varepsilon.
\end{aligned}$$

This shows that the sequence $\{F(t + \tau_m, \varphi)\}$ converges to a unique limiting function $G(t, \varphi)$ uniformly on $]-\infty, \infty[\times S$ for any compact subset S of B. Clearly $G(t, \varphi)$ is continuous on $]-\infty, \infty[\times B$ and belongs to the hull in the usual sense. By the almost-periodicity, there is a number t_m^s with the properties $t_m^s \geq m$ and

$$|F(t + t_m^s, \varphi) - F(t + \tau_m, \varphi)| \leq 1/m \quad \forall (t, \varphi) \in \,]-\infty, \infty[\times K_s,$$

where $t_m^s - \tau_m$ is a $(1/m)$-translation number on K_s. Setting $\sigma_m = t_m^m$, we shall show that $\{F(t + \sigma_m, \varphi)\}$ converges uniformly on $]-\infty, \infty[\times K_s$ for any $s \geq 0$. Then, by the same argument as for $\{F(t + \tau_m, \varphi)\}$, it will be proved that $\{F(t + \sigma_m, \varphi)\}$ converges uniformly on $]-\infty, \infty[\times S$ for any compact subset S of B. We should note that

$$|F(t + \sigma_m, \varphi) - F(t + \tau_m, \varphi)| = |F(t + t_m^m, \varphi) - F(t + \tau_m, \varphi)| \leq 1/m \qquad (2.3.3)$$

on $]-\infty, \infty[\times K_s$ for any $s \leq m$, since $K_s \subset K_m$. On the other hand, it follows from (2.3.1) that there is a number $N_s(\varepsilon)$ such that if $\mu, \nu \geq N_s(\varepsilon)$ then

$$|F(t + \tau_\mu, \varphi) - F(t + \tau_\nu, \varphi)| < \varepsilon \quad \forall (t, \varphi) \in \,]-\infty, \infty[\times K_s.$$

Hence, if $\mu, \nu \geq \max\{N_s(\frac{1}{3}\varepsilon), \frac{1}{3}\varepsilon, s\}$ then we have

$$\begin{aligned}
|F(t + \sigma_\mu, \varphi) - F(t + \sigma_\nu, \varphi)| &\leq |F(t + \sigma_\mu, \varphi) - F(t + \tau_\mu, \varphi)| \\
&\quad + |F(t + \tau_\mu, \varphi) - F(t + \tau_\nu, \varphi)| \\
&\quad + |F(t + \tau_\nu, \varphi) - F(t + \sigma_\nu, \varphi)| < \frac{1}{\mu} + \tfrac{1}{3}\varepsilon + \frac{1}{\nu} \leq \varepsilon
\end{aligned}$$

for all $(t, \varphi) \in \,]-\infty, \infty[\times K_s$; that is, $\{F(t + \sigma_m, \varphi)\}$ converges uniformly on $]-\infty, \infty[\times K_s$ for any s.

Finally, it follows from (2.3.3) that $\{F(t + \sigma_m, \varphi)\}$ converges to the limiting function $G(t, \varphi)$ of the sequence $\{F(t + \tau_m, \varphi)\}$ on $]-\infty, \infty[\times K_s$ for any s. Hence $\{F(t + \sigma_m, \varphi)\}$ converges to $G(t, \varphi)$ on $]-\infty, \infty[\times B$, because $K_\infty = \bigcup K_s$ is dense in B and $G(t, \varphi)$ is continuous on $]-\infty, \infty[\times B$. ∎

This lemma also shows that if $F(t, \varphi)$ is almost-periodic in t uniformly for $\varphi \in B$ then the hull is a compact subset of $C(]-\infty, \infty[\times B, R^n)$. Actually it is compact under the stronger topology, that is, in the sense of uniform convergence on $]-\infty, \infty[\times S$ for any compact subset S of B. In the compact open topology we can prove the following.

Lemma 2.3.2   Let $F(t, \varphi)$ be a continuous R^n-valued function on $[0, \infty[\times B$, where B is a separable Banach space. Then the hull $H(F)$ is compact iff $F(t, \varphi)$ is uniformly continuous in (t, φ) and bounded on $[0, \infty[\times S$ for any compact subset S of B.

Proof   The proof of necessity is not difficult. We now prove sufficiency. Let $\{h_m\}$, $h_m \geq 0$, be a given sequence. Then, by the assumption, the sequence $\{F(t + h_m, \varphi)\}$ is a normal family on $[0, \infty[\times K_s$ for any s, where K_s is the set given in the proof of Lemma 2.3.1. Hence there is a subsequence $\{\tau_m\}$ of $\{h_m\}$ such that the sequence $\{F(t + \tau_m, \varphi)\}$ converges uniformly on $[a, b] \times K_s$ for any s and for any compact interval $[a, b] \subset [0, \infty[$. This assertion corresponds to (2.3.1). Thus, by using the same arguments as in the proof of Lemma 2.3.1, we can prove that $\{F(t + \tau_m, \varphi)\}$ converges uniformly on $[a, b] \times S$ for any compact set $S \subset B$ and for any compact interval $[a, b] \subset [0, \infty[$. Hence the hull $H(F)$ is compact, since $H(F) = T(F)$. ∎

Let S be a compact subset in a phase space B, and let $\alpha > 0$ and $\beta > 0$ be some constants. We set

$$\bigcup(S, \alpha. \beta) \triangleq \{x_t : t \geq 0, x_0 \in S, \|x_s\| \leq \alpha \text{ for } s \in [0, \infty[\text{ and }$$
$$\|x(s_1) - x(s_2)\| \leq \beta|s_1 - s_2| \quad \text{for } 0 \leq s_1, s_2 < \infty\}.$$

Lemma 2.3.3   The closure $\bar{U}(S, \alpha, \beta)$ of the set $U(S, \alpha, \beta)$ is compact in B.

Proof   See Hale and Kato [1, Theorem 3.1 and Lemma 2.1].

Remark 2.3.1   We note that if $F(t, \varphi)$ has a compact hull and if $\Gamma_k \subset B$ are compact sets for $k = 1, 2, \ldots$ then for any $(G^*, \Omega_0) \in H(F)$ there exists a function G such that $(G, \Omega) \in H(F)$ for some $\Omega \supset \Omega_0 \cup \{R \times \bigcup_k \Gamma_k\}$ and $G^*(t, \varphi) = G(t, \varphi)$ on Ω_0.

Consider the system of functional differential equations (2.2.1) and the bounded solution $u(t)$. Then, by noting that $\{u_t | t \geq 0\}^-$ is contained in the compact set $U(\{u_0\}, L)^-$ (see Lemma 2.3.3), we may assume that for any $\Omega \supset I \times \{u_t | t \geq 0\}^-$ we have $(G, \Omega) \in H(F)$, by Remark 2.3.1. Hence it is easily shown that for any $(v, G, \Omega) \in H(u, F)$ the function $v(t)$ is a solution of

$$\frac{dx}{dt} = G(t, x_t) \tag{2.3.4}$$

defined on R_+.

Here and below $H(u)$ (or $H^+(u)$ respectively) and $H(u, F)$ (or $H^+(u, F)$ respectively) are defined similarly to $H(F)$ (or $H^+(F)$ respectively).

Definition 2.3.3 The system (2.2.1) is called *regular* if for any $(G, \Omega) \in H(F)$ any solution $v(t)$ of (2.3.4) is unique for given initial conditions.

Definition 2.3.4 A solution $u(t)$ of (2.2.1) is referred to as *stable in $H(F)$* (or $H^+(F)$ respectively) $(SH(F))$ $(SH^+(F)$ respectively) if and only if for any $\varepsilon > 0$ and any $s \in I$ there exists a $\delta(s, \varepsilon) > 0$ such that for any $s \in I$, $(v, G, \Omega) \in H(u, F)$ (or $H^+(u, F)$ respectively) and $|\varphi - v_s|_B < \delta$ implies the inequality $|x_t(s, \varphi, G) - v_t|_B < \varepsilon$ for all $t \in I_s$.

In the similar manner to Definition 2.2.1, we can formulate definitions of *uniform stability in $H(F)$* (or $H^+(F)$ respectively) $(USH(F))$, *global stability in $H(F)$* (or $H^+(F)$ respectively) $(GSH(F))$, *uniform global stability in $H(F)$* (or $H^+(F)$ respectively) $(UGSH(F))$ and *instability in $H(F)$* (or $H^+(F)$ respectively) $(NSH(F))$.

Definition 2.3.5 A solution $u(t)$ of (2.2.1) is referred to as *attractive in $H(F)$* (or $H^+(F)$ respectively) $(AH(F))$ iff there exist a constant $\Delta > 0$ and a $\tau(s, \varphi, \rho) \in [0, +\infty[$ for any $\rho > 0$ such that if $(v, G, \Omega) \in H(u, F)$ (or $H^+(u, F)$ respectively) and $|\varphi - v_s|_B < \Delta$ then $(t, x_t(s, \varphi, G)) \in \Omega$ for $t \geqslant s$, and the inequality $|x_t(s, \varphi, G) - v_t|_B < \rho$ holds for all $t \in]s + \tau(s, \varphi, \rho), +\infty[$.

In a similar manner to Definition 2.2.2, we can formulate definitions of *φ-uniform attraction in $H(F)$* (or $H^+(F)$ respectively) $(\varphi\text{-}UAH(F))$, *s-uniform attracion in $H(F)$* (or $H^+(F)$ respectively) $(s\text{-}UAH(F))$, *uniform attraction in $H(F)$* (or $H^+(F)$ respectively) $(UAH(F))$ and *global attraction in $H(F)$* (or $H^+(F)$ respectively) $(GAH(F))$.

Definition 2.3.6 A solution $u(t)$ of (2.3.6) is referred to as *asymptotically stable in $H(F)$* (or $H^+(F)$ respectively) $(ASH(F))$ iff it is $SH(F)$ (or $SH^+(F)$ respectively) and $AH(F)$ (or $AH^+(F)$ respectively).

Similarly to (b)–(g) of Definition 2.2.3, one can formulate definitions of *equiasymptotic stability in $H(F)$* (or $H^+(F)$ respectively) $(EAH(F))$, *quasiuniform asymptotic stability in $H(F)$* (or $H^+(F)$ respectively)$(QUASH(F))$, *uniform asymptotic stability in $H(F)$* (or $H^+(F)$ respectively) $(UASH(F))$, *asymptotic global stability in $H(F)$* (or $H^+(F)$ respectively) $(AGSH(F))$ and *exponential stability in $H(F)$* (or $H^+(F)$ respectively) $(ESH(F))$.

Similarly to well-known propositions for ordinary differential equations (see Chapter 1), the following assertions hold for the system (2.1.1).

Proposition 2.3.1 Let the system (2.1.1) be regular. If a solution $u(t)$ is uniformly stable in B (or uniformly asymptotically stable in B respectively) then it is uniformly stable in $H(F)$ (or uniformly asymptotically stable in $H(F)$ respectively).

Proposition 2.3.2 Let the vector function $F(t, \varphi)$ be periodic in t. Then a solution $u(t)$ is uniformly stable in $H(F)$ (or uniformly asymptotically stable in $H(F)$ respectively) if it is uniformly stable in B (or uniformly asymptotically stable in B respectively).

These propositions show that under appropriate conditions, the stability properties of solutions of (2.2.1) inherent to the limiting systems (2.3.1).

2.4　Relationship Between Stability in R^n and Stability in B Space

We state the following results of Hale and Kato [1].

Definition 2.4.1　If in Definitions 2.2.1–2.2.3 substitution of the B-norm $|x_t(s, \varphi) - u_t|_B$ by the norm $\| x(t, s, \varphi) - u(t) \|$ in R^n is admissible then a solution $u(t)$ is said to possess the corresponding type of stability, attraction or asymptotic stability in R^n.

Theorem 2.4.1　Let a solution $u(t)$ of (2.2.1) be

(1) asymptotically stable in B;
(2) equiasymptotically stable in B;
(3) quasiuniformly asymptotically stable in B;
(4) uniformly asymptotically stable in B.

Then this solution is

(1) asymptotically stable in R^n;
(2) equiasymptotically stable in R^n;
(3) quasiuniformly asymptotically stable in R^n;
(4) uniformly asymptotically stable in R^n, respectively.

Proof　By condition (1) (iii) of Assumption 2.1.1 the estimate

$$\| x(t, s, \varphi) \| \leqslant N |x_t|_B, \quad N = \text{const} > 0$$

holds. This estimate and Definitions 2.2.1–2.2.3 provide the proof immediately.

Theorem 2.4.2　Let the zero solution $u(t) = 0$ of (2.2.1) be

(1) uniformly stable in R^n;
(2) asymptotically stable in R^n.

Then it is

(1) uniformly stable in B;
(2) asymptotically stable in B.

Proof　By virtue of the inequality in condition (1) (ii) of Assumption 2.1.1, we have

$$|x_t|_B \leqslant K \sup_{\sigma \leqslant s \leqslant t} \| x(s) \| + M |x_\sigma|_B, \quad t \in I_\sigma \tag{2.4.1}$$

for a solution $x(t)$ defined on I_σ. This inequality and condition (1) of the theorem imply assertion (1).

Now let $u = 0$ be asymptotically stable in R^n. According to the estimate (2.4.1) one can assume that $|x_\sigma|_B \leqslant \delta_0(\sigma)$ implies $|x_t|_B \leqslant c(\sigma)$, $t \geqslant \sigma$, where $c(\sigma) = K + M \delta_0(\sigma)$

depends on σ only through δ_0. For any $\beta > 0$, by condition (1) (ii) of Assumption 2.1.1, we have

$$|x_t|_B \leqslant K \sup_{t-\beta \leqslant s \leqslant t} \|x(s)\| + M(\beta)|x_{t-\beta}|_B, \quad t \geqslant \beta + \sigma. \tag{2.4.2}$$

Hence

$$|x_t|_B \leqslant K \sup_{t-\beta \leqslant s \leqslant t} \|x(s)\| + M(\beta)c(\sigma), \quad t \geqslant \beta + \sigma.$$

Thus, if we take β_0 for any $\varepsilon > 0$ so that $M(\beta)c(\sigma) < \frac{1}{2}\varepsilon$ for all $\beta \geqslant \beta_0$, by the condition $M(\beta) \to 0$ as $\beta \to 0$, then $\|x(t)\| < \varepsilon/2K$ for $t \geqslant \sigma + T$ implies that $|x_t|_B < \varepsilon$ for $t \geqslant \sigma + \beta_0 + T$. This proves the theorem. ∎

Theorem 2.4.3   Let

(1) the vector function $F(t, \varphi)$ be continuous on $R \times B$ and be bounded by a constant on

$$\Omega_0 = \{(t, \varphi): t \geqslant 0, |\varphi - u_t|_B \leqslant \eta\}, \quad \eta > 0;$$

(2) the system (2.2.1) have a bounded solution $u(t)$ defined on $[0, \infty[$,
(3) the vector function $F(t, \varphi)$ have a compact hull $H(F)$.

Then if $u(t)$ is uniformly stable in B and attracting in $H(F)$, it is uniformly asymptotically stable in B.

Proof   Let δ and Δ be the numbers appearing in the definitions of uniform stability and the attractor in $H(F)$, and let

$$\delta_0 = \delta(\min(\eta, \Delta)). \tag{2.4.3}$$

Since we have uniform stability, in order to prove uniformly asymptotic stability it is sufficient to show the existence of T for given ε such that $|x_\sigma - u_\sigma|_B \leqslant \delta_0$ implies $\inf_{\sigma \leqslant t \leqslant \sigma + T} |x_t - u_t|_B < \delta(\varepsilon)$ for any $\sigma \geqslant 0$ and any solution $x(t)$ of the system (2.2.1). Suppose this is not the case. Then there exist sequences $\{\sigma_k\}$, $\sigma_k \geqslant 0$, $\{x^k(t)\}$, solutions of (2.2.1), such that $|x_{\sigma_k}^k - u_{\sigma_k}|_B \leqslant \delta_0$ but $|x_t^k - u_t|_B \geqslant \delta(\varepsilon)$ for $t \in [\sigma_k, \sigma_k + 2k]$. Since $|x_t^k - u_t|_B \leqslant \eta$ for all $t \geqslant \sigma_k$ by (2.4.3), $|\dot{x}^k(t)| \leqslant L$ for an L and all k, $t \geqslant \sigma_k$ by the assumption on F and $|x_t^k|_B \leqslant \eta + \sup_{t \leqslant 0} |u_t|_B < \infty$, which implies the uniform boundedness of $x^k(t + \sigma_k)$ by

$$|\varphi(0)| \leqslant K|\varphi|_B \quad \text{for any } \varphi \in B \text{ and some } K.$$

Therefore for $t_k = \sigma_k + k$ we can assume that $x_{t_k}^k$ and u_{t_k} converge, and hence

$$\Gamma = \text{cl}\{x_t^k, u_t : t \geqslant t_k, k = 1, 2, \ldots\}$$

is a compact subset of B (see Hale, Kato [1, Corollary 3.2]). Hence we can assume that $F(t + t_k, \varphi)$ converges to a $G(t, \varphi)$ uniformly on $[0, A] \times \Gamma$ for any $A > 0$. By the same argument as in the proof of Theorem 3.2 of Hale and Kato [1], one can assume that $u(t + t_k)$ and $x^k(t + t_k)$ converge to $v(t)$ and $y(t)$ respectively, solutions of (2.2.1), and then clearly

$$|y_0 - v_0|_B \leqslant \delta_0 \quad \text{but} \quad |y_t - v_t|_B \geqslant \delta(\varepsilon) \quad \forall t \geqslant 0,$$

which contradicts the attractor in $H(F)$. This completes the proof of Theorem 2.4.3. ■

Corollary 2.4.1   If $F(t, \varphi)$ is continuous and periodic in t and if the zero solution of (2.2.1) is asymptotically stable in B, then it is uniformly asymptotically stable in B.

Proof   It is possible to find an $\eta > 0$ so that $F(t, \varphi)$ is bounded on $R \times \{\varphi : |\varphi|_B \leqslant \eta\}$, because $[0, p] \times \{0\}$ is compact, where $p > 0$ is a period of $F(t, \varphi)$.

Since the zero solution must be unique for the initial-value problem, we can find a $\rho(\varepsilon) > 0$ for any given $\varepsilon > 0$ such that $|x_\sigma|_B \leqslant \rho(\varepsilon), \sigma \in [0, p[$, implies $|x_p|_B \leqslant \varepsilon$ for any solution $x(t)$ of (2.2.1) (cf. Hale and Kato [1, Theorem 2.5]). It follows from this that the zero solution is uniformly stable in B and attracting in $H(F)$, where the associated numbers $\delta(\varepsilon)$ and Δ can be chosen as

$$\delta(\varepsilon) = \rho(\delta(p, \varepsilon)), \quad \Delta = \rho(\delta_0(p))$$

for $\delta(p, \varepsilon)$ and $\delta_0(p)$ appearing in the definition of the asymptotic stability in B at $\sigma = p$.

Thus the rest of the proof follows immediately from Theorem 2.4.3. ■

Here we should note that $H(F) = \{(F(t + s, \varphi), R \times B); s \in [0, p]\}$, which is compact.

Corollary 2.4.2   If the zero solution of periodic system (2.2.1) is asymptotically stable in R^n, then it is uniformly asymptotically stable in R^n.

Proof   According to the assumption, the zero solution is asymptotically stable in B by Theorem 2.4.2. By Corollary 2.4.1, it is uniformly asymptotically stable in B. By Theorem 2.4.1, it is uniformly asymptotically stable in R^n. Thus the proof is complete. ■

2.5   Stability in R^n

Our first aim is to establish a Razumikhin-type Lyapunov theorem as a sufficient condition for stability in R^n. We assume that $F(t, \varphi)$ in (2.2.1) is defined and continuous in the topology induced by $|\cdot|_B$ on $[0, \infty[\times B_H$, $B_H = \{\varphi \in B : |\varphi|_B < H\}$, and that equation (2.2.1) has a zero solution.

We have the following lemmas.

Lemma 2.5.1   If the zero solution of (2.2.1) is unique then there exists a continuous function $L(t, \tau, r)$ such that $L(t, \tau, 0) = 0$ and

$$|x_t(\tau, \varphi)|_B \leqslant L(t, \tau, |\varphi|_B) \quad \text{for } t \geqslant \tau \tag{2.5.1}$$

as long as $x_t(\tau, \varphi) \in B_H$, where $x_t(\tau, \varphi)$ denotes the segment of the solution $x(t, \tau, \varphi)$ of (2.2.1) through (τ, φ). Moreover, if $F(t, \varphi)$ satisfies

$$|F(t, \varphi)| \leqslant L_1 |\varphi|_B$$

for a constant L_1 then $L(t, \tau, r)$ in (2.5.1) can be chosen in the form $L(t, \tau, r) = L(t - \tau)r$;

that is,

$$|x_t(\tau, \varphi)|_B \leqslant L(t - \tau)|\varphi|_B. \tag{2.5.2}$$

The first part of this lemma is nothing other than Theorem 2.5 of Hale and Kato [1], and the second part can be proved is the same way as Theorem 2.2 of Hale and Kato [1].

Lemma 2.5.2   If $F(t, \varphi)$ in (2.2.1) is completely continuous on $[0, \infty[\times B_H$ then for any $\varepsilon \in]0, H[$ and any solution $x(t)$ of (2.2.1) satisfying $(\tau, x_\tau) \in [0, \infty[\times B_\varepsilon$ either $x(t)$ exists for all $t \geqslant \tau$ or there exists a $t_1 > \tau$ for which $x(t)$ exists on $[\tau, t_1]$ and $|x_{t_1}|_B = \varepsilon$.

It is easy to see that our conditions (1)(i)–(iii) on the space B in Assumption 2.1.1 cause no trouble in the proof due to Hale and Kato [1].

From Definitions 2.2.1–2.2.3 we obtain the following.

Definition 2.5.1   The zero solution of (2.2.1) is said to be

(a) *stable in R^n* if for any $\varepsilon > 0$ and any $\tau \geqslant 0$ there exists a $\delta = \delta(\varepsilon, \tau) > 0$ such that

$$|x_\tau|_B < \delta \quad \text{implies} \quad \|x(t)\| < \varepsilon \quad \forall t \geqslant \tau; \tag{2.5.3}$$

(b) *equiasymptotically stable in R^n* if, in addition to stability, for any $\tau \geqslant 0$ there exist a $\delta_0 = \delta_0(\tau) > 0$ and a function $T = T(\varepsilon, \tau)$ of $\varepsilon > 0$ such that

$$|x_\tau|_B < \delta_0 \quad \text{and} \quad t \geqslant \tau + T \quad \text{imply} \quad \|x(t)\| < \varepsilon, \tag{2.5.4}$$

where $x(t)$ denotes any solution of (2.2.1).

If δ, δ_0 and T are independent of τ then the stabilities are called *uniform*, while *exponential stability in R^n* corresponds to the case where there are positive constants α, δ_0 and η such that

$$\|x(t)\| \leqslant \eta e^{-\alpha(t-\tau)}|x_\tau|_B \quad \text{if} \quad |x_\tau|_B < \delta_0 \quad \text{and} \quad t \geqslant \tau. \tag{2.5.5}$$

Remark 2.5.1   In the above it should be understood that each of the relations (2.5.3)–(2.5.5) holds as long as $x(t)$ exists. However, under the complete continuity of $F(t, \varphi)$, Lemma 2.5.2 guarantees that the relation (2.5.3), for example, can be read such that if $|x_\tau|_B < \delta$ then $x(t)$ exists for all $t \geqslant \tau$ and satisfies $\|x(t)\| < \varepsilon$ there.

The following theorem is a simple version of a Lyapunov-type theorem (cf. Driver [1]).

Theorem 2.5.1   Suppose that there exists a continuous real-valued function $V(t, \varphi)$ defined on $[0, \infty[\times B_{H_o}, H \geqslant H_0 > 0$, which satisfies the conditions

$$a(\|\varphi(0)\|) \leqslant V(t, \varphi), \tag{2.5.6}$$

$$V(t, \varphi) \leqslant b(t, |\varphi|_B), \tag{2.5.7}$$

$$\dot{V}_{(2.2.1)}(t, \varphi) \leqslant -c(t, V(t, \varphi)), \tag{2.5.8}$$

where $a(r)$, $b(t, r)$ and $c(t, r)$ are nonnegative, continuous and nondecreasing in $r \geqslant 0$, $a(r) > 0$ for $r > 0$ and $b(t, 0) = 0$.

Then the zero solution of (2.2.1) is stable in R^n.
Moreover, it is asymptotically stable in R^n if, for any $r > 0$ and $t \geq 0$,

$$\int_t^{t+T} c(s, r) \, ds \to \infty \quad \text{as} \quad T \to \infty, \tag{2.5.9}$$

and it is uniformly asymptotically stable in R^n if

$$\text{the divergence in (2.5.9) is uniformly in } t \tag{2.5.10}$$

and if $b(t, r)$ in (2.5.7) can be chosen independent of t, that is,

$$V(t, \varphi) \leq b(|\varphi_B|). \tag{2.5.11}$$

Here,

$$\dot{V}_{(2.2.1)}(t, \varphi) = \sup \overline{\lim_{h \to 0+}} \frac{1}{h} \{ V(t + h, x_{t+h}) - V(t, \varphi) \},$$

where the supremum is taken over the solutions $x(u)$ of (2.2.1) passing through (t, φ).

It will be observed that the segments of the solution may belong to a more restrictive class in B as the time elapses.

This fact may allow us to choose a simpler Lyapunov function.

Example 2.5.1   The zero solution of

$$\dot{x}(t) = - x(t),$$

considered as a functional differential equation on the space $C([-1, 0])$, is certainly exponentially stable in $C([-1, 0])$:

$$|x_t(\tau, \varphi)|_{C([-1,0])} \leq e^{-(t-\tau)} |\varphi|_{C([-1,0])}.$$

It can be seen that a corresponding Lyapunov function $V(t, \varphi)$ can be defined by

$$V(t, \varphi) = \sup_{u \geq 0} |x_{t+u}(t, \varphi)|_{C([-1,0])} e^{cu}$$

for a $c \in (0, 1)$. Hence we have

$$V(t, \varphi) = \max \left\{ \sup_{u \geq 1} e \| \varphi(0) \| e^{(c-1)u}, \sup_{\substack{0 \leq u \leq 1 \\ u-1 \leq s \leq 0}} |\varphi(s)| e^{cu} \right\},$$

which is not so simple as was expected. However, if $x(u)$ is a solution at least on the interval $[t-1, t]$ then

$$V(t, x_t) = |x_t|_{C([-1,0])} = e|x(t)|.$$

Thus the following theorem is expected to be more effective. Such a theorem has been given by Barnea [1] for the uniform stability of an autonomous system with finite delay.

Theorem 2.5.2   Suppose that the condition (2.5.1) is satisfied for the solution of (2.2.1).

Then in Theorem 2.5.1 it is sufficient for $V(t, \varphi)$ to satisfy the condition (2.5.8) in the case that

$$\varphi = x_t \text{ for a function } x(u), \ x(t) \neq 0, \text{ which is a solution of (2.2.1)} \atop \text{at least on the interval } [p(t, V(t, x_t)), t], \qquad (2.5.12)$$

where $p(t, r)$ is a continuous function of $t \geqslant 0$, $r > 0$, increasing in t, nondecreasing in r and $p(t, r) \leqslant t$. Here for the uniform stability we assume that the condition (2.5.2) is satisfied and that $p(t, r)$ in the condition (2.5.12) is of the form

$$p(t, r) = t - q(r) \qquad (2.5.13)$$

which will be referred to as the condition (2.5.12*).

This theorem is a special case of the following one, where $X(\tau) = \{(t, \varphi): \varphi \in B(t - \tau),$ $t \geqslant \tau\}$ and $B(\tau), \tau \geqslant 0$, denotes a subspace of B such that $\varphi \in B(\tau)$ if and only if $\varphi_{-\tau} \in B$ and $\varphi(s)$ is continuous on $[-\tau, 0]$, and we shall introduce a new seminorm $|\cdot|_{B(\tau)}$ in $B(\tau)$ by

$$|\varphi|_{B(\tau)} = \max_{-\tau \leqslant s \leqslant 0} |\varphi_s|_B,$$

which is well defined by the assumptions on B.

Theorem 2.5.3   Suppose that the condition (2.5.1) is satisfied for the solution of (2.2.1) and that for any $\tau \geqslant 0$ there exists a continuous function $V(t, \varphi; \tau)$ defined on $X_{H_0}(\tau) = \{(t, \varphi) \in X(\tau); |\varphi|_{B(t-\tau)} < H_0\}$ that satisfies the conditions (2.5.6)–(2.5.8) under (2.5.12) and

$$V(t, \varphi; \tau) \leqslant b(t, \tau, |\varphi|_{B(t-\tau)}). \qquad (2.5.14)$$

Then the zero solution of (2.2.1) is asymptotically stable in R^n under the condition (2.5.9). Moreover, if (2.5.2),

$$(2.5.14) \text{ with } B \text{ independent of } t \text{ and } \tau, \qquad (2.5.14*)$$

(2.5.10) and (2.5.12*) are assumed then the zero solution of (2.2.1) is uniformly asymptotically stable in R^n.

Proof   Let $x(t)$ be a solution of (2.2.1) starting at $t = \tau$, and let $\varepsilon > 0, \varepsilon < H_0$, be given.
Suppose that $V(\tau, x_\tau; \tau) \leqslant \frac{1}{2} a(\varepsilon)$ and $V(t, x_t; \tau) \geqslant a(\varepsilon)$ for a $t > \tau$, where we may assume that $|x_s|_B < H_0$ for $s \in [\tau, t]$; that is, $|x_t|_{B(t-\tau)} < H_0$. Then there exist t_1 and $t_2, \tau \leqslant t_2 < t_1$, such that

$$t_1 = \inf\{t > \tau: V(t, x_t; \tau) \geqslant a(\varepsilon)\},$$

$$t_2 = \sup\{t < t_1: V(t, x_t; \tau) \geqslant \tfrac{1}{2} a(\varepsilon)\}.$$

The condition (2.5.1) can be read as

$$|x_t|_{B(t-\tau)} \leqslant L(t, \tau, |x_\tau|_B) \quad \text{for } t \geqslant \tau,$$

while the condition (2.5.14) ensures that there exists a continuous function $\eta(\varepsilon, t, \tau) > 0$ such that $|\varphi|_{B(t-\tau)} < \eta(\varepsilon, t, \tau)$ implies $V(t, \varphi; \tau) < \varepsilon$, since $b(t, \tau, 0) = 0$. Since $L(t, \tau, r) \to 0$ as

$r \to 0$, we can choose $\delta = \delta(\varepsilon, \tau) > 0$ such that

$$L(t, \tau, \delta) < \eta(\tfrac{1}{2} a(\varepsilon), t, \tau) \quad \forall t \in [\tau, p_t^{-1}(\tau, \tfrac{1}{2} a(\varepsilon))],$$

where $p_t^{-1}(\tau, r)$ denotes the inverse function of $p(t, r)$ with respect to t for a fixed $r > 0$, which is clearly increasing in τ, nonincreasing in r and satisfies $p_t^{-1}(\tau, r) \geqslant r$. Therefore, if $|x_\tau|_B < \delta$ then $V(t, x_t; \tau) \geqslant \tfrac{1}{2} a(\varepsilon)$ only when

$$t > p_t^{-1}(\tau, \tfrac{1}{2} a(\varepsilon)) \geqslant p_t^{-1}(\tau, V(t, x_t; \tau)),$$

and hence, if $|x_\tau|_B < \delta$,

$$p(t, V(t, x_t; \tau)) > \tau \quad \text{for} \ t \in [t_2, t_1]$$

and, especially, $t_2 > \tau$. Thus, by the assumptions, $V(t, x_t; \tau)$ is nonincreasing on $[t_2, t_1]$, which is a contradiction. This shows that the zero solution of (2.2.1) is stable in R^n. If (2.5.2), (2.5.14*) and (2.5.12*) are satisfied then obviously δ can be chosen independent of τ, since $p_t^{-1}(\tau, r) = \tau + q(r)$.

Next, we shall prove the asymptotic stability. Let $\delta_0(\tau) = \delta(\tfrac{1}{2} H_0, \tau)$ for the δ in the stability, and let $T_1 = T_1(\varepsilon, \tau) \geqslant 0$ be so that

$$\int_\sigma^{\sigma + T_1} c(s, \varepsilon) ds > C(\sigma, \tau) - \varepsilon,$$

where $\sigma = p_t^{-1}(\tau, \varepsilon)$ and

$$C(\sigma, \tau) = \max \left\{ b(\sigma, \tau, r) : r \leqslant \sup_{0 \leqslant s \leqslant \sigma - \tau} [\tfrac{1}{2} K(s) H_0 + M(s) \delta_0(\tau)] \right\}.$$

Let $T(\varepsilon, \tau) = T_1(\varepsilon, \tau) + \sigma(\varepsilon, \tau) - \tau$ and suppose that for a $t_1 > T(\varepsilon, \tau) + \tau$ we have $V(t_1, x_{t_1}; \tau) \geqslant \varepsilon$. Clearly

$$p(t_1, V(t_1, x_{t_1}; \tau)) \geqslant p(\sigma, \varepsilon) = \tau.$$

Let

$$t_2 = \sup \{ t \in [\tau, t_1] : p(t, V(t, x_t; \tau)) \leqslant \tau \},$$

which exists since $p(\tau, r) \leqslant \tau$ for any $r > 0$. Then, by the condition (2.5.8), under (2.5.12), $V(t, x_t; \tau)$ is nonincreasing on $[t_2, t_1]$. Hence we have

$$p(t_2, V(t_2, x_{t_2}; \tau)) \geqslant p(t_2, V(t_1, x_{t_1}; \tau)) \geqslant p(t_2, \varepsilon),$$

which implies $\tau \geqslant p(t_2, \varepsilon)$; that is, $\sigma = p_t^{-1}(\tau, \varepsilon) \geqslant t_2$. Therefore for $t \in [\sigma, t_1]$

$$\dot{V}_{(2.2.1)}(t_1, x_{t_1}; \tau) \leqslant - c(t, V(t, x_t; \tau)), \quad V(t, x_t; \tau) \geqslant \varepsilon,$$

and hence

$$\varepsilon \leqslant V(t_1, x_{t_1}; \tau) \leqslant V(\sigma, x_\sigma; \tau) - \int_\sigma^{t_1} c(s, V(s, x_s; \tau)) ds$$

$$\leqslant V(\sigma, x_\sigma; \tau) - \int_\sigma^{t_1} c(s, \varepsilon) ds.$$

From this, we have

$$\int_\sigma^{t_1} c(s, \varepsilon)ds \leqslant b(\sigma, \tau, |x_\sigma|_{B(\sigma - \tau)}) - \varepsilon \leqslant C(\sigma, \tau) - \varepsilon$$

if $|x_\tau| < \delta_0(\tau)$, because

$$|x_s|_B \leqslant K(s - \tau) \sup_{\tau \leqslant t \leqslant s} \|x(t)\| + M(s - \tau)|x_\tau|_B$$

$$\leqslant \tfrac{1}{2} K(s - \tau) H_0 + M(s - \tau)\delta_0(\tau)$$

for any $s \in [\tau, \sigma]$; that is,

$$|x_\sigma|_{B(\sigma - \tau)} \leqslant \sup_{0 \leqslant s \leqslant \sigma \leqslant -\tau} \{\tfrac{1}{2} K(s) H_0 + M(s)\delta_0(\tau)\}.$$

This contradicts $t_1 \geqslant \sigma + T_1(\varepsilon, \tau)$.

It is not difficult to see that we can choose T together with δ_0 independent of τ under the hypotheses (2.5.2), (2.5.11), (2.5.13) and (2.5.10). ∎

We shall now give a Razumikhin-type theorem for equation (2.2.1). Such theorems have been given by Driver [1], Seifert [2], and Grimmer and Seifert [1]. Here we shall give the following theorem by extending the ideas in Kato [3, 4].

Theorem 2.5.4 Suppose that there is a continuous function $V(t, \varphi)$ defined on $[0, \infty[\times B_{H_0}$ satisfying the conditions (2.5.6)–(2.5.8) and the condition

$$\varphi = x_t \text{ satisfies } V(s, x_s) \leqslant g(V(t, \varphi)) \text{ for any} \qquad (2.5.15)$$
$$s \in [p(t, V(t, \varphi)), t] \text{ in addition to } (2.5.12),$$

where $p(t, r)$ satisfies the same conditions as in (2.5.12) and, moreover, the function $q(t, r) = t - p(t, r)$ is positive, nondecreasing in t and

$$\int_t^{t + T} \frac{ds}{q(p_t^{-1}(s, r), r)} \to \infty \quad \text{as } T \to \infty \qquad (2.5.16)$$

while $g(r)$ is a continuous function of $r > 0$ such that $g(r) > r$ and $g(r)/r$ is nondecreasing. Also, suppose that $c(t, r)$ in (2.5.8) satisfies

$$\sup_{t \geqslant 0} c(t, r) < \infty$$

for any $r \geqslant 0$.

Then, under the conditions (2.5.1) and (2.5.9), the zero solution of (2.2.1) is asymptotically stable in R^n if either $c(t, r)$ in (2.5.8) or $q(t, r)$ in (2.5.15) is independent of t. The latter case corresponds to (2.5.13) and will be referred to as (2.5.15*).

Furthermore, under the conditions (2.5.2), (2.5.11), (2.5.10) and (2.5.15*), the zero solution of (2.2.1) is uniformly asymptotically stable in R^n.

Proof This is done by constructing a Lyapunov function of the type mentioned in Theorem 2.5.3.

Let a, b, c, q, p and g be the functions relating to the conditions (2.5.6), (2.5.7), (2.5.8) and (2.5.15) for $V(t, \varphi)$, and define

$$\alpha(t, r) = \frac{1}{q(p_t^{-1}(t, g^{-1}(\frac{1}{2}r)), g^{-1}(\frac{1}{2}r))} \log \frac{r}{g^{-1}(r)}.$$

Clearly $\alpha(t, r)$ is a continuous function of $t \geqslant 0, r > 0$, $\alpha(t, r) > 0$ for $r > 0$, nondecreasing in r and nonincreasing in t. Put $\alpha(t, 0) = \lim_{r \to 0+} \alpha(t, r)$ and for any $\tau \geqslant 0$ and any $(t, \varphi) \in X_{H_0}(\tau)$ define

$$V(t, \varphi; \tau) = \sup_{\tau - t \leqslant s \leqslant 0} V(t + s, \varphi_s) e^{\alpha(t + s, V(t + s, \varphi_s))s}.$$

Since $\alpha(t, r) \geqslant 0$, this function obviously satisfies the conditions (2.5.6) and (2.5.14) with the same $a(r)$ and

$$b(t, \tau, r) = \sup_{\tau \leqslant s \leqslant t} b(s, r).$$

Now, we shall prove that $V(t, \varphi; \tau)$ satisfies

$$\dot{V}_{(2.2.1)} \leqslant - d(t, V(t, \varphi, \tau))$$

under the condition (2.5.12), where $p(t, r)$ in (2.5.12) is the same one as in (2.5.15) and

$$d(t, r) = \min\{c(t, r), r\alpha(t, r)\}. \tag{2.5.17}$$

Suppose that for a given (t, φ) there exists a solution $x(u)$ of (2.2.1) starting at τ such that $x_t = \varphi$, $p(t, V(t, \varphi; \tau)) \geqslant \tau$. For brevity, set

$$V(u) = V(u, x_u), \qquad W(u) = V(u, x_u; \tau),$$

$$P(s, u) = V(s) e^{\alpha(s, V(s))(s - u)},$$

and we can then find an $s(u) \in [\tau, u]$ such that

$$W(u) = P(s(u), u) \geqslant P(s, u) \quad \text{for } s \in [\tau, u].$$

Clearly we may assume that $s(u)$ is continuous at $u = t$. Let $\{h_k\}$, $h_k \to 0+$, be any sequence. By extracting a subsequence, we may assume that $s(t + h_k) \leqslant t$ for all k or $s(t + h_k) \geqslant t$ for all k.

Case 1   When $s(t + h) \leqslant t$ for any $h \in \{h_k\}$, since $W(t) \geqslant P(s(t + h), t)$,

$$\frac{W(t + h) - W(t)}{h} \leqslant \frac{P(s(t + h), t + h) - P(s(t + h), t)}{h}$$

$$\leqslant W(t + h) \frac{1}{h}(1 - e^{\alpha(s(t + h), V(s(t + h)))h})$$

$$\leqslant - W(t)\alpha(s(t), V(s(t))) + o(1) \leqslant - W(t)\alpha(t, W(t)) + o(1),$$

where we note the monotonicity of $\alpha(t, r)$ in t and r and the fact that $V(s(t)) \geqslant W(t)$.

Case 2   When $t \leqslant s(t+h) \leqslant t+h$ for all $h \in \{h_k\}$, clearly $s(t) = t$, and hence $V(t) = W(t) \geqslant P(s,t)$ for any $s \in [\tau, t]$, which ensures that

$$V(t) \geqslant V(s) e^{-\alpha(s, V(s)) q(t, V(t))} \quad \text{for } s \in [p(t, V(t)), t], \qquad (2.5.18)$$

where we note that $p(t, V(t)) = p(t, W(t)) \geqslant \tau$. If we can prove that for an $s \in [p(t, V(t)), t]$

$$V(t) \geqslant g^{-1}(\tfrac{1}{2} V(s)) \qquad (2.5.19)$$

then we immediately have

$$t \leqslant p_t^{-1}(s, g^{-1}(\tfrac{1}{2} V(s))),$$

and hence, by the definition of $\alpha(t, r)$,

$$\alpha(s, V(s)) q(t, V(t)) \leqslant \log \frac{V(s)}{g^{-1}(V(s))},$$

which implies $V(t) \geqslant g^{-1}(V(s))$ by the relation (2.5.18). Since $V(u)$ is continuous and the relation (2.5.19) holds at $s = t$, this fact also ensures (2.5.19) for all $s \in [p(t, V(t)), t]$ and hence

$$g(V(t)) \geqslant V(s) \quad \forall s \in [p(t, V(t)), t],$$

that is, (t, φ), $\varphi = x_t$, satisfies the condition (2.5.15). By the assumptions, we have $\dot{V}(t) \leqslant -c(t, V(t)) = -c(t, W(t))$. Therefore we have

$$\frac{W(t+h) - W(t)}{h} = V(s(t+h)) \frac{1}{h} (e^{\alpha(s(t+h), V(s(t+h)))(s(t+h)-t-h)} - 1) + \frac{V(s(t+h)) - V(t)}{h}$$

$$\leqslant V(t)\alpha(t, V(t)) \left[\frac{s(t+h) \overset{\cdot}{-} t}{h} - 1 \right] + \dot{V}(t)\frac{s(t+h) - t}{h} + o(1)$$

$$\leqslant -d(t, W(t)) + o(1),$$

where we note that $V(t) = W(t)$ and $(s(t+h) - t)/h \in [0, 1]$. Clearly (2.5.11) and (2.5.12*) for $V(t, \varphi; \tau)$ follow from (2.5.11), and (2.5.15*) respectively for $V(t, \varphi)$. Finally, it is also obvious that $d(t, r)$ in (2.5.17) satisfies the condition (2.5.9) under the assumptions on $c(t, r)$ and $q(t, r)$, and it satisfies the condition (2.5.10) if $c(t, r)$ does and if q is independent of t.   ∎

Example 2.5.2   The zero solution of the equation

$$\dot{x}(t) = -ax(t) + b(t)x(p(t)), \qquad (2.5.20)$$

considered on C_γ for $\gamma > 0$, is asymptotically stable in R^n if $|b(t)| \leqslant \beta < a$ and $t \geqslant p(t) \geqslant \varepsilon t - N$ for an $\varepsilon \in]0, 1]$ and an $N > 0$, because we can apply Theorem 2.5.4 by putting

$$V(t, \varphi) = \varphi^2(0), \qquad g(r) = p^2 r$$

for a $p > 1$ such that $\beta p < a$,

$$p(t, r) = \varepsilon t - N.$$

Remark 2.5.2   Driver [1] has shown that the zero solution of (2.5.20) is asymptotically stable in R^n if $p(t) \to \infty$ as $t \to \infty$, which contains the case where

$$p(t) = \tfrac{1}{2}[(1 + 4t)^{1/2} - 1],$$

but unfortunately our result does not cover this case because $q(t) = t - p(t)$ does not satisfy the relation (2.5.16). However, in Driver [1] the definition of asymptotic stability allows T in (2.5.4) to depend on each solution, while our definition actually requires equiasymptotic stability.

Example 2.5.3   Consider the scalar equation

$$\dot{x}(t) = -ax(t) + \int_{-\infty}^{0} p(t, s, x(t + s))ds, \qquad (2.5.21)$$

and assume that $a > 0$ is a constant and $g(t, s, r)$ is continuous in (t, s, r) and satisfies

$$|g(t, s, r)| \leqslant m(s)|r| \quad \text{for } t \geqslant 0, s \leqslant 0, r \in R.$$

If $m(s)$ satisfies

$$\int_{-\infty}^{0} m(s)ds < a, \qquad \int_{-\infty}^{0} m(s)e^{-\gamma s}ds < \infty$$

for $\gamma \geqslant 0$ then the zero solution of (2.5.21), considered as an equation on $[0, \infty[\times C_\gamma$, is uniformly asymptotically stable in R^1.

Choose a constant $\rho > 1$ and a function $q(r)$ such that

$$\rho \int_{-\infty}^{0} m(s)ds < a$$

and for any $r > 0$

$$2 \int_{-\infty}^{-q(r)} m(s)e^{-\gamma s}ds \leqslant r\left[a - \rho \int_{-\infty}^{0} m(s)ds\right] = c(r).$$

Then $V(t, \varphi) = \varphi^2(0)$ satisfies all the conditions of Theorem 2.5.4 with $g(r) = \rho^2 r$, $p(t, r) = t - q(r)$ and $c(t, r) = c(r)$, while (2.5.2) is satisfied, by Lemma 2.5.1, because

$$\left| -a\varphi(0) + \int_{-\infty}^{0} f(t, s, \varphi(s))ds \right| \leqslant \left[a + \int_{-\infty}^{0} m(s)e^{-\gamma s}ds\right]|\varphi|_{C_\gamma}.$$

Example 2.5.4   Consider the scalar equation

$$\dot{x}(t) = -\int_{-\infty}^{0} f(t, s, x(t + s))ds, \qquad (2.5.22)$$

and assume that $g(t, s, r)$ is a continuous function satisfying

$$0 \leqslant m'(s) \leqslant \frac{f(t, s, r)}{r} \leqslant m(s) \quad (r \neq 0).$$

If $\int_{-\infty}^{0} m(s)e^{-\gamma s}ds < \infty$ for a $\gamma > 0$ and if $\mu(\int_{-\infty}^{0} m(s)ds, 1) > 0$, where

$$\mu(\sigma, \rho) = \int_{-1/\sigma}^{0} m'(s)(1+\sigma s)\,ds + \int_{-2/\sigma}^{-1/\sigma} m(s)(1+\sigma s)ds - \rho \int_{-\infty}^{-2/\sigma} m(s)\,ds,$$

then the zero solution of (2.5.22), considered on the space $[0, \infty[\times C_\gamma$, is uniformly asymptotically stable in R^1. In particular, if $f(t, s, r) = f(t)e^{\alpha s}r$ for a constant $\alpha > 0$ and a continuous function $f(t)$ satisfying

$$\frac{\alpha^2(e^{-\lambda} - e^{-2\lambda})}{\lambda(\lambda - 1 + e^{-\lambda})} < f(t) \leqslant \frac{\alpha^2}{\lambda},$$

with a constant $\lambda > 0$ such that $\lambda - 1 + e^{-2\lambda} > 0$, then the zero solution of (2.5.22), considered on the space $[0, \infty[\times C_\gamma, \gamma < \alpha$, is uniformly asymptotically stable in R^1.

Since $\mu(\sigma, \rho)$ is continuous, we can find constants $\rho > 1$ and $\varepsilon > 0$ such that $\mu(\sigma_0, \rho) > \varepsilon$ for $\sigma_0 = \rho \int_{-\infty}^{0} m(s)\,ds + \varepsilon$. Choose $q(r)$ such that

$$\int_{-\infty}^{-q(r) + 2/\sigma_0} m(s)e^{-\gamma s}ds < \varepsilon r e^{-2\gamma/\sigma_0}.$$

Suppose that the solution $x(u)$ is defined on $[t - q(|x(t)|), t]$ and satisfies $|x(u)| \leqslant \rho|x(t)|$ there. For any $r \in [t - 2/\sigma_0, t]$ we have

$$|\dot{x}(r)| \leqslant \int_{-\infty}^{0} m(s)|x(r+s)|\,ds \leqslant \int_{-\infty}^{r-t} m(t-r+s)|x(t+s)|\,ds$$

$$\leqslant \int_{-q(|x(t)|)}^{r-t} m(t+s-r)\rho|x(t)|\,ds + \int_{-\infty}^{-q(|x(t)|)} m(t+s-r)e^{-\gamma s}ds|x_t|_{C_\gamma}$$

$$\leqslant \rho|x(t)| \int_{-\infty}^{0} m(s)\,ds + \int_{-\infty}^{-q(|x(t)|)+2/\sigma_0} m(s)e^{\gamma(2/\sigma_0 - s)}ds|x_t|_{C_\gamma}$$

$$\leqslant \sigma_0|x(t)| \quad \text{if } |x_t|_{C_\gamma} \leqslant 1,$$

and hence $x(t)x(t+s) \geqslant |x(t)|^2(1 + \sigma_0 s)$ on $s \in [-2/\sigma_0, 0]$. Thus we have

$$\dot{x}(t)x(t) = -x(t) \int_{-\infty}^{0} f(t, s, x(t+s))ds$$

$$\leqslant -|x(t)|^2 \int_{-1/\sigma_0}^{0} m'(s)(1+\sigma_0 s)ds - |x(t)|^2 \int_{-q(|x(t)|)}^{-1/\sigma_0} m(s)\min\{1 + \sigma_0 s, -\rho\}ds$$

$$+ |x(t)| \int_{-\infty}^{-q(|x(t)|)} m(s)|x(t+s)|ds$$

$$\leqslant -\mu(\sigma_0, \rho)|x(t)|^2 + \varepsilon|x(t)|^2$$

if $|x_t|_{C_\gamma} \leqslant 1$, which shows that $V(t, \varphi) = \varphi^2(0)$ satisfies the conditions of Theorem 2.5.4 with $g(r) = \rho^2 r$. Thus we have proved the first part.

For the second part, it is sufficient to show that if

$$m'(s) = \frac{\alpha^2(e^{-\lambda} - e^{-2\lambda})}{\lambda(\lambda - 1 + e^{-\lambda})} e^{\alpha s}, \quad m(s) = \frac{\alpha^2}{\lambda} e^{\alpha s},$$

then $\mu(\int_{-\infty}^{0} m(s)ds, 1) > 0$.   ∎

2.6 Stability under Perturbations

Above, we have discussed stability in R^n, while in Section 2.4 we have discussed relationships between stability in R^n and stability in B. For subsequent presentation of results in this section we need the following lemmas.

2.6.1 Preliminary results

Under the hypotheses on our space B we can reproduce Theorem 6 of Hale and Kato [1] in the following form.

Lemma 2.6.1   The concepts of (uniform) (asymptotic) stability in R^n and B are equivalent. Furthermore, if $M(t)$ in condition (1) (ii) of Assumption 2.1.1 satisfies

$$M(t) \leqslant Me^{-\mu t} \tag{2.6.1}$$

for positive constants M and μ, and if the zero solution of (2.2.1) is exponentially stable in R^n, then it is exponentially stable in B.

Proof   Suppose that the zero solution of (2.2.1) is exponentially stable in R^n; that is, the relation (2.5.5) holds for the solution $x(u)$ of (2.2.1). By the assumptions of the lemma and the condition (1)(ii) of Assumption 2.1.1, we have

$$|x_t|_B \leqslant K \sup_{\sigma \leqslant s \leqslant t} |x(s)| + Me^{-\mu(t-\sigma)}|x_\sigma|_B \tag{2.6.2}$$

for any $\sigma \in [\tau, t]$ if $x(u)$ is a solution of (2.2.1) starting at τ. Putting $\sigma = \tau$ in (2.6.2), we have

$$|x_t|_B \leqslant \{K\eta + M\}|x_\tau|_B \quad \text{if } |x_\tau|_B < \delta_0$$

where δ_0 is as in (2.5.5). Next, putting $\sigma = (\mu t + \alpha \tau)/(\alpha + \mu) \in [\tau, t]$ in (2.6.2), we have

$$|x_t|_B \leqslant K\eta e^{-\alpha(\sigma - \tau)}|x_\tau|_B + Me^{-\mu(t-\sigma)}|x_\sigma|_B$$

$$\leqslant [K\eta e^{-\alpha(\sigma - \tau)} + Me^{-\mu(t-\sigma)}(K\eta + M)]|x_\tau|_B$$

$$\leqslant [K\eta(1 + M) + M^2]|x_\tau|_B e^{-\beta(t-\tau)} \quad \text{if } |x_\tau|_B < \delta_0,$$

which proves exponential stability in B, where $\beta = \alpha\mu/(\alpha + \mu)$, where we note that $\alpha(\sigma - \tau) = \mu(t - \sigma) = \beta(t - \tau)$ for σ in the above.   ∎

We now characterize several properties of V functions in terms of special types of comparison functions.

Definition 2.6.1   A continuous function $a: [0, r_1] \to R_+$ (or a continuous function $a: [0, \infty[\to R_+)$ is said to belong to *class* $\mathcal{K}$, i.e. $a \in \mathcal{K}$, if $a(0) = 0$ and if a is strictly increasing on $[0, r_1]$ (or on $[0, \infty[$).

In the same way as for ordinary differential equations (see e.g. Yoshizawa [2]), it is possible to prove the following lemma (see Sawano [1] for exponential stability).

Lemma 2.6.2   Let

(1) the vector function $F(t, \varphi)$ in (2.2.1) satisfy a Lipschitz condition

$$|F(t, \varphi) - F(t, \psi)| \leqslant L_1 |\varphi - \psi|_B, \tag{2.6.3}$$

with constant L_1;

(2) the zero solution of (2.2.1) be uniformly asymptotically stable in B.

Then there exists a continuous function $V(t, \varphi)$ defined on $[0, \infty[\times B_{H_0}$ for sufficiently small H_0 and satisfying the conditions

$$a(|\varphi|_B) \leqslant V(t, \varphi) \leqslant b(|\varphi|_B), \tag{2.6.4}$$

$$\dot{V}_{(2.2.1)}(t, \varphi) \leqslant -cV(t, \varphi), \tag{2.6.5}$$

$$|V(t, \varphi) - V(t, \psi)| \leqslant L|\varphi - \psi|_B, \tag{2.6.6}$$

where a and b are functions of class $\mathcal{K}$, and c and L are positive constants.

Moreover, if the zero solution of (2.2.1) is exponentially stable in B then $a(r)$ and $b(r)$ in the above can be chosen to be linear in r.

2.6.2   Theorem on stability

Applying these lemmas, we can readily state theorems on the stability of the perturbed system

$$\frac{dx}{dt} = F(t, x_t) + R(t, x_t) \tag{2.6.7}$$

of the system (2.2.1), because if $V(t, \varphi)$ satisfies the relation (2.6.6) then we have

$$\dot{V}_{(2.6.7)}(t, \varphi) \leqslant \dot{V}_{(2.2.1)}(t, \varphi) + LK|R(t, \varphi)|, \tag{2.6.8}$$

where K is as in condition (1) (ii) of Assumption 2.1.1.

Theorem 2.6.1   Suppose that $F(t, \varphi)$ satisfies the condition (2.6.3).

Then if the zero solution of (2.2.1) is uniformly asymptotically stable in R^n, there exists a continuous positive function $\varepsilon(r)$ such that if

$$|R(t, \varphi)| \leqslant \varepsilon(|\varphi|_B), \tag{ }$$

then the zero solution of (2.6.7) is uniformly asymptotically stable in B.

Moreover, if the relation (2.6.2) holds and if the zero solution of (2.2.1) is exponentially stable in R^n then the zero solution of (2.6.7) is exponentially stable in B under the condition

$$|R(t, \varphi)| \leqslant h(t)|\varphi|_B$$

for a small $\eta > 0$ and a continuous function $h(t)$ satisfying

$$\overline{\lim_{t \to \infty}} \int_t^{t+1} h(s)ds < \eta. \tag{2.6.9}$$

Proof   We shall prove the first part of the theorem. Under the assumptions, the zero solution of (2.2.1) is uniformly asymptotically stable in B by Lemma 2.6.1, and hence there is a continuous function $V(t, \varphi)$ as given in Lemma 2.6.2. Therefore, by (2.6.8), we have

$$\dot{V}_{(2.6.7)}(t, \varphi) \leqslant \dot{V}_{(2.2.1)}(t, \varphi) + LK\varepsilon(|\varphi|_B) \leqslant -cV(t, \varphi) + LK\varepsilon(|\varphi|_B).$$

Put

$$\varepsilon(r) = \frac{c}{2LK} a(r).$$

Then clearly we have

$$\dot{V}_{(2.6.7)}(t. \varphi) \leqslant -\tfrac{1}{2} cV(t, \varphi),$$

which guarantees that the zero solution of (2.6.7) is uniformly asymptotically stable in B.

For the second part of the theorem, we note that under the assumptions, there exists a $V(t, \varphi)$ with $a(r) = ar$ and $b(r) = br$ linear in r by Lemmas 2.6.1 and 2.6.2. Therefore we have

$$\dot{V}_{(2.6.7)}(t, x_t) \leqslant -cV(t, x_t) + \frac{LK}{a} h(t)V(t, x_t).$$

Consider the equation

$$\frac{du}{dt} = -cu + \frac{LK}{a} h(t)u, \quad u(\tau) = V(\tau, \varphi),$$

with solution

$$u(t) = u(\tau)\exp\left[-c(t - \tau) + \frac{LK}{a} \int_\tau^t h(s)ds \right].$$

Put $\eta = ac/2LK$ in the relation (2.6.9). We then have

$$\int_\tau^t h(s)ds \leqslant \frac{ac}{2LK}(t - \tau) + N_0$$

for an $N_0 > 0$ and all $t \geqslant \tau$.

Thus, by the comparison theorem, we have

$$|x_t|_B \leqslant \frac{1}{a} V(t, x_t) \leqslant \frac{1}{a} u(t) \leqslant \frac{1}{a} u(\tau)e^{(LK/a)N_0}e^{-(c/2)(t-\tau)} \leqslant N|x_\tau|_B e^{-\alpha(t-\tau)}$$

where

$$u(\tau) = V(\tau, x_\tau) \leqslant b|x_\tau|_B, \quad \alpha = \tfrac{1}{2}c, \quad N = \left(\frac{1}{a}\right)e^{(LK/a)N_0}b,$$

which shows that the zero solution of (2.6.7) is exponentially stable in B.

Example 2.6.1   If $a > 0$ and $ah < \tfrac{1}{2}\pi$, and if

$$|R(t, \varphi)| \leqslant \varepsilon|\varphi|_{C_\gamma}$$

for $\gamma > 0$ and a suitably small $\varepsilon > 0$, then the zero solution of the scalar equation

$$\frac{dx}{dt} = -ax(t - h) + R(t, x_t)$$

is exponentially stable in R^1.

This follows immediately from Theorem 2.6.1, since as seen from Theorem B of Hale [2], the zero solution of the equation

$$\frac{dx}{dt} = -ax(t - h)$$

is exponentially stable in R^1 under the conditions on a and h.

Example 2.6.2   Consider the system

$$\frac{dx}{dt} = A_1 x(t) + A_2 x(t - h) + \int_0^t A(s - t)x(s)ds \qquad (2.6.10)$$

and assume that $h > 0$ is a constant and that A_1, A_2 and $A(s)$ are square matrices satisfying

$$\det\left[\lambda E - A_1 - A_2 e^{-\lambda h} - \int_{-\infty}^0 A(s)e^{\lambda s}ds\right] \neq 0 \quad \text{if } \operatorname{Re}\lambda \geqslant 0, \ \lambda \in C, \qquad (2.6.11)$$

$$\int_{-\infty}^0 |A(s)|e^{-\gamma s}ds < \infty \quad \text{for } \gamma > 0. \qquad (2.6.12)$$

Then the zero solution of (2.6.10) is exponentially stable in R^n.

The system (2.6.10) can be written in the form

$$\frac{dx}{dt} = A_1 x(t) + A_2 x(t - h) + \int_{-\infty}^0 A(s)x(t + s)ds - \int_{-\infty}^{-t} A(s)x(t + s)ds,$$

which is a perturbed system of the autonomous linear system

$$\frac{dx}{dt} = A_1 x(t) + A_2 x(t - h) + \int_{-\infty}^0 A(s)x(t + s)ds. \qquad (2.6.13)$$

The condition (2.6.12) ensures that these systems can be considered as functional differential equations on the space C_γ and that the solution operator induced by (2.6.13) has only point spectra in the half-plane $\{\lambda \in C : \operatorname{Re}\lambda > -\gamma\}$ (see Hale and Kato

[1, Section 5]). On the other hand, the condition (2.6.11) shows that the point spectra of the solution operator have negative real parts, and hence the zero solution of (2.6.13) is exponentially stable in R^n (see Naito [1, Theorem 4.4]). Put

$$h(t) = \int_{-\infty}^{-t} |A(s)| e^{-\gamma s} ds.$$

We then have $h(t) \to 0$ as $t \to \infty$, and

$$\left| \int_{-\infty}^{-t} A(s)\varphi(s) ds \right| \leqslant h(t) |\varphi|_{C_\gamma}.$$

Hence, by applying Theorem 2.6.1, we can see that the zero solution of (2.6.10) is exponentially stable in C_γ and, then, in R^n.

Remark 2.6.1   For $A_2 = 0$ the case in Example 2.6.2 was discussed by Miller [1] under a weaker condition than (2.6.12), but he obtained merely uniformly asymptotic stability in R^n.

2.6.3   Stability under perturbations from $H(F)$

Consider the equation

$$\frac{dx}{dt} = F(t, x_t), \tag{2.6.14}$$

where $F(t, \varphi)$ is defined on $R \times C([-h, 0], R^n)$ and $\|\varphi\| = \sup\{|\varphi(\theta) \cdot \theta \in [-h, 0]\}$. We introduce some notation and definitions. Let $y(t)$ be a solution of the system

$$\frac{dy}{dt} = F(t, y_t) + g(t), \tag{2.6.15}$$

starting at the point (t_0, φ^0) and defined on $[0, \infty[$.

Definition 2.6.2   A solution $y(t)$ of (2.6.14) is said to be *totally stable* if there exists some constant $\gamma(\varepsilon) > 0$ and $\eta(\varepsilon) > 0$ such that if $\|u_{t_0} - \varphi^0\| < \gamma(\varepsilon)$ for any $t_0 \geqslant 0$ and $|g(t)| < \eta(\varepsilon)$, then the inequality $\|u_t - y_t\| < \varepsilon$ is satisfied for all $t \geqslant t_0$.

Let $\bar{C}_{B^*}$ denote the set of functions $\varphi \in C([-h, 0], R^n)$, where $h \geqslant 0$ is a fixed constant, with the property $\|\varphi\| \leqslant B^*$. By Lemma 2.3.1, the hull $H(F)$ is a compact subset of the set $C([0, \infty[\times \bar{C}_{B^*}, R^n)$ if the function $F(t, \varphi)$ is bounded on $[0, \infty[\times \bar{C}_{B^*}$ and uniformly continuous with respect to (t, φ) on $[0, \infty[\times S$ for any compact $S \subset \bar{C}_{B^*}$, since the space $C([-h, 0], R^n)$ is separable.

Further, let $u(t)$ be a solution of (2.6.14) such that

$$\|u_t\| \leqslant B, \quad t \geqslant 0, \quad B < B^*.$$

Since when $(t, \varphi) \in I_0 \times \bar{C}_{B^*}$, $|F(t, \varphi)| \leqslant L$ for a constant $L > 0$, the function $u(t)$ satisfies the Lipschitz condition

$$|u(t) - u(s)| \leqslant L|t - s| \quad \forall t, s \geqslant 0.$$

By Lemma 2.3.1 the hulls $H(u)$ and $H(u, F)$ are compact. Let $(v(t), G(t, \varphi))$ be an element from $H(u, F)$. Then it is clear that the function $v(t)$ is a solution of (2.3.4). For the sake of simplicity, we call $v(t)$ a *solution in the hull* $H(u)$.

Definition 2.6.3   A solution $x(t)$ of (2.3.1) is *uniformly asymptotically stable in the hull* $H(u)$ with respect to the quantities $(\delta(\cdot), \delta_0, T(\cdot))$ if for any $\varepsilon > 0$ and $t_0 \geqslant 0$ and any $(v, G) \in H(u, F)$ the inequality $\| v_t - x_t \| < \varepsilon$ holds for all $t \geqslant t_0$ if $\| v_{t_0} - x_{t_0} \| < \delta(\varepsilon)$, and $\| v_t - x_t \| < \varepsilon$ holds for all $t \geqslant t_0 + T(\varepsilon)$ if $\| v_{t_0} - x_{t_0} \| \leqslant \delta_0$.

We shall now give three lemmas. In these lemmas, $g(t)$ is a continuous function defined on $[0, \infty[$ and $y(t)$ is a solution of the system (2.6.15).

Lemma 2.6.3   Suppose that the solutions in $H(u)$ are unique for the initial conditions. Let ε and T be positive constants.
   Then there exist positive numbers $\delta_1(\varepsilon, T)$ and $\delta_2(\varepsilon, T)$ such that for any $t_0 \geqslant 0$, if

$$\| y_{t_0} - u_{t_0} \| < \delta_1(\varepsilon, T)$$

and if

$$|g(t)| < \delta_2(\varepsilon, T) \quad \text{on } [t_0, t_0 + T]$$

we have

$$\| y_t - u_t \| < \varepsilon \quad \text{on } [t_0, t_0 + T].$$

Proof   This is similar to that of Lemma 6 in Yoshizawa [4].

Lemma 2.6.4   Let the solutions be uniformly stable in $H(u)$.
   Then for any $\varepsilon > 0$ and any $T > 0$ there are positive numbers $\eta_1(\varepsilon)$, independent of T, and $\eta_2(\varepsilon, T)$ such that for any $t_0 \geqslant 0$, if

$$\| y_{t_0} - u_{t_0} \| < \eta_1(\varepsilon)$$

and if

$$|g(t)| < \eta_2(\varepsilon, T) \quad \text{on } [t_0, t_0 + T]$$

we have

$$\| y_t - u_t \| < \varepsilon \quad \text{on } [t_0, t_0 + T].$$

Proof   We can assume that $\varepsilon < B^* - B$. Let K be a compact subset of $\bar{C}_{B^*}$ such that $u_t \in K$ for all $t \geqslant 0$ and that

$$K \supset \{\varphi : \| \varphi \| \leqslant B^*, |\varphi(\theta) - \varphi(\theta')| \leqslant (L + 1)|\theta - \theta'| \text{ on } [-h, 0]\},$$

where L is a bound for $F(t, \varphi)$ on $[0, \infty[\times C_{B^*}$. First of all, we shall show that the conclusion of this lemma is true if $y_{t_0} \in K$, that is, there exists a number $\eta_2'(\varepsilon, T) > 0$ such that for any $t_0 \geqslant 0$, if $\| y_{t_0} - u_{t_0} \| < \frac{1}{2}\delta(\frac{1}{2}\varepsilon)$, $y_{t_0} \in K$, and if $|g(t)| < \eta_2'(\varepsilon, T)$ on $[t_0, t_0 + T]$ then $\| y_t - u_t \| < \varepsilon$ on $[t_0, t_0 + T]$ where $\delta(\cdot)$ is the number appearing in the definition of uniform stability in $H(u)$. Suppose that there is no such $\eta_2'(\varepsilon, T)$. Then there exist sequences $\{t_m\}$, $t_m \geqslant 0$, $\{g_m(t)\}$, $\{y^m(t)\}$, $\{\tau_m\}$, $t_m \leqslant \tau_m \leqslant t_m + T$, such that

$$\sup_{t_m \leqslant t \leqslant t_m + T} |g_m(t)| < \frac{1}{m}, \quad \|y^m_{t_m} - u_{t_m}\| < \tfrac{1}{2}\delta(\tfrac{1}{2}\varepsilon), \quad y^m_{t_m} \in K,$$

$$\|y^m_{\tau_m} - u_{\tau_m}\| = \varepsilon, \quad \|y^m_t - u_t\| < \varepsilon, \quad t \in [t_m, \tau_m],$$

where $y^m(t)$ is a solution, defined on $[t_m, \tau_m]$, of the system

$$\frac{dy}{dt} = F(t, y_t) + g_m(t).$$

Clearly we can assume that the sequences

$$\{(u(t + t_m), F(t + t_m, \varphi))\}, \quad \{\tau_m - t_m\}, \quad \{y^m(t + t_m)\},$$

converge to $(v(t), G(t, \varphi)) \in H(u, F)$, $\sigma \in [0, T]$ and a continuous function $y(t)$ respectively, where it should be noted that $y^m_t \in K$ for all $t \in [t_m, \tau_m]$. From the assumptions, it turns out that $y(t)$ is also a solution of the system

$$\frac{dx}{dt} = G(t, x_t), \quad G(t, \varphi) \in H(F), \qquad (2.6.14^*)$$

defined on $[0, \sigma]$, and clearly

$$\|y_0 - v_0\| \leqslant \tfrac{1}{2}\delta(\tfrac{1}{2}\varepsilon) < \delta(\tfrac{1}{2}\varepsilon), \quad \|y_\sigma - v_\sigma\| = \varepsilon.$$

This contradicts the uniform stability of $v(t)$. Thus there exists an $\eta'_2(\varepsilon, T)$ as mentioned above.

On the other hand, whatever y_{t_o} is, it follows from Lemma 2.6.3 that

$$y_{t_o + h} \in K, \quad \|y_{t_o + h} - u_{t_o + h}\| < \tfrac{1}{2}\delta(\tfrac{1}{2}\varepsilon)$$

for any $t_0 \geqslant 0$, if $\|y_{t_o} - u_{t_o}\| < \delta_1(\tfrac{1}{2}\delta(\tfrac{1}{3}\varepsilon), h)$, $|g(t)| < \min\{\delta_2(\tfrac{1}{2}\delta(\tfrac{1}{2}\varepsilon), h), 1\}$, where δ_1 and δ_2 are as in Lemma 2.6.3. Therefore if we set

$$\eta_1(\varepsilon) = \delta_1(\tfrac{1}{2}\delta(\tfrac{1}{2}\varepsilon), h), \quad \eta_2(\varepsilon, T) = \min\{\delta_2(\tfrac{1}{2}s(\tfrac{1}{2}\varepsilon), h), \eta_2(\varepsilon, T), 1\}$$

then $\eta_1(\varepsilon)$ and $\eta_2(\varepsilon, T)$ satisfy the conditions in of the lemma. This completes the proof. ∎

Lemma 2.6.5 Suppose that the solutions are uniformly asymptotically stable in $H(u)$.
Then there is a triple $(\eta_3, \eta_4(\cdot), \tau(\cdot))$ with the property that for any $\varepsilon > 0$ and any $t_0 \geqslant 0$, if $\|y_{t_o} - u_{t_o}\| < \eta_3$ and if $|g(t)| < \eta_4(\varepsilon)$ on $[t_0, t_0 + \tau(\varepsilon)]$, then the solution $y(t)$ of (2.6.15) is continuable on $[t_0, t_0 + \tau(\varepsilon)]$ and satisfies $\|y_{t_o + \tau(\varepsilon)} - u_{t_o + \tau(\varepsilon)}\| < \varepsilon$.

Proof Suppose that y_{t_o} and $g(t)$ satisfy the conditions

$$\|y_{t_o} - u_{t_o}\| < \eta_1(B^* - B), \quad |g(t)| < \eta_2(B^* - B, \tau(\varepsilon)),$$

where η_1 and η_2 are as given in Lemma 2.6.4. Then, by Lemma 2.6.4, $y(t)$ is continuable to $t_0 + \tau(\varepsilon)$, where $\delta(\cdot)$, δ_0 and $T(\cdot)$ are as in the definition of uniform asymptotic stability in $H(u)$ and set $\tau(\varepsilon) = T(\tfrac{1}{2}\varepsilon) + h$. The other parts of the proof will be completed by the same arguments as in the proof of Lemma 2.6.4. Namely, by putting $\eta'_3 = \min\{\delta_0, \eta_1(B^* - B)\}$, we can find a positive number $\eta'_4(\varepsilon) \leqslant \eta_2(B^* - B, \tau(\varepsilon))$ such that for any $t_0 \geqslant 0$, $\| y_{t_0} - u_{t_0} \| < \eta'_3$ and $|g(t)| < \eta'_4(\varepsilon)$ on $[t_0, t_0 + T(\frac{1}{2}\varepsilon)]$ imply $\| y_{t_0 + T(\varepsilon/2)} - u_{t_0 + T(\varepsilon/2)} \| < \varepsilon$ under the restriction $y_{t_0} \in K$, where K is the compact set given in the proof of Lemma 2.6.4. Thus it turns out that it is sufficient to set

$$\eta_3 = \delta_1(\eta'_3, h), \quad \eta_4(\varepsilon) = \min\{\delta_2(\eta'_3, h), \eta_2(B^* - B, \tau(\varepsilon)), \eta'_4(\varepsilon), 1\},$$

where δ_1 and δ_2 are as in Lemma 2.6.3. ∎

We now have the following theorem.

Theorem 2.6.2 If a solution $u(t)$ of (2.6.14) is uniformly asymptotically stable in the hull then $u(t)$ is totally stable.

Proof Let η_1 and η_2 be as in Lemma 2.6.4 and η_3, η_4 and τ as in Lemma 2.6.5. Setting

$$\rho(\varepsilon) = \min\{\eta_1(\varepsilon), \eta_3\}, \quad \eta(\varepsilon) = \min\{\eta_2(\varepsilon, \tau(\rho(\varepsilon))), \eta_4(\rho(\varepsilon))\},$$

we shall prove that for any $t_0 \geqslant 0$ any solution $y(t)$ of the system (2.6.15) satisfies $\| y_t - u_t \| < \varepsilon$ for all $t \geqslant t_0$, if $\| y_{t_0} - u_{t_0} \| < \rho(\varepsilon)$ and $|g(t)| < \eta(\varepsilon)$ on $[t_0, \infty[$.

It follows from Lemma 2.6.4 that $\| y_t - u_t \| < \varepsilon$ on $[t_0, t_0 + \tau(\rho(\varepsilon))]$, because $\| y_{t_0} - u_{t_0} \| < \eta_1(\varepsilon)$ and $|g(t)| < \eta_2(\varepsilon, \tau(\rho(\varepsilon)))$. Furthermore, since $\| y_{t_0} - u_{t_0} \| < \eta_3$ and $|g(t)| < \eta_4(\rho(\varepsilon))$, we have $\| y_{t_0 + \tau(\rho(\varepsilon))} - u_{t_0 + \tau(\rho(\varepsilon))} \| < \rho(\varepsilon)$ by Lemma 2.6.5. By replacing t_0 by $t_0 + \tau(\rho(\varepsilon))$ and using the same arguments, we can see that $\| y_t - u_t \| < \varepsilon$ on $[t_0 + \tau(\rho(\varepsilon)), t_0 + 2\tau(\rho(\varepsilon))]$ and $\| y_{t_0 + 2\tau(\rho(\varepsilon))} - u_{t_0 + 2\tau(\rho(\varepsilon))} \| < \rho(\varepsilon)$. Thus, by repeating the same arguments, we have $\| y_t - u_t \| < \varepsilon$ for all $t \geqslant t_0$; that is, the solution $u(t)$ is totally stable. ∎

2.6.4 Stability under perturbations from H(F) with respect to a given compact set

We shall give the definition of stability under disturbances from $H(F)$ with respect to a given compact set $K \subset \bar{C}_{B^*}$. Define $\rho(G, P)$ for $G \in H(F)$ and $P \in H(F)$ by

$$\rho(G, P) = \sum_{m=1}^{\infty} \frac{1}{2^m} \frac{\rho_m(G, P)}{1 + \rho_m(G, P)},$$

where $\rho_m(G, P) = \sup\{|G(t, \varphi) - P(t, \varphi)| : (t, \varphi) \in [0, m] \times K\}$. Here we should note that for the space $C([0, \infty[\times K, R^n)$ the metric ρ introduces a topology equivalent to the compact open topology. For convenience, set $\rho^*(G, P, t) = \rho(G^t, P^t)$, where $G^s(t, \varphi) = G(t + s, \varphi) \in T(G)$.

Definition 2.6.4 The solution $u(t)$ of (2.6.14) is said to be *stable under disturbances from H(F) with respect to K* if for any $\varepsilon > 0$ there exist positive numbers $\gamma(\varepsilon)$ and $\eta(\varepsilon)$ such that for any $s \geqslant 0$ and any $T > 0$ the solution $x(t)$ of (2.6.15) satisfies the relation

$$\| u_{t+s} - x_t \| < \varepsilon \quad \forall t \in [0, T]$$

whenever $\| u_s - x_0 \| < \gamma(\varepsilon)$, $x_0 \in K$, $G \in H(F)$ and

$$\rho^*(G, F^s; t) < \eta(\varepsilon) \quad \forall t \in [0, T]. \tag{2.6.16}$$

This definition is a natural generalization of that given by Yoshizawa [4] and Sell [1, 2], because we have the following lemma.

Lemma 2.6.6 Let $F(t, \varphi)$ be almost-periodic in t uniformly for $\varphi \in K$. Then there is a positive continuous function $a(\sigma)$ of $\sigma > 0$, $a(0) = 0$, such that

$$\sigma(G, P) \geqslant \rho^*(G, P; t) \geqslant a(\sigma(G, P)) \tag{2.6.17}$$

for all $t \geqslant 0$ and any $G, P \in H(F)$, where

$$\sigma(G, P) = \sup\{|G(t, \varphi) - P(t, \varphi)| : (t, \varphi) \in] - \infty, \infty[\times K\}.$$

Proof The first part of the inequality (2.6.17) is clear. We shall prove the second part. Since

$$\sigma(G^t, P^t) = \sigma(G, P)$$

for all t and all $G, P \in H(F)$, it is sufficient to show the existence of a function $a(\cdot)$ satisfying

$$\rho(G, P) \geqslant a(\sigma(G, P)).$$

We know that there is an $l(\varepsilon) > 0$ such that any interval of length $l(\varepsilon)$ contains an ε-translation number of F, and we can easily see that an ε-translation number of F is so for any G in $H(F)$. From this, it follows that for any $(t, \varphi) \in] - \infty, \infty[\times K$ we can find an $s \in [0, l(\varepsilon)]$ such that $|G(t, \varphi) - G(s, \varphi)| < \varepsilon$ whatever $G \in H(F)$ is. Put

$$a(\sigma) = \frac{1}{2^{m(\varepsilon)}} \frac{\varepsilon}{1 + \varepsilon} \quad (\varepsilon = \tfrac{1}{3}\sigma),$$

where $m(\varepsilon)$ is an integer greater than $l(\varepsilon)$. Then, if $\sigma(G, P) = 3\varepsilon > 0$, we can show that $\rho_{m(\varepsilon)}(G, P) \geqslant \sigma(G, P) - 2\varepsilon = \varepsilon$. Hence we have

$$\rho(G, P) \geqslant \frac{1}{2^{m(\varepsilon)}} \frac{\rho_{m(\varepsilon)}(G, P)}{1 + \rho_{m(\varepsilon)}(G, P)} \geqslant \frac{1}{2^{m(\varepsilon)}} \frac{\varepsilon}{1 + \varepsilon} = a(3\varepsilon) = a(\sigma(G, P)).$$

By a slight modification, we can make $a(\cdot)$ continuous. This proves the second part of the inequality (2.6.17). ∎

Definition 2.6.5 The bounded solution $u(t)$ of (2.6.14) is said to be *totally stable* if for any $\varepsilon \in 0$ there exist a $\gamma(\varepsilon) > 0$ and an $\eta(\varepsilon) > 0$ such that if

$$\|u_{t_0} - \varphi^0\| < \gamma(\varepsilon), \quad |g(t)| < \eta(\varepsilon)$$

for any $t_0 \geqslant 0$ then a solution $y(t)$ of the system (2.6.14) through (t_0, φ^0) satisfies $\|u_t - y_t\| < \varepsilon$ for all $t \geqslant t_0$.

We shall now prove the following theorems.

Theorem 2.6.3 Let the solution $u(t)$ of the system (2.6.14) be totally stable and K be a compact subset of $\bar{C}_{B^*}$ which is positively invariant for any system in the hull. Then $u(t)$ is stable under disturbances from $H(F)$ with respect to K.

Proof For an $\varepsilon > 0$ suppose that there are a $G \in H(F)$, an $s \geqslant 0$, a $T > 0$ and a solution $x(t)$ of (2.6.14*) such that

$$\|u_{so} - x_0\| < \gamma(\varepsilon), \quad x_0 \in K, \quad \|u_{T+s} - x_T\| = \varepsilon,$$
$$\|u_{t+s} - x_t\| < \varepsilon \quad \text{on } [0, T[$$

and

$$\rho^*(G, F^s; t) < \frac{1}{2} \frac{\eta(\varepsilon)}{1 + \eta(\varepsilon)} \quad \forall t \in [0, T]. \tag{2.6.18}$$

Since

$$\rho^*(G, F^s; t) = \rho(G^t, F^{t+s}) \geqslant \frac{1}{2} \frac{\rho_1(G^t, F^{t+s})}{1 + \rho_1(G^t, F^{t+s})},$$

it follows from (2.6.18) that

$$\eta(\varepsilon) > \rho_1(G^t, F^{t+s}) = \sup\{|G(\tau, \varphi) - F(\tau + s, \varphi)| : (\tau, \varphi) \in [t, t+1] \times K\}$$

for all $t \in [0, T]$. On the other hand, the positive invariance of K implies that $x_t \in K$ as long as $x(t)$ exists. Hence we have

$$|G(t, x_t) - F(t + s, x_t)| < \eta(\varepsilon) \quad \text{on } [0, T]$$

or

$$|G(t - s, x_{t-s}) - F(t, x_{t-s})| < \eta(\varepsilon) \quad \text{on } [s, s + T].$$

Therefore we can easily find a continuous function $g(t)$ that is defined on $[0, \infty[$ and satisfies

$$g(t) = G(t - s, x_{t-s}) - F(t, x_{t-s}) \quad \text{for } t \in [s, s + T]$$

and

$$|g(t)| < \eta(\varepsilon) \quad \forall t \geqslant 0.$$

Clearly $x(t - s)$ is a solution of the system (2.6.15) on $[s, s + T]$, and, by the total stability of $u(t)$, we should have

$$\|u_t - x_{t-s}\| < \varepsilon \quad \text{on } [s, s + T],$$

or

$$\|u_{t+s} - x_t\| < \varepsilon \quad \text{on } [0, T].$$

Thus a contradiction arises, which shows that $u(t)$ is stable under disturbances from $H(F)$ with respect to K. ∎

Theorem 2.6.4 Let $u(t)$ be stable under disturbances from $H(F)$ with respect to K, and let the compact set $K \subset \bar{C}_{B*}$ be that given in the proof of Lemma 2.6.4.
Then the solutions are uniformly stable in $H(u)$.

Proof Let $(\gamma(\cdot), \eta(\cdot))$ be the pair given in Definition 2.6.5. First, we shall prove that for any $\varepsilon > 0$, any $t_0 \geqslant 0$, any $(v, G) \in H(u, F)$ and any solution $x(t)$ of (2.6.14*)

$$\|x_{t_o} - v_{t_o}\| < \tfrac{1}{2}\gamma(\tfrac{1}{2}\varepsilon), x_{t_o} \in K \quad \text{imply} \quad \|x_t - v_t\| < \varepsilon \quad \forall t \geqslant t_0. \tag{2.6.19}$$

If not, then there exist $(v, G) \in H(u, F)$, $\varepsilon > 0$, $t_0 \geqslant 0$ a solution $x(t)$ of (2.6.14) and $\tau > t_0$ satisfying

$$\| x_{t_0} - v_{t_0} \| < \tfrac{1}{2} \gamma(\tfrac{1}{2}\varepsilon), \quad x_{t_0} \in K,$$

and

$$\| x_\tau - v_\tau \| = \varepsilon.$$

By replacing $(v(t + t_0),\ G(t + t_0, \varphi))$ by $(v(t), G(t, \varphi))$, we can assume $t_0 = 0$. Since $(v, G) \in H(u, F)$, there exists an $s \geqslant 0$ such that

$$\rho^*(F^s, G; t) < \eta(\tfrac{1}{2}\varepsilon), \quad \| u_{t+s} - v_t \| < \tfrac{1}{2}\gamma(\tfrac{1}{2}\varepsilon) \quad \text{for } t \in [0, \tau].$$

Since $x_0 \in K$ and

$$\| x_0 - u_s \| \leqslant \| x_0 - v_0 \| + \| v_0 - u_s \| < \gamma(\tfrac{1}{2}\varepsilon),$$

we should have

$$\| x_t - u_{t+s} \| < \tfrac{1}{2}\varepsilon \quad \forall t \in [0, \tau],$$

by the stability under disturbances from $H(F)$ of $u(t)$. Hence

$$\| x_t - v_t \| \leqslant \| x_t - u_{t+s} \| + \| u_{t+s} - v_t \| < \varepsilon \quad \forall t \in [0, \tau],$$

from which a contradiction arises at $t = \tau$. This proves the assertion (2.6.19).

Furthermore, from (2.6.19), we can easily prove that the solutions in $H(u)$ are unique for initial conditions, because $v_t \in K$ for all $v(t) \in H(u)$ and for all $t \geqslant 0$. Therefore, from the fact that $H(v) \subset H(u)$ for every $v \in H(u)$, it follows that the solutions in $H(v)$ are unique for initial conditions. Hence, by Lemma 2.6.3, we can prove that

$$\| x_{t_0 + h} - v_{t_0 + h} \| < \tfrac{1}{2}\gamma(\tfrac{1}{2}\varepsilon), \quad x_{t_0 + h} \in K,$$

whenever

$$\| x_{t_0} - v_{t_0} \| < \delta(\varepsilon) = \delta_1(\tfrac{1}{2}\gamma(\tfrac{1}{2}\varepsilon), h),$$

where δ_1 is as given in Lemma 2.6.3. By modifying the proof of Lemma 2.6.3, it is not difficult to see that δ_1 can be chosen independently of each v in $H(u)$, since $H(u)$ is compact. This completes the proof. ■

The following corollary is an immediate consequence of Theorems 2.6.3 and 2.6.4, where we consider the compact set K given in the proof of Lemma 2.6.4.

Corollary to Theorems 2.6.3 and 2.6.4   If $u(t)$ is totally stable then the solutions are uniformly stable in $H(u)$.

Remark 2.6.2   As is clear from the proofs of Theorems 2.6.3 and 2.6.4, the relation (2.6.16) in the definition can be replaced by the simpler relation

$$| G(t, \varphi) - F(t + s, \varphi) | < \eta(\varepsilon) \quad \forall (t, \varphi) \in [0, T] \times K.$$

However, in this case, the definition would be slightly different from that given by Yoshizawa [4] and Sell [1, 2], even for an almost-periodic system.

2.6.5 A false conjecture

We shall now give an example that shows that the conjecture that the solutions are uniformly asymptotically stable in $H(u)$ if the solution $u(t)$ of (2.2.1) is uniformly asymptotically stable in R^n is not true, even for almost-periodic systems.

Let $f(x)$ be defined by

$$f(x) = \begin{cases} 2\sqrt{|nx - 1|} & \text{for } 2/(2n + 1) \leqslant x \leqslant 2/(2n - 1) \quad (n = 1, 2, \ldots), \\ 0 & \text{for } x = 0, \\ -f(-x) & \text{for } x < 0. \end{cases}$$

Then we can easily see that the zero solution of the system

$$\frac{dx}{dt} = f(x)$$

is not unique for initial conditions. Let $a(t) = \sum_{k=0}^{\infty} a_k(t)$ be an almost-periodic function (see Sell [1, 2]), where $a_0(t) \equiv 1$ and $a_k(t)$ is a periodic function of period 2^k such that

$$a_k(t) = \begin{cases} 0 & \text{for } 0 < t < 2^{k-1}, \\ -2^{-k} & \text{for } 2^{k-1} \leqslant t \leqslant 2^k \quad (k = 1, 2, \ldots) \end{cases}$$

(continuity can be provided by a slight modification), and consider the ordinary differential equation

$$\frac{dx}{dt} = F(t, x), \tag{2.6.20}$$

where

$$F(t, x) = \begin{cases} f(x) - ca(t)\sqrt{x} & \text{for } x \geqslant 0, \\ -F(t, -x) & \text{for } x < 0 \end{cases}$$

for a constant $c > 2\sqrt{2}$.

Since $a(t) > 0$ for all t, $F(t, 1/n) < 0$ for all t and $n = 1, 2, \ldots$, which implies that any solution of (2.6.20) cannot cross the line $x = 1/n$ upwards for $n = 1, 2, \ldots$. Hence the zero solution of (2.6.20) is uniformly stable. Furthermore, from the fact that $a(t) \geqslant \frac{1}{2}$ for $2m < t < 2m + 1$ and $f(x) \leqslant \sqrt{2x}$, it follows that

$$F(t, x) \leqslant (\sqrt{2} - \tfrac{1}{2}c)\sqrt{x} < 0$$

for $x > 0$ and $2m < t < 2m + 1$, and hence we can see that every solution of (2.6.20) starting from a neighbourhood of $x = 0$ tends to zero as $t \to \infty$ and, more precisely, that the zero solution of (2.6.20) is uniformly asymptotically stable. On the other hand, since

$$a(t + 2^m - 1) \to 0 \quad \text{as } m \to \infty \quad \text{for } 0 \leqslant t \leqslant 1,$$

there exists a $G(t, x) \in H(F)$ such that $G(t, x) = f(x)$ for all $t \in [0, 1]$. Therefore the zero solution of the system

$$\frac{dx}{dt} = G(t, x)$$

is not unique for initial conditions, and consequently it is not uniformly stable. Hence the zero solution of (2.6.20) is neither totally stable or stable under disturbances from $H(F)$.

2.7 Stability under Disturbances and Asymptotic Almost-Periodicity of Solutions

Let $f(t)$ be a continuous vector function defined on $I = [0, \infty[$ with values in R^n. The concept of asymptotic almost-periodicity was introduced by Fréchet [1].

Definition 2.7.1 A vector function $f(t)$ is said to be *asymptotically almost-periodic* if it is the sum of a continuous an almost periodic function $p(t)$ and function $g(t)$ defined on $I = [0, \infty[$ which tends to zero as $t \to +\infty$; that is,

$$f(t) = p(t) + g(t).$$

We denote by $C(I \times B, R^n)$ the set of continuous functions defined on $I \times B$ with values in R^n. A sequence $\{F_k\}$ in $C(I \times B, R^n)$ is said to *converge to G Bohr-uniformly on $I \times B$* if F_k converges to G uniformly on $I \times S$ for any compact set S in B as $k \to \infty$. A function $F(t, \varphi) \in C(I \times B, R^n)$ is said to be *positively precompact* if for any sequence $\{t_k\}$ in I such that $t_k \to \infty$ as $k \to \infty$ the sequence $\{F(t + t_k, \varphi)\}$ contains a Bohr-uniformly convergent subsequence. Then, if $F(t, \varphi) \in C(I \times B, R^n)$ and $F(t, \varphi)$ is positively precompact, $F(t, \varphi)$ is asymptotically almost-periodic in t uniformly for $\varphi \in B$ (cf. Yoshizawa [3]). Also, if $F \in C(I \times B, R^n)$ is asymptotically almost-periodic in t uniformly for $\varphi \in B$, $F(t, \varphi)$ is positively precompact.

We shall now consider the functional differential equation

$$\frac{dx}{dt} = F(t, x_t), \tag{2.7.1}$$

where $F(t, \varphi) \in C(I \times B, R^n)$. We assume that

    (i) $F(t, \varphi)$ is positively precompact;
    (ii) for any $H > 0$ there is an $L(H) > 0$ such that $|F(t, \varphi)| \leqslant L(H)$ for all $t \geqslant 0$ and $\varphi \in B$ such that $|\varphi|_B \leqslant H$;
    (iii) equation (2.7.1) has a bounded solution $u(t)$ defined on I such that $|u_t|_B \leqslant c$ for all $t \in I$.

A system

$$\frac{dx}{dt} = G(t, x_t) \tag{2.7.2}$$

is called a *limiting equation of* (2.7.1) when $G \in H(F)$, where $H(F)$ the set of all limit functions G such that $F(t + t_k, \varphi)$ converges to G Bohr-uniformly for some sequence $\{t_k\}$ such that $t_k \to \infty$ as $k \to \infty$.

Under the above assumptions, it is clear that the closure of the set $\{u_t : t \geqslant 0\}$ is contained in the compact set $\bar{U}(\{u_0\}, N_H, L(H))$ and that $G(t, \varphi)$ is almost-periodic in

t uniformly for $\varphi \in B$ if $G \in H(F)$, where $\bar{U}(\{u_0\}, \alpha, \beta)$ in the set given in Lemma 2.3.3. We shall write $(v, G) \in H(u, F)$ when there exists a sequence $\{t_k\}$, $t_k \to \infty$ as $k \to \infty$, such that $F(t + t_k, \varphi) \to G(t, \varphi) \in H(F)$ Bohr-uniformly and $u(t + t_k) \to v(t)$ uniformly on any compact set in I. In this case there exists a subsequence $\{\tau_k\}$ of $\{t_k\}$ such that $u_{\tau_k} \to w \in \bar{U}(\{u_0\}, NH, L(H))$ and $u(t + \tau_k) \to v(t)$ uniformly on any compact interval in I. Since

$$|u(\tau_k) - w(0)| \leqslant N |u_{\tau_k} - w|_B,$$

we have

$$w(0) = v(0).$$

Thus if we let $v_0 = w$ then $v_t \in B$ for all $t \geqslant 0$, and we have

$$|u_{t + \tau_k} - v_t|_B \leqslant K \sup_{0 \leqslant s \leqslant t} |u(s + \tau_k) - v(s)| + M(t) |u_{\tau_k} - v_0|_B. \tag{2.7.3}$$

Thus we can see that $|u_{t + \tau_k} - v_t| \to 0$ uniformly on any compact interval in I as $k \to \infty$. This implies that $v(t)$ is a solution of (2.7.2).

We now have the following lemma (see Hino and Yoshizawa [1]).

Lemm 2.7.1   The solution $u(t)$ of (2.7.1) is totally stable, iff for any $\varepsilon > 0$ there exists a $\delta(\varepsilon) > 0$ such that if $s \geqslant 0$, $|u_s - \psi|_B < \delta(\varepsilon)$ and $k(t)$ is a continuous function which satisfies $|k(t)| < \delta(\varepsilon)$ on $[s, \infty[$ then $|u_t - x_t(s, \psi, F + k)|_B < \varepsilon$ for $t \geqslant s$, where $x(s, \psi, F + k)$ is a solution through (s, ψ) of

$$\frac{dx}{dt} = F(t, x_t) + k(t). \tag{2.7.4}$$

Proof   The necessity is clear. Now consider a solution x through (s, ψ) of

$$\frac{dx}{dt} = F(t, x_t) + R(t, x_t), \tag{2.7.5}$$

where $s \geqslant 0$, $|u_s - \psi|_B < \delta(\varepsilon)$ and $R(t, \varphi)$ is a continuous function such that $|R(t, \varphi)| < \delta(\varepsilon)$ for $t \in [s, \infty[$ and φ such that $|u_t - \varphi|_B \leqslant \varepsilon$ for $t \geqslant s$. Suppose that $|u_\tau - x_\tau|_B = \varepsilon$ for some $\tau > s$ and $|u_t - x_t| < \varepsilon$ on $s \leqslant t < \tau$. Since $|R(t, x_t)| < \delta(\varepsilon)$ on $s \leqslant t \leqslant \tau$, there exists a continuous function as $[s, \infty[$ such that $|k(t)| < \delta(\varepsilon)$ for all $t \geqslant s$ and $k(t) = R(t, x_t)$ on $s \leqslant t \leqslant \tau$. Then x is also a solution of (2.7.4) defined on $[s, \tau]$, and $|u_s - \psi|_B < \delta(\varepsilon)$ and $|k(t)| < \delta(\varepsilon)$ for all $t \geqslant s$. Therefore $|u_t - x_\tau|_B < \varepsilon$, which contradicts $|u_t - x_\tau|_B = \varepsilon$. Thus we have $|u_t - x_t|_B < \varepsilon$ for $t \geqslant s$. This shows that $u(t)$ is totally stable. ∎

Lemma 2.7.2   If a solution $u(t)$ is totally stable then the solutions in $H(u)$ are totally stable.

Proof   By assumption, there exist positive constants $\delta(\varepsilon)$ and $\eta(\varepsilon)$ for a given $\varepsilon > 0$ such that $\tau \geqslant 0$, $|x_\tau - u_\tau|_B < \delta(\varepsilon)$ and $\sup_{t \geqslant \tau} |g(t)| < \eta(\varepsilon)$ implies $|x(t) - u(t)| < \varepsilon$ for all $t \geqslant \tau$ and any solution x of

$$\frac{dx}{dt} = F(t, x_t) + g(t).$$

On the other hand, since $\bar{U}(\{u_0\}, NH, L(H))$ is compact, there exists a $\beta > 0$ such that for any sequence $\{t_k\}, t_k \to \infty$ as $k \to \infty$, the sequence $\{F(t + t_k, \varphi)\}$ contains a subsequence that converges uniformly on $I \times \Omega_\beta, \Omega_\beta = \{\varphi \in B : |\varphi - \psi|_B < \beta$ for some $\psi \in \bar{U}(\{u_0\},$ $NH, L(H))$. In fact, if not, there exists a constant $\varepsilon > 0$ and sequences $\{t_k\}, t_k \to \infty, \{s_k\},$ $s_k \geqslant 0$, and $\{\varphi_k\}, \varphi_k \in \Omega$, for which $F(s_k + t_k, \varphi_k) - G(s_k, \varphi_k)| \geqslant \varepsilon$, where we may assume that $\{F(t + t_k, \varphi)\}$ converges to $G(t, \varphi)$ uniformly on $I \times S$ for any compact $S \subset B$. This yields a contradiction, because $\{\varphi_k\}$ is relatively compact.

Let $(v, G) \in H(u, F)$. Then there exists a $\{t_k\}, t_k \to \infty$ as $k \to \infty$, such that $u(t + t_k) \to v(t)$ uniformly an any compact set of I and $F(t + t_k, \varphi) \to G(t, \varphi)$ uniformly on any $I \times \Omega_\beta$. Let $y(t)$ be a solution of

$$\frac{dy}{dt} = G(t, y_t) + h(t)$$

satisfying $|y_\tau - v_\tau|_B < \delta(\tfrac{1}{2}\varepsilon)$ and $\sup_{t \geqslant \tau} |h(t)| < \eta(\tfrac{1}{2}\varepsilon)$, and assume that

$$|y(s) - v(s)| = \varepsilon, \qquad |y(t) - v(t)| \leqslant \varepsilon \quad t \in [\tau, s]$$

for a $\varepsilon \in (0, \varepsilon_0)$, where $K\varepsilon_0 + \sup_{t \geqslant 0} M(t)\delta(\varepsilon_0) < \beta$. Let k be sufficiently large that

$$|u_{t_k + \tau} - v_\tau|_B < \delta(\tfrac{1}{2}\varepsilon) - |y_\tau - v_\tau|_B,$$

$$|F(t + t_k, \varphi) - G(t, \varphi)| < \eta(\tfrac{1}{2}\varepsilon) - \sup_{t \geqslant \tau} |h(t)| \quad \text{on } I \times \Omega_\beta,$$

$$|u(t + t_k) - v(t)| < \tfrac{1}{2}\varepsilon \quad \text{on } [\tau, s].$$

Then we have $|y_\tau - u_{t_k + \tau}|_B < \delta(\tfrac{1}{2}\varepsilon)$ and $|g(t)| < \eta(\tfrac{1}{2}\varepsilon)$ for $g(t) = h(t) + F(t + t_k, y_t) - G(t, y_t)$. Thus $|y(t + t_k) \leftarrow u(t)| < \tfrac{1}{2}\varepsilon$ for all $t \in [\tau + t_k, s + t_k]$, since $y(t)$ is a solution of

$$\frac{dy}{dt} = F(t + t_k, y_t) + g(t) \quad \text{on } \tau \leqslant t \leqslant s.$$

Therefore we have

$$|y(t) - v(t)| \leqslant |y(t) - u(t + t_k)| + |u(t + t_k) - v(t)| < \varepsilon,$$

which implies a contradiction, and $v(t)$ is totally stable.   ∎

Lemma 2.7.3   If $v \in H(u)$ then $H(v) = H(u)$.

Proof   Let $u(t + t_k) \to v(t)$ for a sequence $\{t_k\}, t_k \to \infty$. If $v(t + s_k) \to w(t)$ then for any k there exist k' and k'' such that

$$|u(t + t_{k'} + s_{k''}) - w(t)| < \frac{1}{k};$$

that is, $w \in H(u)$. Hence $H(v) \subset H(u)$.

Conversely, let $u(t + s_k) \to w(t)$. Here we may assume that $s_k \geqslant 2t_k$. Therefore $s_k - t_k = r_k \to \infty$, and we have $v(t + t_k) - u(t + t_k + r_k) \to 0$, and hence $w \in H(v)$, since $u(t + t_k + r_k) = u(t + s_k)$; that is, $H(u) \subset H(v)$.   ∎

Theorem 2.7.1   If a solution $u(t)$ of (2.7.1) is totally stable, then it is asymptotically almost-periodic in t.

Proof   What we should prove is that for any sequence $\{t_k\}$, $t_k \to \infty$, the sequence $\{u(t + t_k)\}$ contains a subsequence that converges uniformly on I.

Here we may assume that $u(t + t_k) \to v(t)$, $(v, G) \in H(u, F)$, uniformly on any compact set of I. By Lemma 2.7.2, the solution $v(t)$ of (2.7.2) is totally stable, while $u(t + t_k)$ satisfies

$$\frac{dx}{dt} = G(t, x_t) + F(t + t_k, u_{t+t_k}) - G(t, u_{t+t_k}),$$

and hence

$$|u_{t_k} - v_0|_B < \delta(\varepsilon) \text{ implies } |u(t + t_k) - v(t)| < \varepsilon \text{ for all } t \geqslant 0;$$

that is, $\{u(t + t_k)\}$ converges uniformly on I.   ∎

Combining Theorem 2.7.1 and Lemma 2.7.3, we have the following corollary.

Corollary to Theorem 2.7.1   If the system (2.7.1) admits a limiting equation (2.7.2) whose solution $v(t)$, $(v, G) \in H(u, F)$, is totally stable then $u(t)$ is asymptotically almost-periodic in t.

2.8   Eventual Stability under Perturbations

2.8.1   *Definitions and assumptions*

Let T, L and M be normed spaces in the space $C(I, R^n)$ with norms $\|\cdot\|_T$, $\|\cdot\|_L$ and $\|\cdot\|_M$ respectively, defined by

$$\|h\|_T = \sup_{t \geqslant 0} |h(t)|,$$

$$\|h\|_L = \int_0^\infty |h(s)| \, ds,$$

$$\|h\|_M = \sup_{t \geqslant 0} \int_t^{t+1} |h(u)| \, du.$$

We suppose for the system (2.2.1) that the vector function $F(t, \varphi): I \times B \to R^n$ is continuous and that its solution $u(t):]-\infty, \infty[\to R^n$ with initial condition $u_0 \in B$ is uniformly continuous on I and $\sup_{t \geqslant 0} |u_t|_B = H < \infty$.

Together with (2.2.1) we consider the system (2.6.7). Suppose that $R(t, \varphi) \in C(I \times B, R^n)$ and the following conditions are satisfied:

(i) for any $H > 0$ there exists a constant $L(H) > 0$ such that $|F(t, \varphi)| \leqslant L(H)$ for all $t \geqslant 0$ and $\varphi \in B$ such that $|\varphi|_B \leqslant H$;

(ii) the hull $H(F)$ is compact in $C(I \times B, R^n)$;

(iii) $\sup_{0 \leqslant u \leqslant 1} |\int_t^{t+u} R(s, \varphi) \, ds| \to 0$ for each φ from B when $t \to \infty$, and there exist $r > 0$ and functions $b(s)$, with $b(0) = 0$ and $b(s)$ nondecreasing in s, and $h(t)$, with $\int_t^{t+1} |h(s)| \, ds \to 0$ as $t \to \infty$, such that

$$|R(t, \varphi) - R(t, \psi)| \leqslant b(|\varphi - \psi|_B) + h(t)$$

for all $t \geqslant 0$ and all φ and ψ satisfying the conditions $|\varphi|_B \leqslant H$ and $|\psi|_B \leqslant H + r$.

Kato and Yoshizawa [1, 2] proposed the following definitions. (Here and below X denotes one of the spaces $\{0\}$, T, L or M.)

Definition 2.8.1   A solution $u(t)$ of (2.2.1) is said to be

(a) *eventually stable under perturbations from the space X ($E_v S X$)* if for any $\varepsilon > 0$ there exist $\alpha = \alpha(\varepsilon) \geqslant 0$ and $\delta = \delta(\varepsilon) > 0$ such that $|u_s - \varphi|_B < \delta(\varepsilon)$, $s \geqslant \alpha(\varepsilon)$ and $\| R \|_X < \delta(\varepsilon)$ imply $|u_t - x_t(s, \varphi, F + R)|_B < \varepsilon$ for all $t \geqslant s$;

(b) *eventually equiasymptotically stable under perturbations from the space X ($E_v E A S X$)* if it is $E_v S X$ and there exist $\delta_0 > 0$ and some $\alpha_0 \geqslant 0$, and for any $\varepsilon > 0$ and $s \geqslant \alpha_0$ there exist $T(\varepsilon, s) > 0$ and $\gamma(\varepsilon) > 0$ such that $|u_s - \varphi|_B < \delta_0$ and $\| R \|_X < \gamma(\varepsilon)$ imply $|u_t - x_t(s, \varphi, F + R)|_B < \varepsilon$ for all $t \geqslant s + T(\varepsilon, s)$.

The qualification "eventually" (E_v) in (a) and (b) is omitted iff $\alpha(\varepsilon) \equiv 0$ and $\alpha_0 \equiv 0$ respectively.

2.8.2   *Some theorems on eventual stability*

It is not difficult to prove the following assertions (see Hale and Kato [1]; Murakami [1]).

Lemma 2.8.1   Let $x_n(t)$, $n = 1, 2, \ldots$, be functions on $]-\infty, T[$, $0 < T \leqslant \infty$, with values in R^n, satisfying the following conditions:

(1) the sequence $\{x_n(t)\}$ converges to a continuous function as $n \to \infty$ uniformly on any compact subset of $]-\infty, T[$;

(2) there exists an $N > 0$ with the property that for each $L > 0$ there exists an $n_0(L) > 0$ such that $(x_n)_{-L} \in B$, $|(x_n)_{-L}|_B \leqslant N$ and $x_n(t)$ is continuous on $[-L, T]$ for all $n \geqslant n_0(L)$.

Then there exists a function $x(\cdot) : (-\infty, T) \to R^n$ such that $x_0 \in B$, $x(t)$ is continuous on $[0, T[$ and $(x_n)_t \to x_t$ as $n \to \infty$ uniformly on any compact subset of $[0, T[$.

Lemma 2.8.2   Let $x_n(t)$, $n = 1, 2, \ldots$, and $x(t)$ be functions defined on $]-\infty, T[$, $0 < T \leqslant \infty$, with values in R^n, satisfying the following conditions:

(1) $x_0 \in B$, $(x_n)_0 > B$, $x(t)$ and $x_n(t)$ are continuous on $[0, T[$ for $n = 1, 2, \ldots$, and $(x_n)_t \to x_t$ as $n \to \infty$ uniformly on any compact subset of $[0, T[$,

(2) $\sup_{0 \leqslant t \leqslant T} |(x_n)_t|_B \leqslant r_1$ for a constant $r_1 > 0$ and all $n = 1, 2, \ldots$;

(3) $x_n(t)$ is a solution of

$$\frac{dx}{dt} = F_n(t, x_t) + G_n(t, x_t)$$

defined on $[0, T[$ for each $n = 1, 2, \ldots$, where F_n and G_n are continuous functions satisfying

(i) $F_n(t, \varphi) \to F^*(t, \varphi)$ as $n \to \infty$ uniformly on any compact subset of $I \times B$;
(ii) $\sup_{0 \leqslant u \leqslant 1} |\int_t^{t+u} G_n(s, \varphi) ds| \to 0$ as $n \to \infty$ for each φ in B, and

$$|G_n(t, \varphi) - G_n(t, \psi)| \leqslant b(|\varphi - \psi|_B) + h_n(t) \quad \forall t \geqslant 0.$$

with φ and ψ satisfying $|\varphi|_B \leqslant r_1$ and $|\psi|_B \leqslant r_1$, where $b(s)$ is continuous and nondecreasing in $s \geqslant 0$, with $b(0) = 0$, $h_n \in M$, with $\|h_n\|_M \to 0$ as $n \to \infty$.

Then $x(t)$ is a solution of

$$\frac{dx}{dt} = F^*(t, x_t)$$

defind on $[0, T[$.

Lemma 2.8.3   Suppose that $u(t)$ is the unique solution of (2.2.1) for the initial conditions. If $u(t)$ is $E_w SX$ (respectively $E_v EASX$) for (2.2.1), then it is SX (respectively $E_q ASX$) for (2.2.1).

Theorem 2.8.1   If $u(t)$ is attracting in $H(F)$ and if $u(t)$ is $E_v SX$ for (2.6.7) then it is ASX for (2.6.7). Furthermore, if $u(t)$ is SX for (2.6.7) then it is ASX for (2.6.7)

Proof   We shall employ arguments similar to those in Hale and Kato [1] although the proof is cumbersome because of the presence of a perturbation term G. Let δ_1 be the number arising for attraction in $H(F)$ of $u(t)$, and assume that $u(t)$ is $E_v SX$ for (2.2.1). We may assume $\delta_1 < r$, where r is the number given in (iii) in Section 2.8.1. Define $\delta_0 = \delta(\frac{1}{2}\delta_1)$ and $\alpha_0 = \alpha(\frac{1}{2}\delta_1)$, where $\delta(\cdot)$ and $\alpha(\cdot)$ are as given in Definition 2.8.1 (a). For these δ_0 and α_0 and any $\varepsilon > 0$, we shall show that there exist numbers $T(\varepsilon)$ and $\gamma(\varepsilon)$ satisfying the condition in Definition 2.8.1(b). Indeed, suppose that this is not the case. Then there exist an $\varepsilon > 0$, sequences $\{\tau_n\}$, $\tau_n \geqslant \alpha_0$, $\{t_n\}$, $t_n \geqslant \tau_n + 2n$, $\{\varphi_n\} \subset B$, $|\varphi_n - u_{\tau_n}|_B < \delta_0$, $\{p_n\} \subset X$, $\|p_n\|_X < \min(1/n, \delta_0)$, and $\{x(\cdot, \tau_n, \varphi_n, F + R + p_n)\}$ such that

$$|x_{t_n}(\tau_n, \varphi_n, F + R + p_n) - u_{t_n}|_B \geqslant \varepsilon \tag{2.8.1}$$

for all $h = 1, 2, \ldots$,
Since $\tau_n \geqslant \alpha(\frac{1}{2}\delta_1)$, $|\varphi_n - u_{\tau_n}|_B < \delta(\frac{1}{2}\delta_1)$ and $\|p_n\|_X < \delta(\frac{1}{2}\delta_1)$, by the assumption that $u(t)$ is $E_v SX$ for (2.6.7), we have

$$|(x_n)_t - u_t|_B < \tfrac{1}{2}\delta_1 \tag{2.8.2}$$

for all $t \geqslant \tau_n$ and $n = 1, 2, \ldots$, where $x_n(t) = x(t, \tau_n, \varphi_n, F + R + p_n)$, $t \geqslant \tau_n$. Select an integer $n_0 = n_0(\varepsilon)$ such that $\tau_n + n \geqslant \alpha(\varepsilon)$ and $\|p_n\|_X < 1/n < \delta(\varepsilon)$ if $n \geqslant n_0$. Then, by (2.8.1) and the assumption that $u(t)$ is $E_v SX$ for (2.6.7), we have

$$|(x_n)_t - u_t|_B \geqslant \delta(\varepsilon) \quad \forall t \in [\tau_n + n, \tau_n + 2n] \tag{2.8.3}$$

if $n \geqslant n_0$. Taking a subsequence if necessary, we may assume that for a $(v, G) \in H(u, F)$,

$$(u_{\tau_n + n}, F_{\tau_n + n}) \to (v, G) \quad \text{as } n \to \infty. \tag{2.8.4}$$

Now (2.8.2) implies $|(x_n)_s|_B \leqslant |u_s|_B + \frac{1}{2}\delta_1 \leqslant H + \frac{1}{2}\delta_1 = r_1$, and consequently $|F(s,(x_n)_s)| \leqslant L(r_1) = L < \infty$ for all $s \geqslant \tau_n$ by (i) in Section 2.8.1. Let $N > 0$ be given. When $t_1, t_2 \in [-N, N]$, $0 < t_1 - t_2 < 1$ and $n \geqslant 2N$, we have, by (ii),

$$|x_n(t_1 + \tau_n + n) - x_n(t_2 + \tau_n + n)|$$

$$\leqslant \int_{t_2+\tau_n+n}^{t_1+\tau_n+n} |F(s,(x_n)_s)|\,ds + \left| \int_{t_2+\tau_n+n}^{t_1+\tau_n+n} R(s,(x_n)_s)\,ds \right| + \int_{t_2+\tau_n+n}^{t_1+\tau_n+n} |p_n(s)|\,ds$$

$$\leqslant L(t_1 - t_2) + \|p_n\|_X + \int_{t_2+\tau_n+n}^{t_1+\tau_n+n} |R(s,(x_n)_s) - R(s,\varphi)|\,ds + \left| \int_{t_2+\tau_n+n}^{t_1+\tau_n+n} R(s,\varphi)\,ds \right|$$

$$\leqslant [L + b(2r_1)](t_1 - t_2) + \frac{1}{n} + \sup_{t \geqslant n/2} \int_t^{t+1} |h(s)|\,ds + \sup_{0 \leqslant u \leqslant 1, t \geqslant n/2} \left| \int_t^{t+u} R(s,\varphi)\,ds \right|,$$

where φ is an arbitrarily chosen element in B such that $|\varphi|_B \leqslant r_1$. Hence, we can conclude that $\{\bar{x}_n(t)\}$ is uniformly bounded and equicontinuous on $[-N, N]$, where $\bar{x}_n(t) = x_n(t + \tau_n + n)$, $t \geqslant -n$, because

$$\sup_{0 \leqslant u \leqslant 1, t \geqslant n/2} \left| \int_t^{t+u} R(s,\varphi)\,ds \right| \to 0$$

and

$$\sup_{t \geqslant n/2} \int_t^{t+h} |h(s)|\,ds \to 0$$

as $n \to \infty$. Consequently, since $N > 0$ is arbitrary, from Ascoli's theorem and the diagonalization procedure, it follows that there exists a subsequence of $\{\bar{x}_n(t)\}$ that converges to a continuous function as $n \to \infty$ uniformly on any compact subset of R. Here $|(\bar{x}_n)_t|_B = |(x_n)_{t+\tau_n+n}|_B \leqslant r_1$ for all $t \geqslant -n$. Therefore, by Lemma 2.8.1, there exists a function $x(\cdot) : R \to R^n$ such that $x_0 \in B$, $x(t)$ is continuous on I and $(x_n)_t \to x_t$ as $n \to \infty$ uniformly on any compact subset of I. On the other hand, since $\bar{x}_n(t)$ is a solution of

$$\frac{dx}{dt} = F(t + \tau_n + n, x_t) + R(t + \tau_n + n, x_t) + p_n(t + \tau_n + n),$$

by applying Lemma 2.8.2 to $\{\bar{x}_n(t)\}$ and $x(t)$, we can conclude that $x(t)$ is a solution of $dx/dt = G(t, x_t)$ defined on I. Since (2.8.2) and (2.8.3) imply that

$$\delta(\varepsilon) \leqslant |(x_n)_t - u_{t+\tau_n+n}|_B = |(x_n)_{t+\tau_n+n} - u_{t+\tau_n+n}|_B < \frac{1}{2}\delta_1$$

for all $t \in [0, n]$ if $n \geqslant n_0$, on letting $n \to \infty$ in this inequality it follows from (2.8.4) that

$$\delta(\varepsilon) \leqslant |x_t - v_t|_B \leqslant \frac{1}{2}\delta_1 \quad \forall t \in I. \tag{2.8.5}$$

Since $|x_0 - v_0|_B < \delta_1$, we must have $|x_t - v_t|_B \to 0$ as $t \to \infty$, because $u(t)$ is attracting in $H(F)$. This is a contradiction to (2.8.5), which shows that $u(t)$ is $E_V ASX$ for (2.6.7).

Next, suppose that $u(t)$ is SX for (2.6.7). Clearly, $u(t)$ is the unique solution of (2.6.7) for initial conditions. Therefore $u(t)$ is ASX for (2.6.7) by Lemma 2.8.3, since $u(t)$ is $E_V ASX$ for (2.6.7). This completes the proof. ∎

Example 2.8.1 Consider the second-order scalar delay–differential equation

$$\ddot{x}(t) + p(t, x(t), \dot{x}(t))\dot{x}(t) + q(x(t - r(t))) = G(t)x(t - h_1),\tag{2.8.6}$$

and the equivalent system to it,

$$\frac{dx}{dt} = y(t)$$

$$\frac{dy}{dt} = -p(t, x(t), y(t))y(t) - q(x(t))\tag{2.8.7}$$

$$+ G(t)x(t - h_1) + \int_{-r(t)}^{0} \dot{q}(x(t + s))y(t + s)ds,$$

where $0 \leqslant r(t)$, $h_1 \leqslant h < \infty$, and suppose that

$$G(t) \text{ is bounded and } \sup_{0 \leqslant u \leqslant 1} \left| \int_{t}^{t+u} G(s)\,ds \right| \to 0 \quad \text{as } t \to \infty.\tag{2.8.8}$$

Assume for the system (2.8.7) that $p(t, x, y)$ and $r(t)$ are bounded on $I \times$ (any bounded set of $R \times R$) and $p(t, x, y) \geqslant dh + e$, $|\dot{q}(x)| \leqslant L < d$ and $xq(x) > 0$ ($|x| < c$) for some positive constants c, d, e and L.

The system (2.8.7) is considered as an FDE on the space $C([-h, 0])$. We shall show that $(0, 0)$ is $E_v ASX$ for (2.8.7). Indeed, the limiting equations of the system (2.8.7) are of the form

$$\frac{dx}{dt} = y(t)$$

$$\frac{dy}{dt} = -\tilde{p}(t, x(t), y(t))y(t) - \dot{\tilde{q}}(x(t)) + \int_{-\tilde{r}(t)}^{0} \dot{\tilde{q}}(x(t + s))y(t + s)ds,\tag{2.8.9}$$

where $\tilde{p}(t, x, y)$ and $\tilde{r}(t)$ satisfy the same condition as given above for p, and r, with the same constants c, d, e and L. Select a constant δ, $L/d < \delta < 1$, and define a functional $V(\varphi, \psi)$ on $C([-h, 0])$ by

$$V(\varphi, \psi) = \int_{0}^{\varphi(0)} q(s)ds + \tfrac{1}{2}\psi^2(0) + \tfrac{1}{2}d\delta \int_{-h}^{0}\left[\int_{s}^{0}\psi^2(u)du\right]ds.$$

An easy computation shows that $\dot{V}_{(2.8.9)}(\varphi, \psi) \leqslant -e\psi^2(0) \leqslant 0$. Hence, with a few modifications of the arguments in Hale [2; Razumikhin [1, 2]; Theorem 5.2 and pp. 118–126], we conclude that $(0, 0)$ is UAS in $H(F)$.

2.9   Comments and References

2.0   In a number of works (e.g. Burton [1, 3]; Driver [1]; Hale [2]; Krasovsky [1]; Rezvan [1] Razumikhin [1, 2] results on FDEs have been given not only in classic area but also in the context of the problems considered here.

2.1   The source of results in this section is Hale and Kato [1].

2.2   The definitions of stability in a space B and in R^n for systems of FDEs are formulated on the basis of work by Hale and Kato [1] and Hino [2], as was done for ODEs by Grujić, Martynyuk and Ribbens-Pavella [1].

2.3   The construction of limiting systems and the investigation of their properties here follows Hale and Kato [1] as well as Hino [2] and Kato [2].

2.4   In this section we have followed Hale and Kato [1].

2.5   In this section the account follows Kato [4].

2.6   The main results of this section are contained in the work of Hale and Kato [1] and Kato [2, 5].

2.7   The source of the results in this section is Hino and Yoshizawa [1]. See also Fink [1].

2.8   In this section our account follows Hale and Kato [1] and Murakami [1].

3 Limiting Systems and Stability of Motion under Small Forces

3.0 Introduction

This chapter concentrates on the stability analysis of solutions of systems of differential equations involving small parameters. The problem is solved in terms of a new generalization of Lyapunov's direct method for investigation of the stability of motion. This generalization is developed for the following types of systems:

(i) large-scale systems with weakly interacting subsystems;
(ii) oscillating systems;
(iii) systems with nonasymptotically stable unperturbed systems,
(iv) systems that can be investigated via perturbed Lyapunov functions only.

The systems of differential equations considered in this chapter are applicable in the theory of nonlinear oscillations to mechanics, physics, electronics, electrical engineering and biology.

3.1 Preliminary Remarks

3.1.1 Description of the system

We investigate properties of solutions of systems of the form

$$\frac{dx}{dt} = f(t, x) + \mu r(t, x), \tag{3.1.1}$$

where μ is a small parameter. These properties are related to the classical definitions of stability and instability of the equilibrium state $x = 0$ of the system

$$\frac{dx}{dt} = f(t, x), \quad f(t, 0) = 0, \tag{3.1.2}$$

under small persistent perturbations.

The Lyapunov function used to investigate these properties is that corresponding to the unperturbed system (3.1.2). It is assumed that the general solution of the limiting system

$$\frac{dx}{dt} = g(t, x), \quad x(t_0) = x_0, \tag{3.1.3}$$

is known. The method of constructing this limiting system is described below.

Assumption 3.1.1   For the systems (3.1.1)–(3.1.3) the following conditions hold.

(1) $x \in R^n$ and $\mu \in]0, \mu^*[$, with $\mu^* = \text{const} > 0$.
(2) The vector functions $X(t, x)$ and $r(t, x)$ are defined on the domain $\mathscr{D} = R_+ \times \Omega$, with $R_+ = [0, \infty[$ and $\Omega = \{x \in R^n : \|x\| < H = \text{const} > 0\}$, and satisfy conditions guaranteeing the existence and uniqueness of the solution of the Cauchy problem for the system (3.1.1) and, moreover,
  (i) on the domain $\mathscr{D}$ the vector function $f(t, x)$ satisfies a Lipschitz conditions on x, i.e. for all $t \in R_+$, $\|f(t, x) - f(t, x')\| \leqslant L \|x - x'\| \quad \forall(t, x), (t, x') \in \mathscr{D}$, where $L = \text{const} > 0$;
  (ii) for the vector function $r(t, x)$ there exist a summable function $M(t)$ and a constant M_0 such that on any finite interval $[t_1, t_2]$

  $$\|r(t, x)\| \leqslant M(t) \quad \text{on } \mathscr{D},$$

  $$\int_{t_1}^{t_2} M(t)\, dt \leqslant M_0(t_2 - t_1) \tag{3.1.4}$$

(3) The unperturbed system (3.1.2) has a nonasymptotically stable equilibrium state $x = 0$.
(4) The vector function $g(t, x)$ is continuous and satisfies a Lipschitz condition on x with constant L in a domain $R_+ \times \Omega_1$, where $\Omega_1 \subseteq \Omega$, and the general solution $x(t; g) = x(t; t_0, x_0; g)$ of the system (3.1.3) exists for all $(t_0, x_0) \in R_+ \times \Omega_1$ and $t \geqslant t_0$.

We introduce the following notation. Let $V \in C^1(R_+ \times \Omega, R_+)$ and

$$(\text{grad } V)^T = \left(\frac{\partial V}{\partial x_1}, \frac{\partial V}{\partial x_2}, \dots, \frac{\partial V}{\partial x_n} \right),$$

$$\dot{V}(t, x) = \frac{\partial V}{\partial t} + (\text{grad } V)^T f(t, x),$$

$$\varphi(t, x) = (\text{grad } V)^T r(t, x),$$

$$\varphi^0(\theta, t_0, x_0) = \int_{t_0}^{t_0 + \theta} \varphi(t; x(t; t_0, x_0; g))\, dt.$$

B_ρ is an open sphere with radius $\rho > 0$, $\rho < H$, and centre at the origin. As usually positive-definite function $V(t, x)$, allowing an infinitesimally small upper limit in the domain $\mathcal{D}$, there exist functions $a(r)$ and $b(r)$ of class K such that

$$a(\|x\|) \leqslant V(t, x) \leqslant b(\|x\|) \quad \forall (t, x) \in \mathcal{D}.$$

Then we have the following conditions.

(5) For any value of $\rho \in]0, H[$ the limit

$$\lim_{t \to \infty} \|f(t, x) - g(t, x)\| = 0 \tag{3.1.5}$$

holds uniformly on $x \in B_\rho$.

(6) There exist a summable function $F(t)$, a function $\chi(r) \in \mathcal{K}$ and a constant F_0 such that for $t \in R_+$ and any x', x'' from Ω the inequalities

$$|\varphi(t, x') - \varphi(t, x'')| \leqslant F(t)\chi(\|x' - x''\|),$$

$$\int_{t_1}^{t_2} F(t)dt \leqslant F_0(t_2 - t_1) \tag{3.1.6}$$

hold on any finite interval $[t_1, t_2] \subset R_+$.

The following properties of the solutions of the system (3.1.1) are investigated.

Definition 3.1.1   The system (3.1.1) is μ-*stable* if for any $\varepsilon > 0$ an $\eta(\varepsilon) > 0$ and a $\mu_0(\varepsilon) > 0$ can be found such that the condition $(t_0, x_0) \in R_+ \times B_\eta$ implies the inequality $\|x(t; t_0, x_0, \mu)\| < \varepsilon$ for all $t \geqslant t_0$ when $\mu < \mu_0(\varepsilon)$.

Definition 3.1.2   The system (3.1.1) is μ-*stable with respect to some variables* if for any $\varepsilon > 0$ there exist an $\eta(\varepsilon) > 0$ and a $\mu_0(\varepsilon) > 0$ such that the condition $(t_0, x_0) \in R_+ \times B_\eta$ implies the inequality $\|y(t; t_0, x_0, \mu)\| < \varepsilon$ for all $t \geqslant t_0$ when $\mu < \mu_0(\varepsilon)$ and $y(t, \cdot)$ is a subvector of the solution $x(t, \cdot) = (y^T(t, \cdot), z^T(t, \cdot))^T$ of the system (3.1.1).

Definition 3.1.3   The system (3.1.1) is μ-*unstable* if its solution is not μ-stable in terms of Definition 3.1.1.

Remark 3.1.1   In the case $r(t, 0) = 0$ we have the property of stability in the Lyapunov sense for the system obtained by adding the equation $\dot{\mu} = 0$ to the system (3.1.1).

3.1.2   Preliminary results

The following assertions concern the closeness of solutions of the system (3.1.1) to those of the system (3.1.2).

Lemma 3.1.1   If conditions (2) (i) and (ii) of Assumption 3.1.1 are satisfied then the estimate

$$\|x(t) - \bar{x}(t)\| \leqslant \mu M_0 l \exp(Ll)$$

holds for the solution $x(t) = x(t; t_0, x_0, \mu)$ of the system (3.1.1) and the solution $\bar{x}(t) = \bar{x}(t; t_0, x_0)$ of (3.1.2) for all $t \in [t_0, t_0 + l]$, $0 < l < \infty$, as long as these solutions exist.

Proof This is based on the Gronwall–Bellman lemma for the integral equations obtained from (3.1.1) and (3.1.2).

Lemma 3.1.2 Let $g(t, x)$ satisfy a Lipschitz condition on x with constant L and let conditions (3) and (5) of Assumption 3.1.1 hold in the domain $\mathscr{D}$. Then for any number $\lambda > 0$ there exists a time τ_0 such that for solutions $\bar{x}(t) = \bar{x}(t; \tau, x(\tau))$ and $x(t; g) = x(t; \tau, x(\tau); g)$ $\tau > \tau_0$, $x(\tau) \in \Omega$ of the systems (3.1.2) and (3.1.3) the inequality

$$\| \bar{x}(t) - x(t; g) \| < \lambda \gamma \exp(L\gamma) \qquad (3.1.7)$$

holds on any finite interval $[\tau, \tau + \gamma[$ as long as these solutions are defined.

Proof By condition (5), for any $\lambda > 0$ there exists $\tau_0(\lambda) > 0$ such that

$$\| f(t, x) - g(t, x) \| < \lambda \qquad (3.1.8)$$

for $t > \tau_0$ and $x \in B_\rho$. Using (3.1.2) and (3.1.3), we can obtain

$$\| \bar{x}(t) - x(t; g) \| \leqslant \int_\tau^t \| f(s, \bar{x}(s)) - g(s, \bar{x}(s)) \| \, ds + \int_\tau^t \| g(s, \bar{x}(s)) - g(s, x(s; g)) \| \, ds$$

for $\tau > \tau_0$ and $t \in [\tau, \tau + \gamma]$ as long as $\bar{x}(t)$, $x(t; g)$ exist.

$$\| \bar{x}(t) - x(t; g) \| \leqslant L \int_\tau^t \| \bar{x}(s) - x(s; g) \| \, ds + \lambda \gamma, \qquad t \in [\tau, \tau + \gamma].$$

Applying the Gronwall–Bellman inequality, we obtain the estimate (3.1.7).

Lemma 3.1.3 Let all hypotheses of Lemma 3.1.1 be satisfied and, moreover, let the condition $f(t, 0, z) = 0$ hold for the vector function $f(t, y, z)$ for all $t \in R_+$ and $z \in \{z \in R^{n-p} : \|z\| < \infty\}$. Then for solutions $x(t) = x(t; t_0, x_0) = (y^T(t; t_0, x_0), z(t; t_0, x_0)^T)$ of the system (3.1.1) the inequality

$$\| y(t; t_0, x_0) \| \leqslant [\| y_0 \| + \mu M_0 \tau_1] \exp(L\tau_1)$$

holds on any finite interval $[t_0, t_0 + \tau_1]$ for $(t_0, x_0) \in \mathscr{D}$ as long as the solutions exist.

Proof For the system (3.1.1), we have

$$\| y(t; t_0, x_0) \| \leqslant \| y_0 \| + \int_{t_0}^t \| f(s, x(s)) \| \, ds + \mu \int_{t_0}^t \| r(s, x(s)) \| \, ds.$$

in view of the inequality (3.1.4), the relation $f(t, 0, z) = 0$ and the Lipschitz condition for $f(t, y, z)$ on (y, z), we find

$$\| y(t; t_0, x_0) \| \leqslant \| y_0 \| + \mu M_0 (t - t_0) + L \int_{t_0}^t \| y(s; t_0, x_0) \| \, ds.$$

Applying the Gronwall–Bellman lemma to this inequality, we get the estimate (3.1.9).

3.2 A Stability Theorem

Systems with small parameters have been considered in many previous investigations. Below we present results obtained by combining the averaging method and the method of Lyapunov functions on the basis of limiting equations.

3.2.1 General theorem

We establish sufficient conditions for stability of solutions of the system (3.1.1) using the limiting system (3.1.3), which may be simpler (for example it may be integrable even if unperturbed system (3.1.2) is not).

Theorem 3.2.1 For the system (3.1.2) let the following hypotheses hold in the domain $\mathscr{D}$.

(1) There exist a Lyapunov function $V(t, x)$ and functions $a, b \in \mathscr{K}$ such that

$$a(\|x\|) \leqslant V(t, x) \leqslant b(\|x\|), \qquad \dot{V}(t, x) \leqslant 0.$$

(2) Conditions (1)–(6) of Assumption 3.1.1 hold.

(3) For any numbers α and β $(0 < \alpha < \beta < H)$ there exist positive numbers μ^0, δ and l such that for every pair of values of t and $x(t \in R_+, \alpha \leqslant \|x\| \leqslant \beta)$ one of the following conditions holds:

  (i) $\dot{V}(t, x) + \mu \varphi(t, x) \leqslant 0$ for $\mu < \mu^0$;

  (ii) $\varphi^0(\theta, t, x) \leqslant -\delta\theta$ for $\theta > l$.

Then the system (3.1.1) is μ-stable.

Proof Let $\varepsilon \in {]}0, H{[}$ be given. We describe the method of constructing numbers $\eta(\varepsilon)$ and $\mu^0(\varepsilon)$ mentioned in the definition of μ-stability of the solution $x(t, t_0, x_0, \mu)$. By condition (1) of the theorem, all points of the moving surface $V(t, x) = a(\frac{1}{2}\varepsilon)$ will satisfy the condition

$$b^{-1}(a(\tfrac{1}{2}\varepsilon)) \leqslant \|x\| \leqslant \tfrac{1}{2}\varepsilon \tag{3.2.1}$$

for all $t \in R_+$. By condition (3) of the theorem, there exist positive numbers δ and l for the numbers $b^{-1}(a(\frac{1}{2}\varepsilon))$ and $\frac{1}{2}\varepsilon$. From Lemma 3.1.2, there exists a time τ_0 for

$$\lambda = \frac{\min\{\chi^{-1}(\delta/4F_0), \tfrac{1}{2}\varepsilon\}}{2l \exp(2Ll)} \tag{3.2.2}$$

such that the inequality

$$\|\bar{x}(t; \tau, x(\tau)) - x(t; \tau, x(\tau); g)\| \leqslant \min\{\chi^{-1}(\delta/4F_0), \tfrac{1}{2}\varepsilon\} \tag{3.2.3}$$

holds when $\tau > \tau_0$ and $t \in [\tau, \tau + 2l]$ for the solutions $\bar{x}(t; \tau, x(\tau))$ and $x(t; \tau, x(\tau); g)$ of the systems (3.1.2) and (3.1.3), where $V(\tau, x(\tau)) = a(\frac{1}{2}\varepsilon)$. By Lemma 3.1.3, when $p = n$, we obtain the estimate

$$\|x(t; t_0, x_0)\| \leqslant (\|x_0\| + \mu M_0 \tau_1) \exp(L\tau_1) \tag{3.2.4}$$

for all $t_0 \in [0, \tau_1]$ and $t \in [t_0, \tau_1]$ (we set $\tau_1 > \tau_0$) for any solution $x(t) = x(t; t_0, x_0)$ of the system (3.1.1).

We take

$$\mu_1 = \frac{b^{-1}(a(\frac{1}{2}\varepsilon))}{2M_0\tau_1 \exp(L\tau_1)}$$

$$\eta = \frac{b^{-1}(a(\frac{1}{2}\varepsilon))}{2\exp(L\tau_1)}.$$

Then the estimate (3.2.4) implies $\|x(t; t_0, x_0)\| < b^{-1}(a(\frac{1}{2}\varepsilon))$ for $\mu < \mu_1$, $\|x_0\| < \eta$, $t_0 \in [0, \tau_1]$ and $t \in [t_0, t_0 + \tau_1]$. Thus, to prove the theorem, it is sufficient to show that $\|x(t; t_0', x_0')\| < \varepsilon$ for $t_0' \geq \tau_1$, $\|x_0'\| < b^{-1}(a(\frac{1}{2}\varepsilon))$ and all $t \geq t_0'$.

Consider a solution $x(t)$ of (3.1.1) for a small μ as indicated later, assume that it leaves the domain $\|x\| < b^{-1}(a(\frac{1}{2}\varepsilon))$, and let $V(\tau, x(\tau)) = a(\frac{1}{2}\varepsilon)$ at some time $t = \tau$, and let one of conditions (3)(i) or 3(ii) of the theorem be satisfied.

Let condition (3)(i) be satisfied at time τ. Then there exist a μ' for the numbers $b^{-1}(a(\frac{1}{2}\varepsilon))$ and $\frac{1}{2}\varepsilon$ such that the total derivative of the Lyapunov function for the system (3.1.1) is nonpositive at τ for $\mu < \mu'$, and thus the solution $x(t)$ cannot intersect the surface $V(t, x) = a(\frac{1}{2}\varepsilon)$ from below at time τ.

Suppose that condition (3)(ii) is satisfied at time τ. Then there exist numbers δ and l for numbers $b^{-1}(a(\frac{1}{2}\varepsilon))$ and $\frac{1}{2}\varepsilon$ such that

$$\varphi^0(\theta, \tau, x(\tau)) \leq -\delta\theta. \tag{3.2.5}$$

for $\theta > l$. Integrating the expression for the total derivative of the Lyapunov function for the system (3.1.1), we get

$$V(t, x(t)) \leq V(\tau, x(\tau)) + \mu \int_\tau^t \varphi(s, x(s, \tau, x(\tau))) ds \tag{3.2.6}$$

for $t > \tau$. The integral in this inequality can be represented in the form

$$\int_\tau^t \varphi(s, x(s)) ds = \int_\tau^t [\varphi(s, x(s)) - \varphi(s, \bar{x}(s))] ds$$

$$+ \int_\tau^t [\varphi(s, \bar{x}(s)) - \varphi(s, x(s; g))] ds + \int_\tau^t \varphi(s, x(s; g)) ds. \tag{3.2.7}$$

By Lemma 3.1.1, the estimate

$$\|x(t) - \bar{x}(t)\| \leq \mu M_0 2l \exp(2lL) \tag{3.2.8}$$

holds for the solutions $x(t)$ and $\bar{x}(t)$ when $t \in [\tau, \tau + 2l]$. We take $\mu_2 = \frac{1}{4}\varepsilon/M_0 l \exp(2lL)$. Then we find from the inequality (3.2.8) that $\|x(t) - \bar{x}(t)\| < \frac{1}{2}\varepsilon$ for $\mu < \mu_2$ and $t \in [\tau, \tau + 2l]$.

Taking into account that, by condition (1) of the theorem, the unperturbed motion $x = 0$ of the system (3.1.2) is stable, i.e. $\|x(t)\| < \frac{1}{2}\varepsilon$, we get $\|x(t)\| < \varepsilon$ for $t \in [\tau, \tau + 2l]$.

Let

$$\mu_3 = \frac{\chi^{-1}(\delta/4F_0)}{2M_0 l \exp(2Ll)}.$$

Then using condition (6) of Assumption 3.1.1 and the inequality (3.2.8), we find that

$$\int_\tau^t |\varphi(s, x(s)) - \varphi(s, \bar{x}(s))| ds \leqslant \tfrac{1}{4} \delta(t - \tau) \tag{3.2.9}$$

for $\mu < \mu_3$ and $t \in [\tau, \tau + 2l]$. The inequality (3.2.3) shows that the solution $x_g(t)$ will not leave an ε-neighbourhood of $x(t)$ on the interval $[\tau, \tau + 2l]$ and also the domain Ω; thus we may use condition (6) of Assumption 3.1.1 to estimate of the second integral in (3.2.7). In view of the inequality (3.2.3), we get

$$\int_\tau^t |\varphi(s, \bar{x}(s)) - \varphi(s, x(s; g))| ds \leqslant \tfrac{1}{4} \delta(t - \tau) \tag{3.2.10}$$

for $t \in [\tau, \tau + 2l]$.

The expression (3.2.7) and the inequalities (3.2.5), (3.2.9) and (3.2.10) imply the estimate

$$\int_\tau^t \varphi(s, x(s)) ds \leqslant -\tfrac{1}{2} \delta(t - \tau)$$

for $\mu < \mu_4 = \min\{\mu_2, \mu_3\}$ and $t \in]\tau + l, \tau + 2l[$. Thus the integral in the inequality (3.2.6) becomes negative for $t \in]\tau + l, \tau + 2l[$. Since the solution $x(t; \tau, x(\tau))$, having left the surface $V(t, x) = a(\tfrac{1}{2}\varepsilon)$, will remain in the domain $\|x\| < \varepsilon$ for $t \in [\tau, \tau + 2l]$ it will return back to the surface in $[\tau + l, \tau + 2l]$. Thus, repeating the procedure we can see that the solution $x(t)$ will not leave the domain $\|x\| < \varepsilon$ for all $t \geqslant t_0$ when $t_0 \in R_+$, $\|x_0\| < \eta$ and $\mu < \mu_0 = \min\{\mu', \mu_1, \mu_4\}$. Thus the theorem is proved. ∎

Remark 3.2.1 In contrast the Malkin's well-known theorem (Malkin [1]), in Theorem 3.2.1 the requirement on the derivative of the Lyapunov function is weakened (uniform asymptotic stability of the equilibrium state $x = 0$ of the unperturbed system is not required).

Definition 3.2.1 $\varphi_0(t_0, x_0)$ is *definite in the set* $E(V^* = 0)$ if for any numbers α and ε $(0 < \alpha < \varepsilon < H)$ numbers $r(\alpha, \varepsilon)$ and $\delta(\alpha, \varepsilon)$, such that there are positive $\varphi_0(t_0, x_0) < \delta(\alpha, \varepsilon)$ for $\alpha < \|x_0\| < \varepsilon$ and $\rho(x_0, E(V^* = 0)) < r(\alpha, \varepsilon)$ for all $t_0 \in [0, \infty[$. Here $\rho(x, M) = \inf\{\|x - x'\| : x' \in M\}$.

The following assertion holds.

Corollary to Theorem 3.2.1 Suppose that

(1) $f(t, x) = g(t, x)$ in the domain $\mathscr{D}$;
(2) there exists a positive-definite function $V(t, x)$ on $\mathscr{D}$ having infinitesimally small upper limit;
(3) there exists a function V^* in the domain Ω such that along the solutions of the system (3.1.2) the total derivative of the function $V(t, x)$ satisfies

$$\dot{V}(t, x) \leqslant V^*(x) \leqslant 0$$

on $\mathscr{D}$;

(4) there exist summable functions $K(t)$, $F(t)$ and $N(t)$, constants K_0, F_0 and N_0, and a nondecreasing function $\chi(\gamma)$, with $\lim_{\gamma \to 0} \chi(\gamma) = 0$, such that

$$\varphi(t, x) \leqslant N(t), \qquad \int_{t_1}^{t_2} N(t)\, dt \leqslant N_0(t_2 - t_1)$$

for $x \in \Omega \setminus E(V^* = 0)$, $t_1, t_2 \in R_+$, and

$$\| r(t, x) \| \leqslant K(t), \qquad \int_{t_1}^{t_2} K(t)\, dt \leqslant K_0(t_2 - t_1),$$

$$|\varphi(t, x') - \varphi(t, x'')| \leqslant \chi(\| x' - x'' \|) F(t), \qquad \int_{t_1}^{t_2} F(t)\, dt \leqslant F_0(t_2 - t_1);$$

(5) there exists a mean

$$\varphi_0(t_0, x_0) = \lim_{\theta \to \infty} \frac{1}{\theta} \int_{t_0}^{t_0 + \theta} \varphi(s, \bar{x}(s))\, ds$$

uniformly with respect to $(t_0, x_0) \in \mathscr{D}$;
(6) $\varphi_0(t_0, x_0) < 0$ is definite in the set $E(V^* = 0)$.

Then the system of (3.1.1) is μ-stable.

Example 3.2.1   Consider the system

$$\dot{x}_1 = -x_1 + x_2 + \mu(x_1^3 - ax_2^3), \qquad x_1(t_0) = x_{10},$$

$$\dot{x}_2 = \mu[a(x_1 + x_2)\cos t + (x_1^3 - ax_2^3)], \qquad x_2(t_0) = x_{20},$$

where a is a constant. From the generating system

$$\dot{x}_1 = -x_1 + x_2,$$

$$\dot{x}_2 = 0,$$

the total derivative of the function $V = x_2^2 + (x_1 - x_2)^2$ is of the form

$$\dot{V}(x) = -2(x_1 - x_2)^2 = V^*(x) \leqslant 0.$$

Having computed the mean $\varphi_0(t_0, x_0)$ along the solutions

$$x_2(t) = x_{20},$$

$$x_1(t) = x_{20} + (x_{10} - x_{20})\exp[-(t - t_0)]$$

of the generating system, we obtain

$$\varphi_0(x_0) = 2x_{20}^4(1 - a).$$

We have $\varphi(x_0) < 0$ definite in the set

$$E(V^* = 0) = \{x : x_1 = x_2\}$$

for $a > 1$. Thus the solution $(x_1(t), x_2(t))$ of the initial system is μ-stable.

3.3   Stability of Large-Scale Systems with Weakly Interacting Subsystems

3.3.1   Preliminary notes and definitions

We consider a system described by equations of the form

$$\frac{dx_i}{dt} = f_i(t, x_i) + \mu r_i(t, x_1, \ldots, x_m), \tag{3.3.1}$$

where $x_i \in R^{n_i}$, $i = 1, 2, \ldots, m$, $t \in R_+$, $f_i : R_+ \times R^{n_i} \to R^{n_i}$ and $r_i : R_+ \times R^{n_1} \times \cdots \times R^{n_m} \to R^{n_i}$. We assume that $X_i(t, x_i) = 0$ for all $t \in R_+$ iff $x_i = 0$. Putting $\Sigma_i n_i = n$, $x = (x_1^T, \ldots, x_m^T)^T \in R^n$,

$$f(t, x) = (f_1^T(t, x_1), \ldots, f_m^T(t, x_m))^T,$$

$$r(t, x) = (r_1^T(t, x), \ldots, r_m^T(t, x))^T,$$

the system (3.3.1) can be rewritten in the form

$$\frac{dx}{dt} = f(t, x) + \mu r(t, x) \triangleq h(t, x, \mu). \tag{3.3.2}$$

It is clear that $f : R_+ \times R^n \to R^n$, $f : R_+ \times R^n \to R^n$ and $h : R_+ \times R^n \times M \to R^n$, $M = (0, \mu^*)$. We assume that $h(t, x, \mu) = 0$ for all $t \in R_+$ iff $x = 0$. The system (3.3.2) is referred to as a *large-scale system with weakly interacting subsystems*

$$\frac{dx_i}{dt} = f_i(t, x_i), \quad i = 1, 2, \ldots, m. \tag{3.3.3}$$

Each of the systems in (3.3.3) is a free subsystem, $\mu r_i(t, x_1, \ldots, x_m)$ are weak interactions (because of the presence of the small parameter μ). The following interaction functions are of interest in applications:

(i) Functions

$$r_i(t, x_1, \ldots, x_m) = \sum_{j=1, i \neq j}^{m} C_{ij} x_j, \tag{3.3.4}$$

where C_{ij} is a constant $n_i \times n_j$ matrix; hence the system (3.3.1) has the form

$$\frac{dx_i}{dt} = f_i(t, x_i) + \mu \sum_{j=1, i \neq j}^{m} C_{ij} x_j \quad i = 1, 2, \ldots, m. \tag{3.3.5}$$

(ii) Functions

$$r_i(t, x_1, \ldots, x_m) = \sum_{j=1, i \neq j}^{m} g_{ij}(t, x_j), \tag{3.3.6}$$

where $g_{ij} : R_+ \times R^{n_j} \to R^{n_i}$; hence the system (3.3.1) has the form

$$\frac{dx_i}{dt} = f_i(t, x_i) + \mu \sum_{j=1, i \neq j}^{m} g_{ij}(t, x_j), \quad i = 1, 2, \ldots, m. \tag{3.3.7}$$

We suppose that the vector functions $f_i(t, x_i)$ and $r_i(t, x)$ are defined and continuous in domains $\mathscr{D}_i = R_+ \times \Omega_i$ and $\mathscr{D}$ respectively, where $\Omega_i = \{x_i \in R^{n_i} : \|x_i\| < H/m\}$, and satisfy a condition guaranteeing the existence and uniqueness of the solution the Cauchy problem for the systems (3.3.1) and (3.3.3). Moreover, the vector functions $f_i(t, x_i)$ satisfy a Lipschitz condition in x_i, with constant L, and there exist summable functions $M_i(t)$ and constants M_{i0} for the vector functions $R_i(t, x)$ such that the inequalities

$$\|r_i(t, x)\| \leqslant M_i(t), \qquad \int_{t_1}^{t_2} M_i(s) \, ds \leqslant M_{i0}(t_2 - t_1)$$

hold on any finite interval $[t_1, t_2]$ in the domain $\mathscr{D}$.

Each subsystem (3.3.3) is assumed to have a uniformly stable equilibrium state $x_i = 0$, which is ensured by the existence of Lyapunov functions $V_i(t, x_i) \in C^1(R_+ \times D_i, R_+)$ having the corresponding properties, and it is assumed that general solutions

$$x_i(t; g) = x_i(t; t_0, x_{i0}; g) \tag{3.3.8}$$

of the limiting systems exist for $(t_0, x_{i0}) \in \mathscr{D}_i$ and all $t \geqslant t_0$.

Assumption 3.3.1 Corresponding to condition (5) of Assumption 3.1.1, the following holds.

(5*) The limits

$$\lim_{t \to \infty} \| f_i(t, x_i) - g_i(t, x_i) \| = 0, \qquad i = 1, 2, \ldots, m, \tag{3.3.9}$$

hold uniformly on $x_i \in \Omega_i$.

We introduce the following notation:

$$(\operatorname{grad} V_i)^T = \left(\frac{\partial V_i}{\partial x_{i_1}}, \ldots, \frac{\partial V_i}{\partial x_{in_i}} \right),$$

$$\dot{V}_i(t, x_i) = \frac{\partial V_i}{\partial t} + (\operatorname{grad} V_i)^T f_i(t, x_i),$$

$$\dot{V}_{ig}(t, x_i) = \frac{\partial V_i}{\partial t} + (\operatorname{grad} V_i)^T g_i(t, x_i),$$

$$\varphi_i^1(t, x_1, x_2, \ldots, x_m) = (\operatorname{grad} V_i)^T r_i(t, x),$$

$$\varphi_i^{\alpha_i}(t, x_{\alpha_i}, x_{\alpha_i + 1}, \ldots, x_m) = (\operatorname{grad} V_i)^T r_i(t, 0, \ldots, 0, x_{\alpha_i}, x_{\alpha_i + 1}, \ldots, x_m);$$

$$\varphi_{i0}^{\alpha_i}(T, t_0, x_{\alpha_i 0}, x_{\alpha_i + 1 0}, \ldots, x_{m0}) = \int_{t_0}^{t_0 + T} \varphi_i^{\alpha_i}(t, x_{\alpha_i}(t; g), x_{\alpha_i + 1}(t; g), \ldots, x_m(t; g)) \, dt,$$

where

$$\alpha_i \leqslant i, \qquad i = 1, 2, \ldots, m.$$

Corresponding to condition (6) of Assumption 3.1.1, the following holds.

(6*) There exist summable functions $F_i(t)$, functions $\chi_i(\beta) \in K$, $i = 1, 2, \ldots, m$, and constants F_{i0}, $i = 1, 2, \ldots, m$, such that the inequalities

$$|\varphi_i^{\alpha_i}(t, x'_{\alpha_i}, x'_{\alpha_i+1}, \ldots, x'_m) - \varphi_i^{\alpha_i}(t, x''_{\alpha_i}, x''_{\alpha_i+1}, \ldots, x''_m)| \leqslant F_i(t)\chi_i\left(\sum_{k=\alpha_i}^m \|x'_k - x''_k\|\right),$$

$$\int_{t_1}^{t_2} F_i(t)\,dt \leqslant F_{i0}(t_2 - t_1)$$

hold for $t \in R_+$ and $x'_j\, x'' \in$ in Ω, where $[t_1, t_2] \subset R_+$ is a finite interval.

3.3.2 Main result

The following result holds.

Theorem 3.3.1 In the domain $\mathscr{D}$ let the following conditions hold for all $i = 1, 2, \ldots, m$.

(1) There exist functions $V_i \in C^1(R_+ \times \mathscr{D}_i, R_+)$ and $a_i, b_i \in K$ such that

(i) $a_i(\|x_i\|) \leqslant V_i(t, x_i) \leqslant b_i(\|x_i\|)$;
(ii) $\dot{V}_i(t, x_i) \leqslant 0$.

(2) The conditions (5*) and (6*) are satisfied.
(3) There exist positive numbers μ'_i, δ_i and l_i for γ_i and β_i $(0 < \gamma_i < \beta_i < H)$ such that one of the following conditions holds for every pair of values of t and x $(t \in R_+, \gamma_i \leqslant \|x_i\| \leqslant \beta_i, x_p \in \Omega_p$ when $p \neq i)$:

(i) $\dot{V}_i(t, x_i) + \mu\varphi_i(t, x) \leqslant 0$ for $\mu < \mu'_i$;
(ii) $\varphi_{i0}^{\alpha_i}(\theta_i, t, x_{\alpha_i}, x_{\alpha_i+1}, \ldots, x_m) \leqslant -\delta_i\theta_i$ for $\theta_i > l_i$.

Then the equilibrium state $x = 0$ of the system (3.3.1) is μ-stable.

Proof We shall show that numbers $\eta > 0$ and $\mu_0 > 0$ can be found for any $\varepsilon \in]0, H[$ such that $\|x(t; t_0, x_0)\| < \varepsilon$ for all $t \geqslant t_0$ when $\mu < \mu_0$ and $x_0 \in B_\eta$, where $t_0 \in R_+$. To do this, it is sufficient to show that for all $i = 1, 2, \ldots, m$, $\eta_i > 0$ and $\mu_i > 0$ can be found for any given $\varepsilon_i > 0$ $(\sum_{i=1}^m \varepsilon_i \leqslant \varepsilon)$ such that for solutions $x_i(t) = x_i(t; t_0, x_0)$ the inequalities $\|x_i(t)\| < \varepsilon_i$ hold for all $t \geqslant t_0$ when $\|x_{i_0}\| < \eta_i$ and $\mu < \mu_i$, $i = 1, 2, \ldots, m$.

Let $\varepsilon_m = \varepsilon/m$. By condition (1) of the theorem, any point of the moving surface $V_m(t, x_m) = a_m(\frac{1}{2}\varepsilon_m)$ will satisfy the inequality

$$b_m^{-1}(a_m(\tfrac{1}{2}\varepsilon_m)) \leqslant \|x_m\| \leqslant \tfrac{1}{2}\varepsilon_m \tag{3.3.10}$$

for all $t \in R_+$. By Lemma 3.1.2, for

$$\lambda_m = \frac{\min\{\chi_m^{-1}(\delta_m/4F_{m0})/(m + 1 - \alpha_m), \tfrac{1}{2}\varepsilon_m\}}{4l_m \exp(2Ll_m)} \tag{3.3.11}$$

a number τ_{m0} can be chosen such that

$$\|\bar{x}_m(t; \tau, x(\tau)) - x_m(t; \tau, x(\tau); g)\| \leqslant 2\lambda_m l_m \exp(2Ll_m) \tag{3.3.12}$$

for any $\tau > \tau_{m0}$ and $t \in [\tau, \tau + 2l_m]$, where δ_m and l_m are numbers satisfying condition (3) of the theorem. Take

$$\eta_m = \frac{b^{-1}(a_m(\tfrac{1}{2}\varepsilon_m))}{2\exp(L\tau_m)}$$

and

$$\varepsilon_{i-1} = \min\{\chi_i^{-1}(\delta_i/4F_{i0})/(2(\alpha_i - 1)), \varepsilon_i\}. \tag{3.3.13}$$

Then for numbers $b_i^{-1}(a_i(\tfrac{1}{2}\varepsilon_i))$ and $\tfrac{1}{2}\varepsilon_i$, from condition (3) of the theorem, we define numbers δ_i and l_i, and by Lemma 3.1.2 we find τ_{i0} for the numbers

$$\lambda_i = \frac{\min\{\chi_i^{-1}(\delta_i/4F_{i0})/(m+1-\alpha_i), \tfrac{1}{2}\varepsilon_i\}}{2l_i\exp(2l_iL)}$$

such that an inequality of the type (3.3.12) holds for $\tau > \tau_{i0}$ and $t \in [\tau, \tau + 2l_i]$. Here F_{i0} is defined from condition (2) of the theorem. We take

$$\eta_i = \min\left\{\frac{b_i^{-1}(a_i(\tfrac{1}{2}\varepsilon_i))}{2\exp(L\tau_i)}, \eta_{i+1}\right\}, \tag{3.3.14}$$

where $i = m-1, m-2, \ldots, 2, 1$. Suppose that $\tau_0 = \max_i\{\tau_{i0}\}$.

It is easy to see that all conditions of the theorem are satisfied for all $i = 1, 2, \ldots, m$; in the rest of the proof a fixed $q \in [1, m]$ will be considered, assuming that for all $\varepsilon_j > 0$ there exist $\mu_j(\varepsilon_j) > 0$, $\eta_j > 0$, $j = 1, \ldots, q-1$, such that for solutions of the jth subsystems of (3.3.1) with $\|x_{j0}\| < \eta_j$ and $\mu < \mu_j$ the inequalities $\|x_j(t)\| < \varepsilon_j$ hold for all $t \geq t_0$.

Let ε_q be as defined by (3.3.13) for $i = q+1$. Consider an arbitrary solution $x_q(t) = x_q(t; t_0, x_0)$ with initial conditions $t_0 \in R_+$, $x_q(t_0) = x_{q0} \in B_{\eta_q}$, where η_q is defined by (3.3.14) for $i = q$. Using Lemma 3.1.3 and taking

$$\mu_{q1} = \frac{b_q^{-1}(a_q(\tfrac{1}{2}\varepsilon_q))}{2M_{q0}\tau_q\exp(L\tau_q)}, \tag{3.3.15}$$

we obtain an estimate $\|x_q(t; t_0, x_0)\| < b_q^{-1}(a_q(\tfrac{1}{2}\varepsilon_q))$ on the finite interval $[t_0, t_0 + \tau_q]$ for $x_{q0} \in B_{\eta_q}$ and $\mu < \mu_{q1}$, where $\tau_q > \tau_0$. Therefore we must prove that the inequality $\|x_q(t; t_0, x_0)\| < \varepsilon_q$ holds for all $t \geq t_0$ when $x_{q0} \in B_{\eta_q}$ and $\mu < \mu_q$.

Consider a solution $x_q(t; t'_{q0}, x(t'_{q0}))$, assuming that it has left the domain $\|x_q\| < b_q^{-1}(a_q(\tfrac{1}{2}\varepsilon_q))$ and that the condition $V_q(\tau^q, x_q(\tau^q)) = a_q(\tfrac{1}{2}\varepsilon_q)$ is satisfied at some time $t = \tau^q$. Then the inequality

$$b_q^{-1}(a_q(\tfrac{1}{2}\varepsilon_q)) \leq \|x_q(\tau^q)\| < \tfrac{1}{2}\varepsilon_q$$

holds at the point $x_q(\tau^q)$ and one of the conditions (3)(i) or (3)(ii) of the theorem is satisfied at the time $t = \tau^q$.

Case 1 Let condition (3)(i) be satisfied at time $t = \tau^q$, and suppose that there exist μ'_q such that the total derivative of the Lyapunov function $V_q(t, x_q)$ with respect to (3.3.1) is nonpositive for $\mu < \mu'_q$ at this time, and therefore the solution $x_q(t, t'_{q0}, x(t'_{q0}))$ cannot intersect the surface

$$V_q(t, x_q) = a_q(\tfrac{1}{2}\varepsilon_q),$$

at time $t = \tau^q$.

Case 2   Suppose that condition (3)(ii) is satisfied; that is, the inequality

$$\varphi_{q0}^{\alpha q}(\theta, \tau^q, x(\tau^q)) \leqslant -\delta_q \theta_q$$

holds for $\theta_q > l_q$. Integrating the total derivative of the function $V_q(t, x_q)$ with respect to (3.3.1) for $t \geqslant \tau^q$, we have

$$V_q(t, x_q(t)) \leqslant V_q(\tau^q, x_q(\tau^q)) + \mu \int_{\tau^q}^t \varphi_q^1(s, x(s, \tau^q, x(\tau^q)))\, ds$$

as long as $x(s)$ remains in the domain. The integral on the right-hand side of this inequality can be written as

$$J_q = J_{q1} + J_{q2} + J_{q3}, \tag{3.3.16}$$

where

$$J_{q1} = \int_{\tau^q}^t [\varphi_q^1(s, x(s, \tau^q, x(\tau^q))) - \varphi_q^{\alpha q}(s, \bar{x}_{\alpha_q}(s, \tau^q, x_{\alpha_q}(\tau^q)),$$

$$\bar{x}_{\alpha_q+1}(s, \tau^q, x_{\alpha_q+1}(\tau^q)), \ldots, \bar{x}_m(s, \tau^q, x_m(\tau^q)))]\, ds,$$

$$J_{q2} = \int_{\tau^q}^t [\varphi_q^{\alpha q}(s, \bar{x}_{\alpha_q}(s, \tau^q, x_{\alpha q}(\tau^q)), \ldots, \bar{x}_m(s, \tau^q, x_m(\tau^q))$$

$$- \varphi_q^{\alpha q}(s, x_{\alpha_q}(s, \tau^q, x_{\alpha_q}(\tau^q); g), \ldots, x_m(s, \tau^q, x_m^g(\tau^q); g))]\, ds,$$

$$J_{q3} = \int_{\tau^q}^t \varphi_q^{\alpha q}(s, x_{\alpha_q}(s, \tau^q, x_{\alpha_q}(\tau^q); g), \ldots, x_m(s, \tau^q, x_m^g(\tau^q); g))\, ds.$$

By Lemma 3.1.1, the estimate

$$\|x_q(t) - \bar{x}_q(t)\| \leqslant 2\mu M_{q0} l_q \exp(2Ll_q) \tag{3.3.17}$$

holds for solutions $x_q(t) = x_q(t; \tau^q, x(\tau^q))$ and $\bar{x}_q(t) = \bar{x}_q(t; \tau^q, x_q(\tau^q))$ of the qth subsystem of (3.3.1) and (3.3.3) when $t \in [\tau^q, \tau^q + 2l_q]$.

Condition (1) of the theorem implies that the equilibrium state $x = 0$ of the qth subsystem of (3.3.2) is stable, i.e. $\|\bar{x}_q(t)\| < \tfrac{1}{2}\varepsilon_q$. Then we obtain the estimate $\|x_q(t)\| < \varepsilon_q$ for

$$\mu < \mu_{q2} = \frac{\varepsilon_q}{4 M_{q0} l_q \exp(2Ll_q)}.$$

Noting that

$$\varphi_q^{\alpha q}(t, \bar{x}_{\alpha_q}, \bar{x}_{\alpha_q+1}, \ldots, \bar{x}_m) = \varphi_q^1(t, 0, \ldots, 0, \bar{x}_{\alpha_q}, \bar{x}_{\alpha_q+1}, \ldots, \bar{x}_m),$$

it is easy to obtain the estimate

$$J_{q1} \leqslant \int_{\tau^q}^{t} F_q(s)\chi_q\left(\sum_{i=1}^{\alpha_q-1} \|x_i(s)\| + \sum_{k=\alpha_q}^{m} \|x_k(s) - \bar{x}_k(s)\|\right)ds \tag{3.3.18}$$

from condition (2) of the theorem.

Since $\|x_i(t)\| < \varepsilon_i$ for $i = 1, \ldots, \alpha_q - 1$, and $\varepsilon_i \leqslant \varepsilon_{\alpha_q-1}$ and $\alpha_q \leqslant q$, we have

$$\sum_{i=1}^{\alpha_q-1} \|x_i(t)\| \leqslant (\alpha_q - 1)\varepsilon_{q-1} \leqslant \tfrac{1}{2}\chi_q^{-1}(\delta_q/4F_{q0}). \tag{3.3.19}$$

For solutions $x_k(t) = x_k(t, \tau^q, x(\tau^q))$ and $\bar{x}_k(t) = \bar{x}_k(t; \tau^q, x_k(\tau^q))$ of the kth subsystem estimates similar to (3.3.17) can be obtained for $t \in [\tau^k, \tau^k + 2l_k]$, where $k = \alpha_q, \ldots, m$. Summing these over k, we find that

$$\sum_{k=\alpha_q}^{m} \|x_k(t) - \bar{x}_k(t)\| \leqslant 2\mu l_q \exp(2Ll_q) \sum_{k=\alpha_q}^{m} M_{k0}.$$

For

$$\mu < \mu_{q_3} = \frac{\chi_q^{-1}(\delta_q/4F_{q0})}{4l_q \exp(2l_q L) \sum\limits_{k=\alpha_q}^{m} M_{k0}}$$

we obtain

$$\sum_{k=\alpha_q}^{m} \|x_k(t) - \bar{x}_k(t)\| < \tfrac{1}{2}\chi_q^{-1}(\delta_q/4F_{q0}). \tag{3.3.20}$$

Substituting the estimates (3.3.19) and (3.3.20) into (3.3.18), we have

$$J_{q1} \leqslant \tfrac{1}{4}\delta_q(t - \tau^q), \tag{3.3.21}$$

because of condition (3) of the theorem.

By Lemma 3.1.3, for λ_q defined by (3.3.13) for $i = q$ there exists a τ_{q0} such that

$$\|\bar{x}_q(t; \tau^q, x_q(\tau^q)) - x_q(t; \tau^q, x_q(\tau^q); g)\|$$
$$\leqslant \min\{\chi_q^{-1}(\delta_q/4F_{q0})/(m + 1 - \alpha_q), \tfrac{1}{2}\varepsilon_q\}, \tag{3.3.22}$$

for $t \in [\tau^q, \tau^q + 2l_q]$, where $\tau^q > \tau_0 \geqslant \tau_{q0}$. By virtue of the stability of the equilibrium state $x_q = 0$ of the qth subsystem of (3.3.3), the inequality (3.3.22) implies that the solution $x_q^g(t, \tau^q, x_q(\tau^q))$ of the qth subsystem of (3.3.8) will not leave an ε_q-neighborhood on the interval $[\tau^q; \tau^q + 2l_q]$ or the domain Ω_q, so condition (2) of the theorem can be used to estimate J_{q2}. For $t \in [\tau^q, \tau^q + 2l_q]$ we get

$$J_{q2} \leqslant \int_{\tau^q}^{t} F_q(s)\chi_q\left(\sum_{k=\alpha_q}^{m} \|\bar{x}_k(s) - x_k^g(s)\|\right)ds \tag{3.3.23}$$

Similarly to the estimate (3.3.22), we obtain

$$\|\bar{x}_k(t; \tau^q, x_k(\tau^q)) - x_k^g(t; \tau^q, x_k(\tau^q))\| \leqslant \min\{\chi_q^{-1}(\delta_q/4F_{q0})/(m + 1 - \alpha_q), \tfrac{1}{2}\varepsilon_q\}.$$

for $t \in [\tau^q, \tau^q + 2l_q]$, when all $k = \alpha_q, \ldots, m$. Summing these inequalities over all k, we get

$$\sum_{k=\alpha_q}^{m} \| \bar{x}_k(t) - x_k^q(t) \| \leqslant \chi^{-1}(\delta_q/4F_{q0}).$$

Substituting this estimate into (3.3.23), we find that

$$J_{q2} \leqslant \tfrac{1}{4}\delta_q(t - \tau^q) \tag{3.3.24}$$

for $t \in [\tau^q, \tau^q + 2l_q]$.

By condition (3)(ii) of the theorem, we have

$$J_{q3} \leqslant -\delta_q(t - \tau^q) \tag{3.3.25}$$

for $t - \tau^q > l_q$. The inequalities (3.3.21), (3.3.24) and (3.3.25) imply the estimate

$$J_q < -\tfrac{1}{2}\delta(t - \tau^q)$$

for $\mu < \mu_{q0} = \min\{\mu_{q1}, \mu_{q2}, \mu_{q3}\}$ and $t \in]\tau^q + l_q, \tau^q + 2l_q]$. Thus the integral J_q becomes negative at least on the interval $[\tau^q + l_q, \tau^q + 2l_q]$. Hence if the solution $x_q(t)$ leaves the domain bounded by the surface $V_q(t, x_q) = a_q(\tfrac{1}{2}\varepsilon_q)$ then, owing to the choice of the small parameter μ, it will remain in the domain $\|x_q\| < \varepsilon$ for $t \in [\tau^q, \tau^q + 2l_q]$ and will return to the domain bounded by surface $V_q(t, x_q) = a_q(\tfrac{1}{2}\varepsilon_q)$ at some time contained in the interval.

Cases 1 and 2 show that the solution $x_q(t) = x_q(t; t_0, x_0)$ will not leave the domain $\|x_q\| < \varepsilon_q$ for all $t \in [t_0, \tau^q + 2l_q]$ when $t_0 \in R_+$, $\|x_{q0}\| < \eta_q$ and $\mu < \mu_2 = \min\{\mu_q', \mu_{q0}\}$.

Now suppose that for all $i \in [1, m]$, ε_i and t_0 there exist η_i and μ_i such that $\|x_i(t; t_0, x_0)\| < \varepsilon_i$ for $t \geqslant t_0$ when $\|x_{i0}\| < \eta_i$ and $\mu < \mu_i$. Then we may state that, given $\varepsilon > 0$ and $t_0 \in R_+$, there exist $\eta > 0$ and $\mu_0 > 0$ such that for all $t \geqslant t_0$ the inequality $\|x(t; t_0, x_0)\| < \varepsilon$ holds when $\|x_0\| < \eta$ and $\mu < \mu_0$. For this, it is sufficient to take $\eta = \eta_1$, and $\mu_0 = \min\{\mu_1, \mu_2, \ldots, \mu_m\}$.

Thus the proof of Theorem 3.3.1 is complete.

Remark 3.3.1 When it is not assumed that the interconnection functions $r_i(t, x_1, \ldots, x_m)$ satisfy $r_i(t, x_1, \ldots, x_m) = 0$ for $x_i = 0$, in order that Theorem 3.3.1 be valid, it additionally necessary that

$$\varphi_i^1(t, x) \leqslant N_i(t), \qquad \int_{t_1}^{t_2} N_i(t)\,dt \leqslant N_{i0}(t_2 - t_1)$$

on any finite interval $[t_1, t_2]$ for $t \in R_+$, $x_i \in \Omega \backslash E(\dot{V}_i = 0)$, $x_k \in \Omega_k$, $k \neq i$, where $N_{i0} = \text{const} > 0$.

Corollary to Theorem 3.3.1 Let the following conditions be satisfied in the domain $\mathscr{D}$ for the system (3.3.1).

(1) There exist positive-definite functions $V_i(t, x_i)$, $i = 1, 2, \ldots, m$, having infinitesimally small upper bounds and

$$\dot{V}_i(t, x_i) \leqslant V_i^*(x_i) \leqslant 0, \qquad i = 1, 2, \ldots, m,$$

where $V_i^*(x_i)$ are definite and continuous in the domains Ω_i.

(2) $f_i(t, x_i) = g_i(t, x_i)$ in $\mathcal{D}$.
(3) There exist summable functions $M_i(t)$ and $F_i(t)$, constants M_{i0} and F_{i0} and functions $\chi_i(\beta) \in \mathcal{K}$ such that

$$|\varphi_i^{\alpha_1}(t, x') - \varphi_i^{\alpha_1}(t, x'')| \leqslant F_i(t)\chi_i\left(\sum_{k=1}^{m} \|x'_k - x''_k\|\right),$$

$$\int_{t_1}^{t_2} F_i(t)\, dt \leqslant F_{i0}(t_2 - t_1),$$

$$\|r_i(t, x)\| \leqslant M_i(t), \qquad \int_{t_1}^{t_2} M_i(t)\, dt \leqslant M_{i0}(t_2 - t_1), \qquad i = 1, 2, \ldots, m$$

on any finite interval $[t_1, t_2]$.
(4) There exist mean values

$$\varphi_{i0}^{\alpha_i}(t_0, x_{\alpha_i 0}, \ldots, x_{m0}) = \lim_{\theta \to \infty} \frac{1}{\theta} \int_{t_0}^{t_0 + \theta} \varphi_i^{\alpha_i}(t, \bar{x}_{\alpha_i}(t), \ldots, \bar{x}_m(t))\, dt$$

uniformly on $t_0, x_{\alpha_i 0}, x_{\alpha_i + 1\, 0}, \ldots, x_{m0}$.
(5) There exist numbers $r_i(\eta_i, \varepsilon_i)$ and $\delta_i(\eta_i, \varepsilon_i)$ for any η_i and $\varepsilon_i(0 < \eta_i < \varepsilon_i < H)$ such that

$$\varphi_{i0}^{\alpha_i}(t_0, x_{\alpha_i 0}, \ldots, x_{m0}) < -\delta$$

for $\eta_i \leqslant \|x_i\| \leqslant \varepsilon_i$, $\rho(x_{i0}, E(V_i^* = 0)) < r_i$ and all $t_0 \in R_+$, $x_{j0} \in \Omega_j$, $j = \alpha_i, \ldots, i - 1$, $i + 1, \ldots, m$.

Then the system (3.3.1) is μ-stable uniformly in t_0.

Example 3.3.1   We consider the system

$$\frac{dx_1}{dt} = -x_1 + x_2^2 + \mu x_2 y_1,$$

$$\frac{dx_2}{dt} = \mu(x_2 y_2 - x_1 x_2 + x_1 z + x_1^2 \sin t),$$

$$\frac{dy_1}{dt} = -y_1 + y_2 + \mu x_2^2, \tag{3.3.26}$$

$$\frac{dy_2}{dt} = \mu(-x_2^2 y_2 z - y_1 y_2 + (y_1 + y_2)\cos t),$$

$$\frac{dz}{dt} = \mu(-z^3 + y_1 y_2 - x_1^2 z + z^2 \cos t).$$

We have for $\mu = 0$

$$\frac{dx_1}{dt} = -x_1 + x_2^2, \qquad \frac{dx_2}{dt} = 0,$$

$$\frac{dy_1}{dt} = -y_1 + y_2, \qquad \frac{dy_2}{dt} = 0, \tag{3.3.27}$$

$$\frac{dz}{dt} = 0.$$

The general solution of the system (3.3.27) is

$$\bar{x}_1(t) = x_{20}^2 + (x_{10} - x_{20}^2)\exp[-(t-t_0)], \quad \bar{x}_2(t) = x_{20},$$
$$\bar{y}_1(t) = y_{20} + (y_{10} - y_{20})\exp[-(t-t_0)], \quad \bar{y}_2(t) = y_{20},$$
$$\bar{z}(t) = z_0.$$

The investigation of the system (3.3.26) begins with the second subsystem in (3.3.27), for which the Lyapunov function is taken in the form

$$V_1(y) = y_2^2 + (y_1 - y_2)^2. \tag{3.3.28}$$

The total derivative of this function along the solutions of the system (3.3.27) will satisfy

$$\dot{V}_1(y) = -2(y_1 - y_2)^2 = V_1^*(y) \leqslant 0. \tag{3.3.29}$$

The mean value φ_{10}^1 for this subsystem is

$$\varphi_{10}^1(y_0, z_0, x_0) = -2y_{20}^2 - 2y_{20}^2 x_{20}^2 z_0^2. \tag{3.3.30}$$

The mean φ_{10}^1 is obviously less than zero on the set

$$E(V_1^* = 0) = \{y : y_1 = y_2\}$$

for any x_{20} and z_0.

For the third subsystem in (3.3.27) we take the function

$$V_2(z) = z^2.$$

Its total derivative is zero, by virtue of the subsystem. Setting $y_1 = 0$ and $y_2 = 0$, we compute the mean value for this subsystem:

$$\varphi_{20}^2(z_0, x_0) = -2(z_0^4 + z_0^2 x_{20}^4). \tag{3.3.31}$$

The mean value φ_{20}^2 is negative-definite with respect to z_0 for any x_{20}.

For the first subsystem in (3.3.27) we take the function

$$V_3(x) = x_2^2 + (x_1 - x_2^2)^2.$$

It can easily be shown that

$$\dot{V}_3(x) = -2(x_1 - x_2^2)^2 = V_3^*(x) \leqslant 0.$$

Then we get for $y_1 = 0$, $y_2 = 0$ and $z = 0$

$$\varphi_{30}^3(x_0) = -2x_{20}^4. \tag{3.3.32}$$

In view of (3.3.30)–(3.3.32), it is clear that all the conditions of the corollary to Theorem 3.3.1 are satisfied, and the system (3.3.26) is μ-stable.

3.4 Instability

It is rather difficult as a rule, to find the general solution of the system (3.1.2). Here we show that the general solution of the limiting system (3.1.3) can be used instead of the general solution of the unperturbed (3.1.2).

We consider a Lyapunov function $V(t,x) \in C^1(R_+ \times \mathscr{D}, R_+)$ and introduce the domain $\mathscr{D}_\tau = J_\tau \times \Omega$, with $J_\tau =]\tau, +\infty[, \tau \in R$.

Theorem 3.4.1   In the domain $\mathscr{D}_\tau$ let the following conditions hold.

(1) There exist a function $V(t,x)$ that is bounded and for which

$$\dot{V}(t,x) \geqslant 0, \qquad \lim_{t \to +\infty} \dot{V}(t,x;g) \geqslant 0$$

in the domain $V > 0$.

(2) Conditions (4)–(6) of Assumption 3.1.1 are satisfied in the domain $V > 0$.

(3) There exist positive numbers μ', γ and δ, l for any arbitrary small $\alpha > 0$ such that one of the following inequalities holds for every pair of values of t and x satisfying the inequality $V(t,x) > \alpha$:

   (i) $\dot{V}(t,x) + \mu\varphi(t,x) \geqslant \gamma$ for $\mu < \mu'$;
   (ii) $\varphi^0(\theta, t, x) \geqslant \delta\theta$ for $\theta > l$.

Then the system (3.1.1) is μ-unstable.

Proof   We set $\varepsilon \in]0, H[$. Let $\eta (0 < \eta < \varepsilon)$ and $\mu_0 \in]0, \mu^*[$ be arbitrary small numbers. We take $x_\tau \in \{x \in \Omega : \|x\| \leqslant \eta\}$, such that $V(\tau, x_\tau) > 0$. Then there exists a number $\alpha > 0$ such that $V(\tau, x_\tau) > 2\alpha$. By condition (1) of the theorem, a constant $h > 0$ can be found such that the inequality

$$|V(t,x)| < h \tag{3.4.1}$$

holds in the domain $V > 0$ if $\|x\| < \varepsilon$.

There exist numbers μ', γ and δ, l for α satisfying condition (3) of the theorem. By Lemma 3.1.2, there exists a time $\tau_0' > \tau$ for

$$\lambda = \frac{\chi^{-1}(\delta/4F_0)}{(2l + l')\exp[L(2l + l')]} \tag{3.4.2}$$

such that the inequality

$$\|\bar{x}(t) - x_g(t)\| < \chi^{-1}(\delta/4F_0), \tag{3.4.3}$$

with $\bar{x}(t) = \bar{x}(t; \tau_1, x(\tau_1))$ and $x(t;g) = x(t; \tau_1 x(\tau_1); g)$,   holds   on   the   interval $[\tau_1, \tau_1 + 2l + l']$ for $\tau_1 \geqslant \tau_0'$, where $l' = (h - \alpha)/\gamma$.

Since $\lim_{t \to \infty} \dot{V}(t,x;g) = c \geqslant 0$ for a constant c, there exists a $\tau_0'' > \tau$ for any $a > 0$ such that the inequality

$$|\dot{V}(t,x;g) - c| \leqslant a \tag{3.4.4}$$

holds for all $t \geqslant \tau_0''$. Hence

$$\dot{V}(t,x;g) \geqslant -|c - a|. \tag{3.4.5}$$

We take a such that

$$|c - a| = \frac{\alpha}{2(2l + l')}. \tag{3.4.6}$$

Let $\tau_0 = \max\{\tau_0', \tau_0''\}$. Choose $\mu_1 > 0$ such that the condition

$$\mu_1 \left| \int_\tau^t \varphi(s, x(s)) \, ds \right| \leqslant \alpha \tag{3.4.7}$$

holds for $t \in [\tau, \tau_0]$, where $x(s) = x(s; \tau, x_\tau)$ is a solution of the system (3.1.1).

It is possible to choose $\mu_1 > 0$ satisfying (3.4.7). In fact, since for the function $V(t, x) \in C^1(R_+ \times \mathscr{D}, R_+)$ the components of the vector $(\text{grad } V)^T$ are bounded on the finite interval $[\tau, \tau_0]$ and in view of condition (2)(ii) of Assumption 3.1.1 for $t \in [\tau, \tau_0]$, we can obtain

$$\left\| \int_\tau^t \varphi(s, x) \, ds \right\| \leqslant b M_0 (\tau_0 - \tau). \tag{3.4.8}$$

For $\mu_1 = L/b M_0(\tau_0 - \tau)$ the inequality (3.4.8) holds, where $\|\text{grad } V\| \leqslant b$ for $t \in [\tau, \tau_0]$. In view of condition (1) of the theorem, we get

$$V(t, x) \geqslant V(\tau, x_\tau) + \mu \int_\tau^t \varphi(s, x(s)) \, ds$$

for $t \in [\tau, \tau_0]$ and $\mu < \mu_1$, i.e. the solution $x(t)$ of the system (3.1.1) will not leave the domain $V > 0$ on the interval $[\tau, \tau_0]$.

Suppose that the inequalities

$$\|x(\tau_2)\| < \varepsilon, \quad V(\tau_2, x(\tau_2)) > \alpha$$

hold at some time $\tau_2 \geqslant \tau_0$ for the solution $x(t; \tau, x_\tau)$.

Case 1   Let conditions (1), (2) and (3)(i) of the theorem hold. We shall estimate the variation of the function $V(t, x)$ along the solution $x(t) = x(t; \tau_2, x(\tau_2))$ of (3.1.1). In the case that the solution $x(t)$ does not leave the domain $V > 0$ until the inequality $\|x(t)\| < \varepsilon$ holds, the inequalities $\|x(t)\| < \varepsilon$ and $V(t, x(t)) > \alpha$ and condition (3)(i) cannot remain valid simultaneously for all $x(t)$. In fact, suppose the contrary. By condition (3)(i), there exist positive numbers μ' and γ for α such that the inequality

$$\dot{V}(t, x) \geqslant \gamma$$

holds when $\mu < \mu'$ for any pair of (t, x) satisfying $V(t, x) > \alpha$. Integrating this inequality on the interval $[\tau_2, \tau_2 + l']$, we obtain

$$V(t, x(t)) \geqslant V(\tau_2, x(\tau_2)) + \gamma(t - \tau_2).$$

Taking into account that $V(\tau_2, x(\tau_2)) > \alpha$, we have for $t = \tau_2 + l'$ that $V(t, x(t)) > h$. This contradicts the boundedness of the function $V(t, x)$ in the domain $V > 0$ for $\|x(t)\| < \varepsilon$. This contradiction implies that there exists a time $\tau_3 \in [\tau_2, \tau_2 + l']$ when one of the conditions $\|x(t)\| < \varepsilon$, $V(t, x(t)) > \alpha$ or (3)(i) does not hold. Violation of the first condition leads to instability.

Suppose that condition (3)(i) is violated. Then condition (3)(ii) holds, since $x(t)$ belongs to the domain $V > 0$.

Case 2   Let the inequalities $\|x(\tau_3)\| < \varepsilon$, $V(\tau_3, x(\tau_3)) > \alpha$ and condition (3)(ii) of the theorem hold. Integrating the expression for the total derivative of the Lyapunov function by virtue of (3.1.1) and in view of condition (1) of the theorem, we get

$$V(t, x(t)) \geqslant V(\tau_3, x(\tau_3)) + \mu \int_{\tau_3}^t \varphi(s, x(s)) \, ds. \tag{3.4.9}$$

The integral in this inequality is estimated as

$$\int_{\tau_3}^t \varphi(s, x(s)) \, ds \geqslant \int_{\tau_3}^t \varphi(s, x(s; g)) \, ds - J_1 - J_2, \tag{3.4.10}$$

where

$$J_1 = \int_{\tau_3}^t |\varphi(s, x(s)) - \varphi(s, \bar{x}(s))| \, ds,$$

$$J_2 = \int_{\tau_3}^t |\varphi(s, \bar{x}(s)) - \varphi(s, x(s; g))| \, ds,$$

and $x(s) = x(s; \tau_3, x(\tau_3))$, $\bar{x}(s) = \bar{x}(s; \tau_3, x(\tau_3))$ and $x(s; g) = x(s; \tau_3, x(\tau_3); g)$ are solutions of the systems (3.1.1), (.3.1.2) and (3.1.3) respectively. The integrals J_1 and J_2 are estimated as in the previous section, but on the interval $[\tau_3, \tau_3 + 2l + l']$. Then for

$$\mu < \mu_2 = \frac{\chi^{-1}(\delta/4F_0)}{4M_0(2l + l') \exp[L(2l + l')]}$$

and $t \in [\tau_3, \tau_3 + 2l + l']$ we get

$$J_1 < \tfrac{1}{4}\delta(t - \tau_3), \tag{3.4.11}$$

$$J_2 < \tfrac{1}{4}\delta(t - \tau_3). \tag{3.4.12}$$

In view of the estimates (3.4.10)–(3.4.12) and the fact that $V(\tau_3, x(\tau_3)) > \alpha$, we obtain from (3.4.9)

$$V(t, x(t)) > \alpha + \mu \int_{\tau_3}^t \varphi(s, x(s; g)) \, ds - \tfrac{1}{2}\mu \, \delta(t - \tau_3).$$

By condition (3)(ii), we find

$$\int_{\tau_3}^t \varphi(s, x(s; g)) \, ds = \varphi_0(t - \tau_3, \tau_3, x(\tau_3)) > \delta(t - \tau_3)$$

for $t - \tau_3 > l$. Hence we have

$$V(t, x(t)) > \alpha + \tfrac{1}{2}\mu\delta(t - \tau_3) \tag{3.4.13}$$

for $t \in]\tau_3 + l, \tau_3 + 2l + l']$.

We take μ_3 such that inequality

$$\mu_3 \left| \int_{\tau_3}^t \varphi(s, x(s)) \, ds \right| \leqslant \frac{1}{2}\alpha$$

holds for $t \in [\tau_3, \tau_3 + 2l + l']$. Then the inequality

$$V(t, x(t)) > \tfrac{1}{2}\alpha.$$

holds on the interval $[\tau_3, \tau_3 + 2l + l']$ for $\mu < \mu_3$. Therefore the solution $x(t; \tau_3, x(\tau_3))$ will not leave the domain $V > 0$. Let $\bar{\mu} = \min\{\mu', \mu_1, \mu_2, \mu_3\}$. Consider a sequence of times $t_i = t_{i-1} + 2l + l'$, where $i = 1, 2, \ldots$. If the conditions of Case 1 hold, then according to the above, there exists a time $t_0' \in \,]t_0, t_0 + l']$ for $x(t; t_0, x_0)$ when $\mu < \mu_0$ at which the conditions of Case 2 hold, i.e. the estimate (3.4.13) is valid for $t \in \,]t_0 + l, t_0 + 2l + l']$, and we get

$$V(t_1, x(t_1)) > \alpha + \tfrac{1}{2}\mu\delta(2l + l') \tag{3.4.14}$$

for $t_1 = t_0 + 2l + l'$.

In Case 2 we also get the estimate (3.4.14) at time t_1. Similar cases will occur on all the later intervals $[t_k, t_{k+1}]$, i.e. $V(t, x(t))$ increases on each interval at least for $\mu\delta(l + l'/2) > 0$ (for $0 < \mu < \mu_0$). Let m be the smallest integer satisfying the condition $m \geqslant (u - \alpha)/\mu\delta(l + \tfrac{1}{2}l')$. Suppose that the solution $x(t; t_0, x_0)$ lies in the domain $\|x\| < \varepsilon$ for $t \in [t_0, t_m]$, but in view of the above, we get $V(t_m, x(t_m)) \geqslant h$ for time t_m, which contradicts the boundedness of the function $V(t, x)$ in the domain $V > 0$. The contradiction thus obtained proves that there exists a time $t^* \in [t_0, t_m]$ such that the solution $x(t; t_0, x_0)$ of (3.1.1) leaves the domain at this time. Thus Theorem 3.4.1 is proved. ∎

Remark 3.4.1   Owing to the system (3.1.1), the derivative of the Lyapunov function $V(t, x)$ can be of variable sign in the domain $V > 0$, in contrast to Chetayev's theorem on instability (Chetayev [1, 2]).

Remark 3.4.2   In contrast to Gorshin's theorem on instability under persistent perturbations (Gorshin [1]) the instability of the equilibrium state $x = 0$ of the unperturbed system (3.1.2) is not required in Theorem 3.4.1.

Corollary to Theorem 3.4.1   Let there exist a function $V(t, x)$ that is bounded and satisfies the following conditions in the domain $V > 0$.

(1) $\dot{V}(t, x) \geqslant 0$ with respect to the system (3.1.2).
(2) In the domain $\mathscr{D}$ the relation $f(t, x) = g(t, x)$ holds.
(3) There exists a summable function $N(t)$ and constant N_0 such that for $x \in \{V > 0\} \backslash E(\dot{V} = 0)$ the estimates

$$\varphi(t, x) \geqslant -N(t), \qquad \int_{t_1}^{t_2} N(t)\,dt \leqslant N_0(t_2 - t_1)$$

hold and condition (2)(ii) of Assumption 3.1.1 are satisfied in $\mathscr{D}$.
(4) The mean $\varphi_0(t_0, x_0)$ exists in the domain $V > 0$.
(5) The mean $\varphi_0(t_0, x_0)$ is greater than zero in the sets $E(\dot{V} = 0)$ of the domain $V > 0$.

Then the system (3.1.1) is μ-unstable.

3.5   Instability Investigation of Large-Scale Systems with Weakly Interacting Subsystems

For instability of the equilibrium state $x = 0$ of (3.3.1) (when it is assumed that $r_i(t, x) = 0$ for $x = 0$), it is sufficient to find a subsystem among the set of interacting subsystems (3.3.1) where solutions are μ-unstable.

Theorem 3.5.1   Let the following conditions be satisfied in the domain $\mathscr{D}$.

(1) For some kth subsystem of (3.3.1) $(1 \leqslant k \leqslant m)$ there exists a function $V_k(t, x_k)$ that is bounded and for which $\dot{V}_k(t, x) \geqslant 0$ and $\lim \dot{V}_{kg}(t, x) \geqslant 0$ on the domain $V_k > 0$.

(2) Conditions (5*) and (6*) of Assumption 3.3.1 hold in the domain $V_k > 0$.

(3) For an arbitrarily small number $\alpha > 0$ there exist numbers $\mu_k', \gamma_k > 0$, $\delta_k > 0$ and $l_k > 0$ such that for every pair of values (t, x_k) satisfying $V_k(t, x_k) > \alpha$ one of the following conditions holds for $x_i \in \Omega_i$, where $i \neq k$:

   (i) $\dot{V}_k(t, x) + \mu \varphi_k(t, x_k) \geqslant \gamma_k$ for $\mu < \mu_k'$;

   (ii) $\varphi_{k_0}^{\alpha_k}(\theta, t, x_{d_k}, x_{\alpha_k + 1}, \ldots, x_m) \geqslant \delta_k \theta$ for $\theta > l_k$.

(4) For the $1, 2, \ldots, (k-1)$th subsystems of (3.3.1) the following conditions hold:

   (i) $\dot{V}_j(t, x_j) > 2\gamma(\alpha) > 0$ for (t_0, x_{j0});

   (ii) $\displaystyle\int_{t_0}^{t_0+t} (\text{grad } V_j)^T r_j(t, \bar{x}_1, \ldots, \bar{x}_k, \ldots, 0)\, dt \geqslant \delta(\alpha)t$ for $t > l(\alpha)$,

   where γ, δ and l are some positive numbers depending on $\alpha > 0$ only.

(5) There exist functions $V_{k+1}(t, x_{k+1}), \ldots, V_m(t, x_m)$ satisfying the conditions of Theorem 3.3.1.

Then the system (3.3.1) is μ-unstable.

Proof   We set $\varepsilon \in]0, H[$. Let $\eta (0 < \eta < \varepsilon)$ be an arbitrarily small number. We take a point x_{k0} belonging to the domain $V_k > 0$, i.e. $V_k(t_0, x_{k0}) > 0$. Then a number $\alpha > 0$ can be chosen such that $V_k(t_0, x_{k0}) > 2\alpha$. By condition (1) of the theorem, there exist a constant $h > 0$ such that the inequality $V_k(t, x_k) < h$ holds for $\|x\| < \varepsilon$ in the domain $V_k > 0$. For the given α there exist numbers $\mu_k', \gamma_k, \delta_k$ and l_k as mentioned in condition (3). By virtue of Lemma 3.1.2, for

$$\lambda = \frac{\chi_k^{-1}(\delta_k/4F_{k0})}{(m + 1 - \alpha_k)(2l_k + l_k')\exp\left[L(2l_k + l_k')\right]} \qquad (3.5.1)$$

there exists a time $\tau_0' > t_0$ such that for $\tau > \tau_0'$ the inequality

$$\|\bar{x}_k(t, \tau, x_k(\tau)) - x_k(t; \tau, x_k(\tau); g)\| < \frac{\chi_k^{-1}(\delta_k/4F_{k0})}{m + 1 - \alpha_k}$$

holds on the finite interval $[\tau, \tau + 2l_k + l_k']$, where $l_k' = (h - \alpha)/\gamma_k$.

Similarly to the proof of Theorem 3.4.3, it can be shown that for any number $a > 0$, $\tau_0'' \geqslant t_0$ can be chosen such that for all $t \geqslant \tau_0''$ the inequality $\dot{V}_k(t, x_k; g) \geqslant -|c_k - a|$ holds where

$$\lim_{t \to \infty} \dot{V}_k(t, x_k; g) = c_k = \text{const} \geqslant 0.$$

The number a can be taken such that the solution $x_k(t; g) = x_k(t; t_0, x_{k0}; g)$ of the kth subsystem of (3.3.8) will not leave the domain $V_k > 0$ on the finite interval $[\tau_0'', \tau_0'' + \sigma(a)]$.

Let $\tau_0 = \max\{\tau_0', \tau_0''\}$ and $\mu_0 \in]0, \mu^*[$. We take $\mu_{k1} > 0$ such that for $t \in [t_0, \tau_0]$ the inequality

$$\mu_{k1} \leqslant \frac{1}{\alpha}\left|\int_{t_0}^{t} \varphi_k(s, x(s))\, ds\right|$$

holds, where $x(t) = x(t; t_0, x_0)$ is a solution of (3.3.1). Suppose that at some time $\tau_2 \geqslant t_0$ the inequalities $\| x_k(\tau_2) \| < \varepsilon$ and $V_k(\tau_2, x_k(\tau_2)) > \alpha$ hold. We consider two possible cases.

Case 1   Let conditions (1), (2), (3)(i) and (4) of the theorem hold. Then the solution $x_k(t)$ will not leave the domain $V_k > 0$ until the inequality $\| x_k(t) \| < \varepsilon$ is satisfied. For $x_k(t)$ the inequality $\| x_k(t) \| < \varepsilon$ and hypothesis (3)(i) cannot hold simultaneously for time l'_k. In fact, supposing the contrary, we have

$$\dot{V}_k(t, x_k) \geqslant \gamma_k \quad \text{for} \quad \mu < \mu'_k$$

by condition (3)(ii) for $V_k(t, x_k) > \alpha$. Integrating this inequality for $t \in [\tau_2, \tau_2 + l'_k]$, we get

$$V_k(t, x_k(t)) \geqslant V_k(\tau_2, x_k(\tau_2)) + \gamma(t - \tau_2).$$

Hence for $t = \tau_2 + l'_k$ we have

$$V_k(\tau_2 + l'_k, x_k(\tau_2 + l'_k)) > h.$$

This contradicts the boundedness of the function $V_k(t, x_k)$ in the domain $V_k > 0$. The contradiction implies that there exists a time $\tau_3 \in [\tau_2, \tau_2 + l']$ at which either of the conditions $\| x_k(t) \| < \varepsilon$ or (3)(i) is violated. Violation of the former leads to instability. Let condition (3)(i) be violated. Then since for $\| x_k \| < \varepsilon$ the vector x_k belongs to the domain $V_k > 0$ and condition (3)(ii) of the Theorem is satisfied.

Case 2   Suppose that $V_k(\tau_3, x_k(\tau_3)) > \alpha$ and condition (3)(ii) of the Theorem is satisfied. Integrating the total derivative of function V_k by virtue of (3.3.1) and in view of condition (1) of the theorem, we obtain

$$V_k(t, x_k(t)) \geqslant \alpha + \mu \int_{\tau_3}^{t} \varphi_k(s, x(s)) ds, \tag{3.5.2}$$

where $x(t) = x(t; \tau_3, x(\tau_3))$. For the integral in this inequality the relation

$$\int_{\tau_3}^{t} \varphi_k(s, x(s)) ds \geqslant J_{k3} - |J_{k1}| - |J_{k2}|, \tag{3.5.3}$$

can be obtained, where J_{k1}, J_{k2} and J_{k3} are defined similarly to J_{q1}, J_{q2} and J_{q3} in (3.3.16). Estimating integrals J_{k1} and J_{k2} as in the previous section but on the interval $[\tau_3, \tau_3 + 2l_k + l'_k]$, we get

$$|J_{k1}| < \tfrac{1}{4} \delta_k(t - \tau_3) \tag{3.5.4}$$

for

$$\mu < \mu_{k_2} = \frac{\chi_k^{-1}(\delta_k / 4 F_{k0})}{4 M_{k0}(2l_k + l'_k) \exp[L(2l_k + l'_k)]}$$

$$|J_{k2}| < \tfrac{1}{4} \delta_k(t - \tau_3). \tag{3.5.5}$$

Condition (3)(ii) implies

$$J_{k3} \geqslant \delta_k(t - \tau_3) \quad \text{for} \quad t - \tau_3 > l_k. \tag{3.5.6}$$

Substituting the estmates (3.5.3)–(3.5.6) into the inequality (3.5.2), we obtain

$$V_k(t, x_k(t)) > \alpha + \tfrac{1}{2} \mu \delta_k(t - \tau_3).$$
(3.5.7)

Let μ_{k3} be a small positive number such that for $\mu < \mu_{k_3}$ the inequality

$$\mu \left| \int_{\tau_3}^{t} \varphi_k(s, x(s)) \, ds \right| \leqslant \tfrac{1}{2} \alpha$$

holds for all $t \in [\tau_3, \tau_3 + 2l_k + l'_k]$. Then the solution $x_k(t)$ will not leave the domain $V_k > 0$ on the interval $[\tau_3, \tau_3 + 2l_k + l'_k]$.

Define $\mu_{k0} = \min\{\mu'_k, \mu_{k1}, \mu_{k2}, \mu_{k3}, \mu_0\}$. Similarly to the proof of Theorem 3.3.1, it can be shown that on each interval $[t_i, t_{i+1}]$ (where t_i is defined as in the proof of Theorem 3.4.1) the function $V_k(t, x_k(t))$ increases, at least for $\mu \delta_k(l_k + \tfrac{1}{2} l'_k) > 0$, for $\mu \in \,]0, \mu_{k0}[$. Let p_1 be the smallest integer p satisfying the condition $p \geqslant (h - \alpha)/\mu \delta_k(l_k + \tfrac{1}{2} l'_k)$ such that for $t \in [t_0, t_p]$ the solution $x_k(t; t_0, x_{k0})$ lies in the domain $\| x_k \| < \varepsilon$. In view of the above, we get

$$V_k(t_{p_1}, x_k(t_{p_1})) \geqslant h$$

for time t_p, which contradicts the boundedness of the function $V_k(t, x_k)$ in the domain $V_k > 0$.

Hence there exists a time $t^* \in [t_0, t_p]$ such that the solution $x_k(t)$ of the kth subsystem of (3.3.1) leaves the domain $\| x_k \| < \varepsilon$. Since $\| x_k(t) \| \leqslant \| x(t) \|$ for all $t \in R_+$, the inequality $\| x(t) \| \geqslant \varepsilon$ holds for the solution $x(t)$ of (3.3.1) at this time. Thus Theorem 3.5.1 is proved. ∎

3.6   Investigation of Stability with Respect to Some Variables

If the integral $\varphi_0(T, t_0, x_0)$ does not satisfy conditions (3)(ii) of Theorems 3.2.1 and 3.4.1, it is reasonable to use the perturbed Lyapunov function

$$V(t, x, \mu) = V_0(t, x) + U(t, x, \mu),$$
(3.6.1)

where $V_0(t, x)$ is a Lyapunov function for the system (3.1.2) and the function $U \in C^{(1)}(R_+ \times \mathscr{D} \times M, R)$ is defined on $R_+ \times \mathscr{D} \times M$, where $M = \,]0, \mu^*[$. The following assumptions are made on the vector functions $X(t, x)$ and $g(t, x)$. The vector x can be represented in the form $x = (y^T, z^T)^T$, and it is assumed that $X: R_+ \times \Omega_1 \times \Pi \to R^n$, where $\Omega_1 = \{y \in R^p : \| y \| < H\}$, $\Pi = \{z \in R^{n-p} : \| z \| < \infty\}$ and $\mathscr{D}_1 = R_+ \times \Omega_1 \times \Pi$. The vector function $X(t, y, z)$ satisfies a Lipschitz condition in $(y^T, z^T)^T$, with a constant L_1, and $f(t, 0, z) = 0$ for all $t \in R_+$ and $z \in \Pi$. The vector function $g(t, x)$ satisfies a Lipschitz conditions, with a constant L_2. We introduce the following notation. Along the solutions of (3.1.1), the total derivative of the function (3.6.1) is

$$\dot{V}(t, x, \mu) = \frac{\partial V_0}{\partial t} + (\mathrm{grad}\, V_0)^T f(t, x) + (\mathrm{grad}\, V_0)^T \mu r$$

$$+ \frac{\partial U}{\partial t} + (\mathrm{grad}\, U)^T f(t, x) + (\mathrm{grad}\, U)^T \mu r.$$

Suppose that there exists a function $\theta_1(\mu)$ of class $\mathcal{K}$ such that

$$(\text{grad } V_0)^T \mu r + \frac{\partial U}{\partial t} + (\text{grad } U)^T f + (\text{grad } U)^T \mu r = \theta_1(\mu)\psi(t, x, \mu).$$

for a continuous function $\psi(t, x, \mu)$.
Then we have

$$\dot{V}(t, x, \mu) = \dot{V}_0(t, x) + \theta_1(\mu)\psi(t, x, \mu), \tag{3.6.2}$$

where

$$\dot{V}_0(t, x) = \frac{\partial V_0}{\partial t} + (\text{grad } V_0)^T f(t, x).$$

Let

$$\psi_0(T, t_0, x_0, \mu) = \int_{t_0}^{t_0 + T} \psi(s, x(s; t_0, x_0; g), \mu) \, ds, \tag{3.6.3}$$

where the integral is computed along solutions $x(t; g)$ of the limiting system (3.1.3).

The problem of stability of solutions of (3.2.1) with respect to the variables y is solved in this section.

Theorem 3.6.1   Let the following conditions be satisfied on the domain D_1.

(1) There exists a function $V_0(t, y, z) \in C^{(1)}(\mathcal{D}_1, R_+)$ and functions $a, b \in \mathcal{K}$ such that
   (i) $a(\| y \|) \leqslant V_0(t, y, z) \leqslant b(\| y \|)$;
   (ii) $\dot{V}_0(t, x) \leqslant 0$, $x = (y^T, z^T)^T$.
(2) Condition (5) of Assumption 3.1.1 is satisfied and there exist summable functions $M(t), F(t), \chi_1(r) \in \mathcal{K}$ and constants M_0 and F_0 such that

$$\| R(t, x) \| \leqslant M(t), \qquad \int_{t_1}^{t_2} M(t) \, dt \leqslant M_0(t_2 - t_1),$$

$$|\psi(t, x', \mu) - \psi(t, x'', \mu)| \leqslant F(t)\chi_1(\| x' - x'' \|)_0;$$

$$\int_{t_1}^{t_2} F(t) \, dt \leqslant F_0(t_2 - t_1)$$

on any finite interval $[t_1, t_2] \subset R_+$.
(3) There exists a function $\chi_2(r) \in \mathcal{K}$ such that the estimate $| U(t, x, \mu)| \leqslant \chi_2(\mu)$ holds for $(t, x) \in \mathcal{D}_1$ and $\mu < \mu_0$, with $\mu_0 \in]0, \mu^*[$.
(4) For any numbers α and β $(0 < \alpha < \beta < H)$ there exist positive numbers μ', δ and l such that for any (t, y, z), $\alpha \leqslant \| y \| \leqslant \beta$ and $z \in \Pi$ one of the following conditions holds:

   (i) $\dot{V}_0(t, x) + \theta_1(\mu)\psi(t, x, \mu) \leqslant 0$ for $\mu < \mu'$;
   (ii) $\psi_0(0, t_0, x_0, \mu) \leqslant -\delta\theta$ for $\theta > l$.

(5) Every solution of (3.1.1) exists in the future.

Then the system (3.1.1) is μ-stable with respect to the variables y.

Proof Let $\varepsilon \in]0, H[$ be given. By condition (1) of the theorem, all points of either of the moving surface

$$V_0(t, y, z) = a(\tfrac{1}{2}\varepsilon) \quad \text{or} \quad V_0(t, y, z) = a(\tfrac{1}{2}\varepsilon) - 2\sigma, \tag{3.6.4}$$

where σ satisfies the inequality $0 < \sigma < \tfrac{1}{2} a(\tfrac{1}{2}\varepsilon)$, will satisfy the condition

$$b^{-1}(a(\tfrac{1}{2}\varepsilon) - 2\sigma) \leqslant \|y\| \leqslant \tfrac{1}{2}\varepsilon \tag{3.6.5}$$

for all $t \in R_+$ and $z \in \Pi$. Now consider the moving surface

$$V(t, y, z, \mu) = a(\tfrac{1}{2}\varepsilon) - \sigma. \tag{3.6.6}$$

Take $\mu_1 = \chi_2^{-1}(\sigma)$. Then, by condition (3) of the theorem, we have $|U(t, y, z, \mu)| < \sigma$ for $\mu < \mu_1$, and all points of the moving surface (3.6.6) as well as each of the surfaces (3.6.4) will satisfy the inequality (3.6.5) for all $t \in R_+$ and $z \in \Pi$ when $\mu < \mu_1$. By condition (4) of the Theorem, for numbers $b^{-1}(a(\tfrac{1}{2}\varepsilon) - 2\sigma)$ and $\tfrac{1}{2}\varepsilon$ there exist positive numbers δ and l. Similarly to the proof of Theorem 3.2.1, for any $\lambda > 0$ there exists a time τ_0 such that when $\tau > \tau_0$ and $t \in [\tau, \tau + 2l]$ for solutions $\bar{x}(t; \tau, x(\tau))$ and $x_g(t; \tau, x(\tau))$ of (3.1.1) and (3.1.3), the inequality

$$\|\bar{x}(t) - x(t; g)\| \leqslant \lambda 2l \exp(2l L_2) \tag{3.6.7}$$

holds for $x(\tau)$ such that $V(\tau, x(\tau), \mu) = a(\tfrac{1}{2}\varepsilon) - \sigma$.

Let

$$\lambda = \frac{\min\{\chi_1^{-1}(\delta/4F_0), \tfrac{1}{4}\varepsilon\}}{[2l \exp(2L_2 l)]}. \tag{3.6.8}$$

Since the conditions of the theorem include all assumptions of Lemma 3.1.3, for any solution $x(t) = (y^T(t; t_0, x_0), z^T(t; t_0, x_0))^T$ for all $t_0 \in [0, \tau_1]$ and $t \in [t_0, \tau_1]$ (we set $\tau_1 > \tau_0$), we obtain the estimate

$$\|y(t; t_0, x_0)\| \leqslant (\|y_0\| + \mu M_0 \tau_1) \exp(L_1 \tau_1). \tag{3.6.9}$$

We take

$$\mu_2 = \frac{b^{-1}(a(\tfrac{1}{2}\varepsilon) - 2\sigma)}{2M_0 \tau_1 \exp(L_1 \tau_1)},$$

$$\eta = \frac{b^{-1}(a(\tfrac{1}{2}\varepsilon) - 2\sigma)}{2\exp(L_1 \tau_1)}.$$

The estimate (3.6.9) implies that

$$\|y(t; t_0, x_0)\| < b^{-1}(a(\tfrac{1}{2}\varepsilon) - 2\sigma)$$

for $\|y_0\| < \eta$, $z_0 \in \Pi$, $t \in [0, \tau_1]$, $t \in [t_0, \tau_1]$ and $\mu < \mu_2$. Thus to prove the theorem, it is sufficient to show that $\|y(t; t_0', x_0')\| < \varepsilon$ for $t_0' \geqslant \tau_1, t \geqslant t_0'$ and $\|y_0'\| < b^{-1}(a(\tfrac{1}{2}\varepsilon) - 2\sigma)$, $z_0 \in \Pi$. Consider the solution $x(t) = x(t; t_0', x_0')$ of (3.1.1), and assume that it leaves the domain $\|y\| < b^{-1}(a(\tfrac{1}{2}\varepsilon) - 2\sigma)$ and the condition

$$V(\tau, x(\tau), \mu) = a(\tfrac{1}{2}\varepsilon) - \sigma$$

is satisfied at time τ. Then the inequality (3.6.5) and one of conditions (4)(i) or (4)(ii) of the theorem holds at the point $(\tau, x(\tau))$. Now we consider two cases.

Case 1   Let condition (4)(i) be satisfied at τ, i.e. for number $b^{-1}(a(\frac{1}{2}\varepsilon) - 2\sigma)$ and $\frac{1}{2}\varepsilon$ there exists μ' such that the total derivative of the perturbed Lyapunov function is nonpositive at τ for $\mu < \mu'$, and therefore the solution $x(t)$ can not intersect the surface (3.6.6) at time τ.

Case 2   Suppose that condition (4)(ii) is satisfied at time τ, i.e. the inequality

$$\psi_0(t, \tau, x(\tau), \mu) \leqslant -\delta(t - \tau) \tag{3.6.10}$$

holds for all $t > \tau + l$. Integrating the expression (3.6.2) for $t \geqslant \tau$, we get

$$V(t, x(t), \mu) \leqslant a(\tfrac{1}{2}\varepsilon) - \sigma + \theta_1(\mu) \int_\tau^t \psi(s, x(s), \mu) \, ds. \tag{3.6.11}$$

The integral in this inequality can be represented in the form

$$\int_\tau^t \psi(s, x(s), \mu) \, ds = \int_\tau^t [\psi(s, x(s), \mu) - \psi(s, \bar{x}(s), \mu)] \, ds$$

$$+ \int_\tau^t [\psi(s, \bar{x}(s), \mu) - \psi(s, x(s; g), \mu)] \, ds$$

$$+ \int_\tau^t \psi(s, x(s; g), \mu) \, ds$$

$$= I_1 + I_2 + I_3,$$

where $\bar{x}(t) = \bar{x}(t; \tau, x(\tau))$ and $x(t; g) = x(t; \tau, x(\tau); g)$ are solutions of (3.1.2) and (3.1.3).

The value of parameter μ can be chosen such that the solution $x(t; \tau, x(\tau))$ does not leave the domain $\|y\| < \varepsilon$ on the interval $[\tau, \tau + 2l]$. By Lemma 3.1.1, for solutions $x(t) = x(t; \tau, x(\tau))$ and $\bar{x}(t) = \bar{x}(t; \tau, x(\tau))$ of (3.1.1) and (3.1.2) the estimate

$$\|x(t) - \bar{x}(t)\| \leqslant \mu M_0 2l \exp(2L_2 l)$$

holds on the interval $[\tau, \tau + 2l]$. Since

$$\|y(t) - \bar{y}(t)\| \leqslant \|x(t) - \bar{x}(t)\|,$$

we have

$$\|y(t) - \bar{y}(t)\| \leqslant \mu M_0 2l \exp(2L_2 l).$$

We set $\mu < \mu_3 = \varepsilon/[4 M_0 \exp(2L_2 l)]$. Then the latter inequality gives $\|y(t) - \bar{y}(t)\| < \frac{1}{2}\varepsilon$. By condition (1) of the theorem, the system (3.1.2) is y-stable. From the two latter inequalities we have $\|y(t)\| < \varepsilon$ for $t \in [\tau, \tau + 2l]$ and $\mu < \mu_3$.

We estimate the integral I_1. Let

$$\mu_4 = \frac{\chi^{-1}(\delta/(4F_0))}{2 M_0 l \exp(2L_2 l)}. \tag{3.6.12}$$

Then for all $\mu < \mu_4$ and $t \in [\tau, \tau + 2l]$, by virtue of condition (2) of the theorem, we find that

$$I_1 < \tfrac{1}{4}\delta(t - \tau). \tag{3.6.13}$$

As was shown in Section 3.2, the solution $x_g(t)$ of the limiting system (3.1.3) will not leave the domain Ω on the interval $[\tau, \tau + 2l]$, and therefore it cannot leave the domain $\mathcal{D}_1 \times \Pi$. In view of the inequality (3.6.7), the value of λ and condition (2) of the theorem, we obtain

$$I_2 < \tfrac{1}{4}\delta(t - \tau) \tag{3.6.14}$$

for $t \in [\tau, \tau + 2l]$. From the expression (3.6.12) and in view of the inequalities (3.6.10), (3.6.13) and (3.6.14), we obtain the estimate

$$\int_\tau^t \psi(s, x(s), \mu)ds < -\tfrac{1}{2}\delta(t - \tau) \tag{3.6.15}$$

for $\mu < \mu_4$ on the interval $[\tau, \tau + 2l]$. Thus the integral in (3.6.11) becomes negative, at least for $t \in \,]\tau + l, \tau + 2l[$ when $\mu < \mu_4$, and this means that the solution $x(t; \tau, x(\tau))$, having left the surface $V(t, x, \mu) = a(\tfrac{1}{2}\varepsilon) - 2\sigma$, remains in the domain $\|y\| < \varepsilon$, owing to the choice of μ_3, for $\mu < \mu_3$ and $t \in [\tau, \tau + 2l]$, and will return to the domain bounded by the surface $V(t, x, \mu) = a(\tfrac{1}{2}\varepsilon) - 2\sigma$ at some time within the interval $]\tau, \tau + l[$.

Owing to the choice of η, Cases 1 and 2 imply that the solution $x(t; t_0, x_0)$ for $t_0 \in R_+$, $\|x_0\| < \eta$, $\mu < \mu_0 = \min\{\mu', \mu_1, \mu_2, \mu_3, \mu_4\}$ for all $t \geqslant t_0$ will not leave the domain $\|y\| < \varepsilon$, and the numbers μ_0 and η can be taken independent of t_0. Thus Theorem 3.6.1 is proved. ∎

Remark 3.6.1   In Theorem 3.6.1 the perturbed Lyapunov function (3.6.1) and its derivative can be of variable sign owing to (3.1.1), in contrast to Rumyantzev's theorem on stability with respect to some variables (see Rumyantzev, Oziraner [1]).

3.7   Estimation of Stability Intervals on Time-Variable Sets

Basic theorems on the stability of unperturbed motion of the system (3.1.2) on time-variable sets were derived by Grujič, Martynyuk and Ribbens-Pavella [1]. In this chapter we investigate time intervals for which the solution does not leave some sets that are time-variable in the general case. The main theorem presented here is due to Martynyuk and Chernetzkaya [1].

3.7.1   Preliminary results

For the systems (3.1.1) and (3.1.2) the following assertion holds.

Theorem 3.7.1   Let the following conditions be satisfied.

(1) There exists a function $V(t, x)$, $V \in C^1(I \times \Omega, R_+)$ and function $a, b \in \mathcal{K}$ for which
   (i) $a(\|x\|) \leqslant V(t, x) \leqslant b(\|x\|)$ $\forall (t, x) \in I \times \Omega$;
   (ii) $\dfrac{\partial V}{\partial t} + (\text{grad } V)^T f(t, x) \leqslant 0$.

(2) for any ε, ρ $(0 < \rho < \varepsilon < H)$ in the domain $\rho \leqslant \|x\| < \varepsilon$ the function $\varphi(t, x)$ is definite and there exists a function $c \in \mathcal{K}$ such that $|\varphi(t, x)| \leqslant c(\|x\|)$.

Then for any $\varepsilon_0 \in]\rho, \varepsilon[$ there exists an $\eta(\varepsilon_0) > 0$ such that $\| x(t; t_0, x_0) \| \leqslant \varepsilon$ for all $t \in [t_0, t_0 + T(\mu)]$ when $\| x_0 \| \leqslant \eta(\varepsilon_0)$, where

$$T(\mu) = \frac{1}{\mu} \int_{\varepsilon_0}^{\varepsilon} \frac{da(s)}{c(s)}, \qquad \eta(\varepsilon_0) = b^{-1}(a(\varepsilon_0)).$$

Example 3.7.1   Consider in the space R^{2n} a system with weakly connected oscillators:

$$\frac{dx_{2i-1}}{dt} = \omega_i x_{2i}, \quad \frac{dx_{2i}}{dt} = -\omega_i x_{2i-1} + \mu \omega_i \cos v_i t \prod_{\substack{j=1 \\ j \neq 2i}}^{2n} x_j \prod_{j=1}^{n} \omega_j, \quad i = 1, 2, \ldots, n \quad (3.7.1)$$

with initial conditions

$$x_i(t_0) = x_i^0, \quad i = 1, 2, \ldots, 2n.$$

The function V for the degenerate system corresponding to (3.7.1) has the form $V(x) = \sum_{i=1}^{2n} x_i^2$. It is easily verified that condition (1)(ii) of Theorem 3.7.1 is satisfied and

$$\varphi(t, x) = 2 \prod_{j=1}^{2} \omega_j \prod_{j=1}^{2n} x_j \sum_{i=1}^{n} \omega_i \cos v_i t. \qquad (3.7.2)$$

In the space R^{2n} the norm of a vector x is defined by

$$\| x \| = \max_i |x_i|, \quad i = 1, 2, \ldots, 2n.$$

Here the functions $a, b, c \in \mathcal{K}$ have the forms

$$a(r) = r^2, \quad b(r) = 2nr^2, \quad c(r) = 2 \prod_{j=1}^{n} \omega_j \sum_{i=1}^{n} \omega_i r^{2n}.$$

From Theorem 3.7.1, we conclude that the solution $x(t; t_0, x_0)$ of system (3.1.1) under the initial conditions on (t_0, x_0) given by

$$\| x_0 \| < \varepsilon_0 (2n)^{-1/2}, \quad \varepsilon_0 < \varepsilon, \quad t_0 \in R$$

will the domain $\| x \| < \varepsilon$ for all $t \in [t_0, t_0 + T(\mu)]$, where $T(\mu)$ is defined by

$$T(\mu) = \begin{cases} (\mu \omega^2)^{-1} \ln \dfrac{\varepsilon}{\varepsilon_0} & \text{for } n = 1, \\[3mm] \left(\dfrac{1}{\varepsilon_0^{2n-2}} - \dfrac{1}{\varepsilon^{2n-2}} \right) \left[2(n-1)\mu \sum_{i=1}^{n} \omega_i \prod_{j=1}^{n} \omega_j \right]^{-1} & \text{for } n \neq 1. \end{cases}$$

3.7.2   Theorem on estimation of stability intervals

Together with the system (3.1.2), the limiting system

$$\frac{dx}{dt} = g(t, x), \quad g(t, 0) = 0 \qquad (3.7.3)$$

is considered.

Assumption 3.7.1   Corresponding to condition (5) of Assumption 3.1.1, the following holds.

(5**) There exists a limit

$$\lim_{t \to \beta} \| f(t,x) - g(t,x) \| = 0, \quad 0 < \beta < \tau \leqslant +\infty,$$

uniformly on $x \in S(t)$, where $S(t)$ is a time-variable set.

Remark 3.7.1 If $\tau = \infty$ and $S(t) = \Omega$ then the limiting system (3.1.3) coincides with (3.7.3).

Suppose that the unperturbed motion $x = 0$ of the system (3.7.3) is nonasymptotically stable and for the system there exists a function $V(t,x)$ with the corresponding properties.

Theorem 3.7.2 Let the following condition be satisfied.

(1) There exists a function $V(t,x)$ and functions $a, b \in \mathcal{K}$ $a'(s) > 0$, such that
    (i) $a(\|x\|) \leqslant V(t,x) \leqslant b(\|x\|)$ $\forall (t,x) \in I \times \Omega$;
    (ii) $dV^0/dt \leqslant 0$ $\forall (t,x) \in I \times \Omega$.
(2) For any $\rho (0 < \rho < H)$ in the domain $\rho \leqslant \|x\| < H, H < +\infty$ the function $\varphi(t,x)$ is definite and there exist a function $c \in \mathcal{K}$ and a continuous nonnegative function $\varkappa(t)$ for all $t \in I$ such that $|\varphi(t,x)| \leqslant c(\|x\|)\varkappa(t)$.
(3) There exist a function $\omega \in \mathcal{K}$ and a continuous nonnegative function $\lambda(t)$ such that

$$|(\operatorname{grad} V)^T (f - g)| \leqslant \omega(\|x\|)\lambda(t) \quad \forall (t,x) \in I \times \Omega.$$

Then for any $\varepsilon_0 > \rho$ there exists an $\eta(\varepsilon_0) > 0$ such that any solution of the system (3.1.2) starting at $t = t_0$ in the domain $\|x_0\| \leqslant \eta$ will not leave the set $B_s = \{x : \|x\| < s(t)\}$ on the interval of existence of positive solutions of the differential equation

$$\frac{ds}{dt} = \frac{\omega(s)}{a'(s)} \lambda(t) + \frac{c(s)}{a'(s)} \varkappa(t) \tag{3.7.4}$$

for the initial value $s(t_0) = \varepsilon_0 > 0$ such that $s(t) < H$.

Proof Let $\varepsilon_0 \in]\rho, H[$. For the function $s(t)$ with value $s \in [\varepsilon_0, H]$ when $t \in \mathcal{T} = [t_0, t_0 + \tau[$, where τ is a finite number or ∞, we consider a time-variable set $B_s = \{x : \|x\| \leqslant s(t)\}$ and a movable surface $S_s = \{x : V(t,x) = a(s(t))\}$. Continuity of $V(t,x)$ implies that the surface S_s is closed. From condition (1)(i), $S_s \subset B_s$. In fact, for $\|x\| = s(t)$ for all $t \in I_1 \subset \mathcal{T}_\tau$ we have $V(t,x) \geqslant a(s(t))$ and owing to the monotonicity of the function $a(s)$, the equilibrium $V(t,x) = a(s)$ implies as $x \in B_s$. Hence $S_s \subset B_s$, and for any s_1 and s_2 such that $\varepsilon_0 < s_1 < s_2 < H$ the surface S_{s_1} is enclosed in S_{s_2}. So, for any $\varepsilon \in]\varepsilon_0, H[$ and $\eta = b^{-1}(a(\varepsilon_0))$. By condition (1)(i), this set B_η lies within the surface S_{ε_0}. Let the solution $x(t; t_0, x_0)$ starting on the set B_η at time $t = t_\rho$ intersect the surface S_{ε_0} at $t^* \geqslant t_0$ before leaving B_ε. We denote the time at which the solution $x(t_s, t_0, x_0)$ reaches the surface S_s at t_ε. Consider the behaviour of the function $V(t,x)$ along part of the trajectory with starting point at (t_s, x_s) and final point at (t_{s+ds}, x_{s+ds}) on the surface

S_{s+ds}. We find from the expression for the function $V(t, x)$ that the total derivative due to the system (3.1.1) is

$$V(t_{s+ds}, x_{s+ds}) \leqslant V(t_s, x_s) + \int_{t_s}^{t_s+ds} |(\text{grad } V)^T(f-g)| dt + \int_{t_s}^{t_s+ds} |\varphi(t, x)| dt.$$

In view of the definition of the surface S_s, when conditions (2) and (3) are satisfied, we obtain

$$a(s+ds) - a(s) = a'(s)ds \leqslant \omega(s)\lambda(t)dt + c(s)\varkappa(t)dt.$$

Since the function $a(\cdot)$ is strictly monotonically increasing, $a'(s) > 0$ and

$$\frac{ds}{dt} \leqslant \frac{\omega(s)}{a'(s)}\lambda(t) + \frac{c(s)}{a'(s)}\varkappa(t). \tag{3.7.5}$$

With the same notation for the variables and the transition to (3.7.4) from (3.7.5), we find that for all values of t for which $s(t)$ is positive and satisfies (3.7.4), the solution $x(t; t_0, x_0)$ will remain in the time-variable set B_s.

Remark 3.7.2   If $g(t, x) \equiv f(t, x)$ then the nonlinear equation (3.7.4) becomes

$$\frac{ds}{dt} = \frac{c(s)}{a'(s)}\varkappa(t).$$

Hence

$$\int_{\varepsilon_0}^{\varepsilon} \frac{a'(s)ds}{c(s)} = \int_{t_0}^{t} \varkappa(s)ds. \tag{3.7.6}$$

If in condition (2) we set $|\varphi(t, x)| \leqslant \mu c(\|x\|)$, i.e. $\varkappa(t) = 1$, then (3.7.6) implies the estimate of T from Theorem 3.7.1.

3.7.3   Some corollaries to Theorem 3.7.2

We shall make the following assumption about the functions $f(t, x)$, $g(t, x)$ and $r(t, x)$ and the Lyapunov function $V(t, x)$.

Assumption 3.7.2   There exist nonnegative functions $\lambda_1(t)$ and $\varkappa_1(t)$, positive constants a and b, and numbers r_1, r_2 and r_3 such that

  (i) $\|f(t, x) - g(t, x)\| \leqslant \lambda_1(t)\|x\|^{r_1}$;
  (ii) $\|r(t, x)\| \leqslant \varkappa(t)\|x\|^{r_2}$;
  (iii) $a\|x\|^{r_3} \leqslant V(t, x) \leqslant b\|x\|^{r_3}$.

Corollaries to Theorem 3.7.2   In conditions (1)(i) (2) and (3) of Theorem 3.7.2 let Assumption 3.7.2 be taken into account. Then equation (3.7.4) becomes

$$\frac{ds}{dt} = \lambda(t)s^{r_1} + \varkappa(t)s^{r_2}, \tag{3.7.7}$$

$$s(t_0) = \left(\frac{b}{a}\right)^{1/r_3} \|x_0\|, \tag{3.7.8}$$

where $\lambda(t)$ and $\varkappa(t)$ are defined by the inequalities

$$|(\text{grad } V)^T (f - g)| \leqslant a r_3 \lambda(t) \| x \|^{r_1 + r_3 - 1},$$

$$|\varphi(t, x)| \leqslant a r_3 \varkappa(t) \| x \|^{r_2 + r_3 - 1},$$

and the solution $x(t; t_0, x_0)$ remains on the set B_s for all values of t for which the Cauchy problem (3.7.7), (3.7.8) has a positive solution $s(t; t_0, s_0)$.

Proof This is by direct substitution of the functions under consideration into (3.7.4) and in view of the fact that $a(s) = a s^{r_3}$.

We consider some cases of the problem (3.7.7), (3.7.8).

Case A When the right-hand sides of the limiting system (3.7.3) and of (3.1.3) coincide, we have $\lambda(t) \equiv 0$, and (3.7.7) becomes

$$\frac{ds}{dt} = \varkappa(t) s^{r_2}, \tag{3.7.9}$$

where $s(t_0)$ is defined by (3.7.8). Hence for $r_2 = 1$

$$s(t) = s(t_0) \exp\left[\int_{t_0}^{t} \varkappa(s) ds \right],$$

and

$$s(t) = \left[\left(\frac{b}{a} \right)^{q/r_3} \| x_0 \|^q + q \int_{t_0}^{t} \varkappa(s) ds \right]^{1/q}, \qquad q = 1 - r_2,$$

for $r_2 \neq 1$. Here the value of τ in the interval $[t_0, t_0 + \tau[$ is defined as

$$\tau = \sup\{ t \in \mathcal{T}_0 : s(t) \in \mathring{R}_+ \}, \qquad \mathring{R}_+ =]0, + \infty[. \tag{3.7.10}$$

Remark 3.7.3 A similar result is obtained for $R(t, x) \equiv 0$ in (3.1.1) or for $r_1 = r_2$.

Case B When the system (3.1.2) is linear, i.e. $f(t, x) = A(t)x$, where $A(t)$ is an $n \times n$ matrix with elements that we continuous and bounded on $\mathcal{T}_0$, we have $g(t, x) = A^0(t)x$, $r_1 = 1$, and equation (3.7.5) becomes

$$\frac{ds}{dt} = \lambda(t) s + \varkappa(t) s^{r_2}, \tag{3.7.11}$$

where $s(t_0)$ is defined by (3.7.8). By means of the substitution $s = \eta^{1/(1 - r_2)}$, equation (3.7.11) is reduced to a linear one, and can be integrated in quadratures. We have

$$s(t) = E(t) \left\{ \left[\left(\frac{b}{a} \right)^{1/2} \| x_0 \| \right]^q + q \int_{t_0}^{t} \varkappa(s) E(-qs) ds \right\}^{1/q}, \tag{3.7.12}$$

where

$$E(t) = \exp\left[\int_{t_0}^{t} \lambda(s) ds \right].$$

The value of τ in the interval where $\|x(t)\| \leqslant s(t)$, i.e. $x(t; t_0, x_0) \in B_s$, is estimated according to (3.7.10), in view of (3.7.12).

Remark 3.7.4   Since the unperturbed system (3.1.2) is linear, the Lyapunov function $V(t, x)$ is taken as a quadratic form, and thus $r_3 = 2$.

Example 3.7.2   Consider the system

$$\frac{dx_1}{dt} = \alpha \frac{\beta - t}{t} x_1 + \omega x_2,$$

$$\frac{dx_2}{dt} = -\omega x_1 + \mu \omega x_1 \cos^2 vt, \tag{3.7.13}$$

where μ is a small parameter, and α, ω and v are positive constants.

For the system (we set $\mu = 0$)

$$\frac{dx_1}{dt} = \frac{\alpha(\beta - t)}{t} x_1 + \omega x_2,$$

$$\frac{dx_2}{dt} = -\omega x_1 \tag{3.7.14}$$

the total derivative of the Lyapunov function $V = x_1^2 + x_2^2$ is nonpositive, and therefore the state $x_1 = x_2 = 0$ of (3.7.14) is unstable. Thus Theorem 3.7.1 does not apply to the system (3.7.13). Since

$$A^0 = \lim_{t \to \infty} \begin{bmatrix} \alpha(\beta - t)/t & \omega \\ -\omega & 0 \end{bmatrix} = \begin{bmatrix} -\alpha & \omega \\ -\omega & 0 \end{bmatrix},$$

and the state $x_1 = x_2 = 0$ of the corresponding limiting system is stable (nonasymptotically), Theorem 3.7.2 is applicable to the system.

Taking as before the norm of a vector $x \in R^2$ as

$$\|x\| = \max\{|x_1|, |x_2|\},$$

the constants $r_1 = r_2 = 1$, $r_3 = 2$, $a = 1$ and $b = 2$. The functions $\varkappa(t)$ and $\lambda(t)$ are

$$\varkappa(t) = \tfrac{1}{2} \mu \omega (1 + \cos 2t), \quad \lambda(t) = \alpha\left(\frac{\beta}{t} - 1\right).$$

Applying Theorem 3.7.2, we find that in Case A $\|x(t)\| < s(t)$ for all $t \in [t_0, t_0 + \tau[, t_0 > 0,$ where

$$s(t) = \sqrt{2} \|x_0\| \left(\frac{t}{t_0}\right)^{\alpha\beta} \exp[(\tfrac{1}{2}\mu\omega - \alpha)(t - t_0) + \tfrac{1}{4}\mu\omega(\sin 2t - \sin 2t_0)]$$

and τ is defined by (3.7.10).

3.8 Some Applications of the General Theorem to Mechanical Systems

Example 3.8.1   We investigate the stability of solutions of the system

$$\dot{x}_1 = -x_1 + [1 - p(t)]x_2 + \mu(x_1^3 - ax_2^3),$$
$$\dot{x}_2 = -p(t)(x_2 - x_1) + \mu[a(x_1 + x_2)\cos t + (x_1^3 - ax_2^3)], \qquad (3.8.1)$$

where $a = \text{const} > 1$, and $p(t)$ is a continuous function such that $0 \leqslant p(t) \leqslant m = \text{const}$ and $\lim_{t \to \infty} p(t) = 0$. We apply Theorem 3.2.1. The Lyapunov function is taken in the form

$$V = \tfrac{1}{2}[x_2^2 + (x_1 - x_2)^2].$$

The derivative of the Lyapunov function with respect to the unperturbed system

$$\dot{x}_1 = -x_1 + [1 - p(t)]x_2,$$
$$\dot{x}_2 = p(t)(x_1 - x_2) \qquad (3.8.2)$$

has the form

$$\dot{V}(t, x_1, x_2) = -[1 + p(t)](x_2 - x_1)^2 \leqslant 0.$$

The integral $\varphi_0(T, t_0, x_0)$ computed along the solutions

$$x_{1g} = x_{20} + (x_{10} - x_{20})\exp[-(t - t_0)],$$
$$x_{2g} = x_{20}$$

of the limiting system

$$\dot{x}_1 = -x_1 + x_2,$$
$$\dot{x}_2 = 0$$

will satisfy condition (3)(ii) of Theorem 3.2.1 for $t_0 \geqslant 0$ and

$$x_0 \in E = \{x_0 : |x_{20}| \leqslant 2|x_{10}|, |x_{10}| \leqslant 2|x_{20}|\},$$

since there exists a negative mean for $a > 1$, $t_0 \geqslant 0$ and $x_0 \in E$. Since $\dot{V}(t, x)$ is negative, condition (3)(i) of Theorem 3.2.1 is satisfied in some other neighbourhood of zero. Thus the system (3.8.1) are μ-stable.

Example 3.8.2   We consider a mechanical system of pendulum type. Suppose that the system is subjected to linear nonstationary resistance and elastic extend forces. Moreover, the system is subjected to the influence of small nonlinear forces, and its motion system is described by the equation

$$\ddot{y} + a(t)\dot{y} + b(t)y + \mu[c(t)g(\dot{y}) + d(t)q(y)] = 0, \qquad (3.8.3)$$

where μ is a small parameter, $a(t) \geqslant 0$, $b(t) > 0$, $c(t)$ and $d(t)$ are continuous functions, with $\lim_{t \to \infty} a(t) = 0, \lim_{t \to \infty} b(t) = b_0 > 0, db/dt \leqslant 0, 0 < c_1 \leqslant c(t) \leqslant c_2, 0 < d_1 \leqslant d(t) \leqslant d_2$, $\lim_{t \to \infty} c(t) = c_0$ and $\lim_{t \to \infty} d(t) = d_0$, and $q(y)$ and $q(\dot{y})$ are analytic functions of the variables y and $\dot{y}$ respectively, and furthermore $g^{(2i+1)}(\dot{y})|_{\dot{y}=0} \geqslant 0$. Equation (3.8.3) is equivalent to the system

$$\dot{y} = x,$$
$$\dot{x} = -a(t)x - b(t)y - \mu[c(t)g(x) + d(t)q(y)].$$

For the unperturbed system

$$\dot{y} = x,$$
$$\dot{x} = -a(t)x - b(t)y \tag{3.8.4}$$

there exists a Lyapunov function

$$V = \tfrac{1}{2}[x^2 + b(t)y^2],$$

where derivative is constantly negative owing to (3.8.4):

$$\dot{V}(t,x,y) = -a(t)x^2 + \frac{1}{2}\frac{db}{dt}y^2.$$

The limiting system

$$\dot{y} = x,$$
$$\dot{x} = -b_0 y$$

has a general solution of the form

$$x_g(t; t_0, x_0, y_0) = \sqrt{b_0}A \cos[\sqrt{b_0}(t - t_0) + \alpha],$$
$$y_g(t; t_0, x_0, y_0) = A \sin[\sqrt{b_0}(t - t_0) + \alpha],$$

where

$$A = \sqrt{\frac{x_0^2 + b_0 y_0^2}{b_0}}, \qquad \cos\alpha = \frac{x_0}{A\sqrt{b_0}}, \qquad \sin\alpha = \frac{y_0}{A}.$$

Since the function

$$\varphi(t,x,y) = -x[c(t)g(x) + d(t)q(y)]$$

has a negative-definite mean value computed along the solution of the limiting system, condition (3)(i) of Theorem 3.2.1 is satisfied. The fact that the other hypotheses are also satisfied is easily verified. Thus, Theorem 3.2.1, the system (3.8.3) are μ-stable.

Remark 3.8.1   In the above examples the solutions of the unperturbed systems (3.8.2) and (3.8.4) in general do not reduce to quadratures, and cannot be expressed in terms of elementary functions in closed form. This prevents application of Theorem 3.2.1 when the solutions of the unperturbed systems are used.

Example 3.8.3   We investigate the stability of the system

$$\dot{x}_1 = -x_1 + \left(1 - \frac{1}{2e^t}\right)x_2 + \mu(yz_1^2 - x_1 z_2^2 - x_1^3),$$

$$\dot{x}_2 = \frac{1}{2e^t}(x_1 - x_2) + \mu[(x_1 + x_2)\sin t - x_2^3 - z_2^2 x_2 y^2],$$

$$\dot{y} = -p(t)(y + y^3) + \mu(y^5 \sin t - y^3 - x_1 x_2 y^4 - z_2^2 y), \tag{3.8.5}$$

$$\dot{z}_1 = -z_1 + \frac{1}{1+t^2}z_2 + \mu(z_2^2 x_1 - z_1^3 e^{4t} + z_1 x_2^3 y + z_1 z_2^5),$$

$$\dot{z}_2 = -\frac{1}{1+t^2}(z_1 + z_2^3) + \mu(z_1^4 y + z_1^2 x_2^4 \cos t - x_2^2 z_2 - z_1^4 z_2 - z_2^3),$$

where $p(t)$ is a continuous function, with $0 \leqslant p(t) \leqslant M = \text{const}$, $\lim_{t \to \infty} p(t) = 0$. For $\mu = 0$ the system (3.8.5) can be decomposed into three independent systems:

$$\dot{x}_1 = -x_1 + \left(1 - \frac{1}{2e^t}\right)x_2,$$

$$\dot{x}_2 = \frac{1}{2e^t}(x_1 - x_2); \tag{3.8.6}$$

$$\dot{y} = -p(t)(y + y^3); \tag{3.8.7}$$

$$\dot{z}_1 = -z_1 + \frac{1}{1 + t^2}z_2,$$

$$\dot{z}_2 = -\frac{1}{1 + t^2}(z_1 + z_2^3). \tag{3.8.8}$$

With the help of Theorem 3.3.1, the stability of the solutions of (3.8.7) can be shown. For the systems (3.8.6)–(3.8.8) we take functions

$$V_1(x_1, x_2) = \tfrac{1}{2}[x_2^2 + (x_1 - x_2)^2],$$

$$V_2(y) = \tfrac{1}{2}y^2,$$

$$V_3(z_1, z_2) = \tfrac{1}{2}(z_1^2 + z_2^2).$$

The time derivatives of these functions have the forms

$$\dot{V}_1(x_1, x_2) = -\left(1 + \frac{1}{2e^t}\right)(x_1 - x_2)^2 \leqslant 0,$$

$$\dot{V}_2(y) = -p(t)(1 + y^2)y^2 \leqslant 0,$$

$$\dot{V}_3(z_1, z_2) = -\left(z_1^2 + \frac{z_2^4}{1 + t^2}\right) \leqslant 0$$

along the solutions of the corresponding systems (3.8.6)–(3.8.8). The limiting systems corresponding to (3.8.6)–(3.8.8) are

$$\dot{x}_1 = x_2 - x_1,$$

$$\dot{x}_2 = 0; \tag{3.8.9}$$

$$\dot{y} = 0; \tag{3.8.10}$$

$$\dot{z}_1 = -z_1,$$

$$\dot{z}_2 = 0. \tag{3.8.11}$$

The general solutions of (3.8.9)–(3.8.11) have the forms

$$x_{1g}(t) = x_{20} + (x_{20} - x_{10})e^{-(t - t_0)},$$

$$x_{2g}(t) = x_{20},$$

$$y_g(t) = y_0,$$

$$z_{1g}(t) = z_{10}e^{-(t-t_0)},$$

$$z_{2g}(t) = z_{20}.$$

The integrals

$$\varphi_{10}^1(\theta, t_0, x_{10}, x_{20}, y_0, z_{10}, z_{20}),$$

$$\varphi_{20}^2(\theta, t_0, y_0, z_{10}, z_{20}),$$

$$\varphi_{30}^3(\theta, t_0, z_{10}, z_{20})$$

computed along the solutions of (3.8.9)–(3.8.11) satisfy condition (3)(ii) of Theorem 3.3.1 for $t_0 \geqslant 0$ and $x_0 = (x_{10}, x_{20}, y_0, z_{10}, z_{20})$ in some neighbourhood of zero, where $|x_{10}| \leqslant 2|x_{20}|$ and $|x_{20}| \leqslant 2|x_{10}|$, since there exist mean values

$$\lim_{\theta \to \infty} \theta^{-1}\varphi_{10}^1 = -x_{20}^2(x_{20}^2 + y_0^2 z_{20}^2),$$

$$\lim_{\theta \to \infty} \theta^{-1}\varphi_{20}^2 = -y_0^2(y_0^2 + z_{20}^2),$$

$$\lim_{\theta \to \infty} \theta^{-1}\varphi_{30}^3 = -z_{10}^4 e^{4t_0} - z_{20}^4$$

that are negative for these values of t_0 and x_0. In some other neighborhood of zero condition (4)(i) is satisfied owing to the negativeness of the derivatives $\dot{V}_1$, $\dot{V}_2$ and $\dot{V}_3$. By Theorem 3.3.1, the solutions of (3.8.5) are μ-stable.

3.9   Comments and References

3.0   Many effective methods have been developed for the analysis of nonautonomous equations involving small parameters. The basis and application of these methods to a variety of problems have been the subjects of many investigations (see e.g. Bogolyubov, Mitropolsky [1]; Volosov [1]; Grebenikov, Ryabov [1, 2]: Martynyuk [5]; Chetaev [1], Hale [1]). In the Institute of Mechanics of the National Ukrainian Academy of Sciences research into developing the method of Lyapunov functions in terms of limiting equations has been carried out under the direction of Martynyuk (see Karimzhanov [1–3]; Karimzhanov, Kosolapov [1]; Martynyuk, Karimzhanov [1–3]). These developments have led to general results that appear to be more effective in applications than those proposed by Hapaev [1].

3.1   Condition (5) of Assumption 3.1.1 has been introduced in a generalization of the approach proposed by Chetayev [1] for the investigation of characteristic numbers of linear nonstationary systems.

3.2   The results presented in this section correspond to those obtained by Karimzhanov and Kosolapov [1] and Martynyuk and Karimzhanov [1]. The corollary to Theorem 3.2.1 was proved by Kosolapov [1] as an independent result without the use of limiting equations.

3.3   The technique for large-scale systems, generalized by Sage [1] and Šiljak [1] was developed for the investigation of weakly connected systems by Martynyuk [1–5]. Weakly connected systems were later investigated by many researchers (see Karimzhanov [3]; Kosolapov [1]; Chernetskaya [1]). This section is based on results of Martynyuk and Karimzhanov [1].

3.4   The results of this section were first presented by Karimzhanov [3].

3.5   The theorem on instability of large-scale systems with weakly interacting subsystems in terms of limiting equations is due to Martynyuk and Karimzhanov [2, 3].

3.6   The investigation of stability with respect to some variables in terms of limiting equations follows Karimzhanov [3].

3.7   The results of this section are adapted from those of Martynyuk and Chernetzkaya [1].

3.8   Some examples in this section were investigated by Karimzhanov, Kosolapov, Martynyuk and Obolensky at the Institute of Mechanics of the National Ukranian Academy of Sciences in 1983–1989.

4 Stability Analysis of Solutions of ODEs (Continued)

4.0 Introduction

In this chapter we proceed with the analysis of systems of ordinary differential equations undertaken in Chapter 1. We shall investigate nonautonomous systems of differential equations whose right-hand sides satisfy the conditions for existence of limiting differential equation systems (see Chapter 1).

Through a combination of the comparison technique with the method of Lyapunov functions and the method of limiting equations, results are proved on the limiting behaviour of solutions, and asymptotic stability and instability, including stability with respect to some variables.

4.1 Preliminary Results

We consider the system of differential equations

$$\frac{dx}{dt} = f(t, x), \quad x(t_0) = x_0, \tag{4.1.1}$$

where $f \in C(R_+ \times R^n, R^n)$ is the function described in Chapter 1. Assume that equation (4.1.1) satisfies the condition of positive precompactness, and its solution $x(t) = x(t; t_0, x_0)$ cannot be continuable so that to differ from x by the definition interval. Such a solution $x(t)$ will be referred to as *noncontinuable*. We denote by $\Omega^+(x(t))$ the set of limiting points of the noncontinuable solution $x(t) = x(t; t_0, x_0)$ of the system (4.1.1).

Along with the system (4.1.1), we consider the comparison equation

$$\frac{dw}{dt} = g(t, w, x), \quad w(t_0) = w_0 \geqslant 0, \tag{4.1.2}$$

where $g \in C(R_+ \times R_+ \times R^n, R)$, $g(t, w, x)$ is nondecreasing relative to w and satisfies the condition of positive precompactness. Let W be an open subset in R_+ and $g_n(t, w, x) = g(t + t_n, w, x)$, where $t_n \to +\infty$ as $n \to \infty$.

We shall also consider the set

$$\Omega(g) = \{g^0 \in C(R_+ \times W \times R^n, R): \{t_n\} \subset R_+, t_n \to +\infty \text{ as } n \to \infty,$$

$$g_n(t, w, x) \to g^0(t, w, x) \text{ in the compact open topology}\}.$$

Together with the comparison equation (4.1.2), we shall consider the limiting comparison equation

$$\frac{dw}{dt} = g^0(t, w, x), \qquad w(t_0) = w_0 \geqslant 0, \tag{4.1.3}$$

where $g^0 \in \Omega(g)$.

We recall that the zero solution of the comparison equation (4.1.2) (or (4.1.3)) is stable in the Lyapunov sense if for given $t_0 \in R_+$ and $\varepsilon > 0$ there exists a $\delta^* = \delta^*(t_0, \varepsilon^*) > 0$ such that the conditions $\|x(t)\| < +\infty$ and $w_0 < \delta^*$ imply the inequality $w(t; t_0, w_0, x_0) < \varepsilon^*$ for all $t \geqslant t_0$.

We shall now prove some assertions about properties of the zero solution of equations (4.1.1)–(4.1.3).

Lemma 4.1.1   Let

(1) the function $g(t, 0, x) = g^0(t, 0, x) = 0$ for all $t \in R_+$ be nondecreasing relative to w for all $(t, x) \in R_+ \times R^n$;
(2) there exist an integrable function $\alpha(t)$ and a constant p such that

$$|g(t, w, x) - g^0(t, w, x)| \leqslant \alpha(t) \quad \forall w \in W, x \in R^n$$

and

$$\int_{t_0}^t \alpha(s)\, ds \leqslant p < +\infty \quad \forall t \geqslant t_0;$$

(3) for $g^0 \in \Omega(g)$ the following inequality hold:

$$g^0(t, w, x) \leqslant 0 \quad \forall(t, w, x) \in R_+ \times W \times R^n.$$

Then the solution $w = 0$ of the comparison equation (4.1.2) is stable in the Lyapunov sense.

Proof   On the basis of equations (4.1.2) and (4.1.3), we have

$$\frac{dw}{dt} = g^0(t, w, x) + [g(t, w, x) - g^0(t, w, x)].$$

Therefore, in view of conditions (2) and (3), we have

$$w(t; t_0, w_0) \leqslant w_0 + \int_{t_0}^t \alpha(s)\, ds < \varepsilon^* \quad \forall t \geqslant t_0$$

when $w_0 < \delta^*$.

Theorem 4.1.1   Assume that

(1) for system (4.1.1) there exists a function $V \in C(R_+ \times R^n, R_+)$, with $V(t,x)$ is locally Lipschitzian in x and

$$a(\|x\|) \leqslant V(t,x) \quad \forall (t,x) \in R_+ \times R^n,$$

where $a \in \mathscr{K}$;

(2) for $(t,x) \in R_+ \times R^n$ the inequality

$$D^+ V(t,x) \leqslant g(t, V(t,x), x)$$

is satisfied, where $g \in C(R_+ \times R_+ \times R^n, R)$, $g(t,0,0) = 0$ for all $t \in R_+$;

(3) conditions (1)–(3) of Lemma 4.1.1 are satisfied.

Then the zero solution of the system (4.1.1) is stable in the Lyapunov sense.

Proof   If conditions (1) and (2) of the theorem are satisfied then equation (4.1.2) is a comparison equation for (4.1.1). As a consequence of Theorem 3.1.5 by Lakshmikantham, Leela and Martynyuk [1] for the maximal solution of the comparison equation (4.1.2) $r(t; t_0, x_0, w_0)$ we have

$$V(t, x(t)) \leqslant r(t; t_0, x_0, w_0), \quad t \geqslant t_0. \tag{4.1.4}$$

By the conditions of Lemma 4.1.1, the zero solution $w = 0$ of equation (4.1.2) is stable. Therefore for any $t_0 \in R_+$ and $\varepsilon > 0$ a $\delta^* = \delta^*(t_0, \varepsilon) > 0$ can be found such that $w(t; t_0, x_0, w_0) < a(\varepsilon)$ when $w_0 < \delta^*$ and $\|x(t)\| < +\infty$. The continuity of the function $V(t,x)$ implies that a $\delta(t_0, \varepsilon) > 0$ can be found such that the inequality $\|x_0\| < \delta$ implies the inequality $u_0 = V(t_0, x_0) < \delta^*$. Now we have from condition (1) of the theorem and the estimate (4.1.4)

$$a(\|x\|) \leqslant V(t, x(t)) \leqslant r(t; t_0, x_0, w_0) < a(\varepsilon).$$

Hence $\|x(t; t_0, x_0)\| < \varepsilon$ for all $t \geqslant t_0$. This proves the theorem.   ∎

Definition 4.1.1   We shall call (G, g^0) a *limiting pair of functions* for the system (4.1.1) and the comparison equation (4.1.2) if $G(t,x)$ and $g^0(t,w,x)$ are limiting functions for $f(t,x)$ and $g(t,w,x)$ respectively for the same sequence $t_n \to +\infty$.

Given a limiting pair (G, g^0), we denote by $M^+(G, g^0)$ the set of noncontinuable solutions of the limiting system

$$\frac{dx}{dt} = G(t,x), \tag{4.1.5}$$

where $G \in \Omega(f)$, that lie on the set $g^0(t,w,x) = 0$ for all $t \in R_+$, $x \in D$ and $w \in W$ over their whole interval of definition. We denote by $M_*^+(G, g^0)$ the union with respect to all couples (G, g^0), i.e.

$$M_*^+(G, g^0) = \bigcup M^+(G, g^0). \tag{4.1.6}$$

Lemma 4.1.2   Let

(1) there exist a function $V \in C(R_+ \times R^n, R_+)$, with $V(t,x)$ locally Lipschitzian in x and $V(t,x) \geqslant 0$ for all $(t,x) \in R_+ \times D$;

(2) conditions (2) and (3) of Theorem 4.1.1 be satisfied;
(3) there exist at least one limiting couple (G, g^0) such that the set $M^+(G, g^0)$ is nonempty for all $(t, x) \in R_+ \times D$.

Then for every solution $x(t)$ of the system (4.1.1) defined on $[t_0, \infty[$ the set of its limit points satisfies the inclusion

$$\Omega^+(x(t)) \cap D \subset M_*^+(G, g^0).$$

Proof If $\|x(t)\| \to +\infty$ or $\Omega^+ \subset \partial D$ then the assertion of the lemma follows immediately. When condition (3) of the lemma is satisfied, we have for the solution $x(t; t_0, x_0) \Omega^+(x(t)) \cap D \neq \varnothing$, and there exists an $x_0^* \in \Omega^+ \cap D$ such that $x(t_n, t_0, x_0) \to x_0^*$ for some sequence $t_n \to +\infty$ as $n \to \infty$. The function $V(t)$ is bounded from below along the solution $x(t; t_0, x_0)$ and, according to condition (1) of the lemma and conditions (2)–(3) of Theorem 4.1.1, satisfies the inequality

$$V(t, x(t; t_0, x_0)) \leqslant r(t) \tag{4.1.7}$$

and the condition

$$V(t, x(t; t_0, x_0)) \to \sigma \quad \text{as } t \to +\infty. \tag{4.1.8}$$

Here $r(t)$ is a maximal solution of the comparison equation (4.1.2) and $\sigma = \text{const} > 0$.

When condition (3) of the lemma is satisfied, we take from the sequence $\{t_n\}$ a subsequence τ_k such that

$$f_k(t, x) = f(t + \tau_k, x) \to G(t, x), \tag{4.1.9}$$

$$g_k(t, w, x) = g(t + \tau_k, w, x) \to g^0(t, w, x) \tag{4.1.10}$$

simultaneously. It is also clear that the sequence $x_k(t) = x(t + \tau_k, t_0, x_0)$ converges to the solution $\psi(t) = \psi(t; t_0, x_0^*)$ of the limiting system (4.1.5), and the sequence $r_k(t) = r(t + \tau_k, t_0, w_0)$ converges to the solution $r(t)$ of the comparison equation (4.1.3). Moreover, the convergence will be uniform over every interval $[0, \alpha] \subseteq J_1 \cap J_2$, where J_1 and J_2 are the intervals of existance of solutions of the system (4.1.5) and equation (4.1.3). By condition (2) of the lemma, we have

$$V(t + \tau_k) - V(\tau_k) \leqslant \int_{t_0}^t g_k(s, V(s), x_k(s)) \, ds. \tag{4.1.11}$$

Taking the limit in the inequality (4.1.11) as $\tau_k \to +\infty$, and in view of the uniform convergence (4.1.9), (4.1.10) and the relations (4.1.7) and (4.1.8), we obtain

$$\sigma - \sigma \leqslant \int_0^t g^0(\tau, \sigma, \psi(\tau)) \, d\tau \leqslant 0.$$

Hence it follows that $g^0(t, \sigma, \psi(t)) \equiv 0$ for all $t \in J_1 \cap J_2$ and $x_0^* \in M^+(G, g^0)$. Then $\Omega^+(x(t)) \subset M_*^+(G, g^0)$. Thus the lemma is proved. ∎

4.2 Theorems on Stability and Uniform Asymptotic Stability

Utilizing the preliminary results, it is easy to prove some assertions on the stability of the solution $x = 0$ of the system (4.1.1).

Theorem 4.2.1   Assume that

(1) there exists a function $V \in C(R_+ \times R^n, R_+)$, with $V(t, x)$ locally Lipschitzian in x and

$$a(\|x\|) \leqslant V(t, x) \quad \forall(t, x) \in R_+ \times D, \quad D \subseteq R^n,$$

where $a \in \mathcal{K}$;

(2) for $(t, x) \in R_+ \times D$ the inequality

$$D^+ V(t, x) \leqslant g(t, V(t, x), x)$$

holds, where $g \in C(R_+ \times R_+ \times R^n, R)$, with $g(t, w, x)$ a function nondecreasing in w satisfying the condition of positive precompactness;

(3) all conditions of Lemma 4.1.1 are satisfied;

(4) for any limiting pair (G, g^0) the set $\{g^0(t, w, x) = 0\}$ as a subset of $R_+ \times R^n$ for all $t \in R_+$ does not contain the solutions of the system

$$\frac{dx}{dt} = G(t, x), \quad x(t_0) = x_0^*$$

except for $x = 0$.

Then the zero solution of the system (4.1.1) is asymptotically stable.

Proof   Conditions (1)–(3) of the theorem imply that the solution $x = 0$ of the system (4.1.1) is stable in the Lyapunov sense. Moreover, we find from condition (1) of the theorem an estimate of the domain of attraction of the solution $x(t)$ as

$$V_{t,\alpha}^{-1} = \{x \in D : \sup V(t, x) \leqslant a(\alpha), \alpha < H < +\infty\}. \tag{4.2.1}$$

Let $M_*^+(G, g^0)$ be the union with respect to all pairs (G, g^0) of the subsets $M^+(G, g^0) \subset \{g^0(t, w, x) = 0\}$ invariant relative to solutions of the system

$$\frac{dx}{dt} = G(t, x), \quad x(t_0) = x_0^*.$$

By Lemma 4.1.2 for any solution $x(t) = x(t; t_0, x_0)$, when $x_0 \in V_{t_0, \alpha}^{-1}$, the inclusion

$$\Omega^+(x(t)) \subset M_*^+(G, g^0)$$

holds. According to condition (4) of the theorem, we have $M^+(G, g^0) \equiv \{x = 0\}$, and therefore $M_*^+(G, g^0) \equiv \{x = 0\}$. Hence it follows that $\Omega^+(x(t)) \equiv \{x = 0\}$ and $\lim_{t \to +\infty} x(t) = 0$. This proves the theorem.

Corollary to Theorem 4.2.1 (Andreev [1])   Suppose that

(1) condition (1) of Theorem 4.2.1 is satisfied;

(2) the function $g(t, V(t, x), x) \equiv -W(t, x) \leqslant 0$, where $W(t, x)$ satisfies a Lipschitz condition in t and x on every compact $[t_0, t_0 + \tau] \times D_1, D_1 \subset D, \tau > 0, t_0 \geqslant 0$;

(3) for any limiting pair

$$f_k(t, x) = f(t + t_k, x) \to \Phi(t, x)$$
$$W_k(t, x) = W(t + t_k, x) \to \omega(t, x)$$

the set $\{\omega(t, x) = 0\}$ does not contain the solutions of the system

$$\frac{dx}{dt} = \Phi(t, x), \qquad x(t_0) = x_0$$

except for $x = 0$.

Then the zero solution of the system (4.1.1) is asymptotically stable.

Theorem 4.2.2   Assume that

(1) there exists a function $V \in C(R_+ \times R^n, R_+)$, with $V(t, x)$ locally Lipschitzian in (t, x) and

$$a(\|x\|) \leqslant V(t, x) \leqslant b(\|x\|),$$

where $a, b \in \mathscr{K}$;
(2) for $(t, x) \in R_+ \times D$ the inequality

$$D^+ V(t, x) \leqslant g(t, V(t, x), x)$$

holds, where $g \in C(R_+ \times R_+ \times R^n, R)$ is quasimonotonically nondecreasing in w for every $\|x(t)\| < k < +\infty$ and $t \in R_+$;
(3) the solution $w = 0$ of equation (4.1.2) is uniformly stable;
(4) for $(t, w, x) \in R_+ \times R_+ \times D$ the inequality

$$g^0(t, w, x) \leqslant 0$$

holds;
(5) for every limiting pair (G, g^0) the set $\{g^0(t, w, x) = 0\}$ does not contain the solutions of the system (4.1.5) except for $x = 0$.

Then the zero solution of the system (4.1.1) is uniformly asymptotically stable.

Proof   By conditions (1)–(3) and the comparison principle, the solution $x = 0$ of the system (4.1.1) is uniformly stable. The assertion of the theorem will be proved if the point $x = 0$ is shown to be attractive for every limiting system from the domain $D_0 \subset D$. As in Theorem 4.2.1, we denote by $x = \psi(t)(\psi(t_0) = x_0 \in D_0)$ the solution of the limiting system (4.1.5) and by $w = \zeta(t)$ the solution of the limiting comparison equation (4.1.2). According to condition (5) of the theorem, we take from the sequence $\{t_n\}$ a subsequence $t_k \to \infty$ as $k \to \infty$ such that

$$V_k(t, x) = V(t + t_k, x) \to \lambda^0(t, x), \tag{4.2.2}$$

$$g_k(t, w, x) = g(t + t_k, w, x) \to g^0(t, w, x), \tag{4.2.3}$$

$$f_k(t, x) = f(t + t_k, w, x) \to G(t, x). \tag{4.2.4}$$

Moreover, the convergence in (4.2.2)–(4.2.4) is uniform over every compact set $[t_0, t_0 + \tau] \times D_2, D_2 \subset D_0$.
According to condition (1) of the theorem, we have

$$a(\|x\|) \leqslant \lambda^0(t, x) \leqslant b(\|x\|) \quad \forall (t, x) \in R_+ \times D_2. \tag{4.2.5}$$

Consider the sequence of solutions $x = x_k(t)$, $t \geq t_0$, and $w = w_k(t)$, $t \geq t_0$, of the equations

$$\frac{dx}{dt} = f_k(t, x), \quad x_k(t_0) = x_0,$$

$$\frac{dw}{dt} = g_k(t, w, x), \quad w_k(t_0) = w_0 \geq 0.$$

The convergence in (4.2.3) and (4.2.4) implies that the sequences $x_k(t)$ and $w_k(t)$ will converge uniformly over every interval $[t_0, t_0 + \tau]$ to the functions $\psi(t)$ and $\zeta(t)$, $0 < \tau < +\infty$. Besides, the functions $x_k(t)$ will be solutions of the system (4.1.1) with the initial conditions $x(t_k + t) = x_0$, and $w_k(t)$ will be solutions of equation (4.1.2) with the initial conditions $w(t_k + t) = w_0 \geq 0$. In view of condition (2) of the theorem, we have

$$V_k(t, x_k(t)) - V_k(t_0, x_0) \leq \int_{t_0}^{t} g_k(\tau, V_k(\tau, x_k(\tau)), x_k(\tau)) \, d\tau. \tag{4.2.6}$$

Hence, going to the limit and in view of condition (4) of the theorem, we get

$$\lambda^0(t, \psi(t)) - \lambda^0(t_0, x_0) \leq \int_{t_0}^{t} g^0(\tau, \lambda^0(\tau, \psi(\tau)), \psi(\tau)) \, d\tau \leq 0. \tag{4.2.7}$$

The estimate (4.2.5) and the inequality (4.2.7) imply that the solution $x = 0$ of the system

$$\frac{dx}{dt} = G(t, x), \quad x(t_0) = x_0, \tag{4.2.8}$$

is stable, and any of its solutions starting in D_0 remains in the domain D_1.

Furthermore, for the functions $\lambda^0(t, x)$, $g^0(t, w, x)$ and $G(t, x)$ we take from the sequence $\{t_n\}$ a subsequence $t_s \to \infty$ as $s \to \infty$ such that

$$\lambda_s^0(t, x) = \lambda^0(t + t_s, x) \to \lambda^*(t, x),$$
$$g_s^0(t, w, x) = g^0(t + t_s, w, x) \to g^*(t, w, x),$$
$$G_s(t, x) = G(t + t_s, x) \to G^*(t, x).$$

Using the functions G^* and g^*, we construct the limiting system

$$\frac{dx}{dt} = G^*(t, x), \quad x(t_0) = x_0, \tag{4.2.9}$$

and the limiting comparison equation

$$\frac{dw}{dt} = g^*(t, w, x). \tag{4.2.10}$$

The system (4.2.9) and equation (4.2.10) are limiting for the system (4.1.1) and equation (4.1.2). Therefore, by condition (5) of the theorem, if (G^*, g^*) is a limiting pair for the pair (G, g^0), the set $\{g^*(t, w, x) = 0\}$ does not contain the solutions of the system (4.2.9) except for $x = 0$. It then follows that the solution $x = 0$ of the system (4.2.9) is asymptotically stable. Since the zero solution of the system (4.1.1) is uniformly stable and $x = 0$ is the point of attraction of the solutions to any limiting system, the zero solution of the system (4.1.1) is uniformly asymptotically stable. ∎

Corollary to Theorem 4.2.2 (Andreev [1])   Assume that

(1) condition (1) of Theorem 4.2.2 is satisfied;
(2) for $(t, x) \in R_+ \times D$ the inequality

$$D^+ V(t, x) \leqslant -W(t, x) \leqslant 0$$

   holds, where $W(t, x)$ is the function described in the corollary to Theorem 4.2.1;
(3) condition (3) of the corollary to Theorem 4.2.1 is satisfied.

Then the zero solution of the system (4.1.1) is uniformly asymptotically stable.

4.3   An Instability Theorem

We shall continue with our investigation of the system (4.1.1) via the approach described above. We show that the following assertion holds.

Theorem 4.3.1   Suppose that

(1) there exists a function $V \in C(R_+ \times R^n, R_+)$, with $V(t, x)$ precompact locally Lipschitzian in x and such that

$$V(t, x) \leqslant a(\| x \|) \quad \forall (t, x) \in R_+ \times D; \tag{4.3.1}$$

(2) for any $\delta > 0$ and $t_0 \in R_+$ an $x_0 \in B(0, \delta)$ can be found such that $V(t_0, x_0) > 0$;
(3) for $(t, x) \in R_+ \times D$ the inequality

$$D^+ V(t, x) \geqslant F(t, V(t, x), x) \tag{4.3.2}$$

   holds, where $F \in C(R_+ \times R_+ \times R^n, R)$ is precompact on $R_+ \times W \times W_1$ and $F(t, 0, 0) = 0$ for all $t \in R_+$;
(4) the solution $u = 0$ of the limiting equation

$$\frac{du}{dt} = F^0(t, u, x), \quad u(t_0) = u_0 \geqslant 0, \tag{4.3.3}$$

   is unstable;
(5) there exists a limiting pair of functions (G, F^0) such that the set $\{F^0(t, u, x) = 0\}$ as a subset of $R_+ \times R^n$ for all $t \in R_+$ does not contain the solutions of the limiting system

$$\frac{dx}{dt} = G(t, x), \quad x(t_0) = x_0, \tag{4.3.4}$$

   except for $x = 0$.

Then the zero solution of the system (4.1.1) is unstable.

Proof   The fact that the solution $u = 0$ of equation (4.3.3) is unstable implies that an $\varepsilon^* > 0$ and a $t_0 \in R_+$ can be found such that for all $\delta^* > 0$ there exists a u_0 with $0 \leqslant u_0 \leqslant \delta^*$ and $t^* \geqslant t_0$ for which $u^-(t; t_0, u_0) \geqslant \varepsilon^*$, where $u^-(t; t_0, u_0)$ is a minimal solution of equation (4.3.3).

We take $\varepsilon > 0$ such that $a(\varepsilon) \leqslant \varepsilon^*$. Given $\delta > 0$, we take a $\delta^* > 0$ such that for all u_0 with $0 \leqslant u_0 \leqslant \delta^*$ an x_0 with $\|x_0\| < \delta$ can be found such that $u_0 \leqslant V(t_0, x_0)$. This results from condition (2) of the theorem.

According to the precompactness of the functions $V(t, x)$, $f(t, x)$ and $F(t, V, x)$, we shall take from the sequence $\{t_n\}$ a subsequence $t_k \to \infty$ as $k \to \infty$ such that

$$V_k(t, x) = V(t + t_k, x) \to \rho(t, x),$$
$$f_k(t, x) = f(t + t_k, x) \to G(t, x),$$
$$F_k(t, w, x) = F(t + t_k, w, x) \to F^0(t, w, x).$$

By condition (3) of the theorem, we have the estimate

$$V_k(t, x_k(t)) - V_k(t_0, x_0) \geqslant \int_{t_0}^t F_k(\tau, V(\tau, x_k(\tau)), x_k(\tau)) \, d\tau$$

which yields, on taking the limit as $t_k \to +\infty$,

$$\rho(t, \psi(t)) \geqslant u_0 + \int_{t_0}^t F^0(\tau, \rho(\tau, \psi(\tau)), \psi(\tau)) \, d\tau \geqslant u^-(t; t_0, u_0), \qquad (4.3.5)$$

where $\psi(t)$ is a solution of the limiting system (4.3.4). Since $\rho(t, x) \leqslant a(\|x\|)$, we then find from (4.3.5) that

$$a(\|x(t)\|) \geqslant \rho(t, x(t)) \geqslant u^-(t; t_0, u_0), \qquad t \geqslant t_0. \qquad (4.3.6)$$

We now fix u_0, with $0 \leqslant u_0 \leqslant \delta^*$, and $t \geqslant t_0$, $t \in R_+$, such that $u^-(t, \cdot) \geqslant \varepsilon^*$. If x_0 is chosen as above and $t \notin J(t_0, x_0)$, the interval of existence of the solution $x(t) = x(t; t_0, x_0)$ of the system (4.1.1), then the theorem is proved, since the solution cannot cease to exist without leaving the domain $B(0, \varepsilon)$. Let $t \in J(t_0, x_0)$; then we find from estimate (4.3.6) that $\|x(t)\| \geqslant \varepsilon$. Condition (5) ensures the absence of solutions other than $x = 0$ to the system (4.1.1) and equation (4.3.3). Thus the solution $x = 0$ of the system (4.1.1) is unstable. ∎

Corollary 1 to Theorem 4.3.1 (Andreev [1]) Suppose that

(1) there exists a function $V \in C(R_+ \times R^n, R_+)$, with $V(t, x)$ locally Lipschitzian in x, for which conditions (1) and (2) of Theorem 4.3.1 are satisfied;
(2) for $(t, x) \in R_+ \times D$ the inequality

$$D^+ V(t, x) \geqslant W(t, x) \geqslant 0,$$

holds, where $W(t, x)$ is the function described in the corollary to Theorem 4.2.1;
(3) there exists a limiting pair of functions (G, W^0) for the functions $f(t, x)$ and $W(t, x)$ such that the set $\{W^0(t, x) = 0\}$ does not contain the solutions of system (4.3.4) except for $x = 0$.

Then the zero solution of the system (4.1.1) is unstable.

We shall consider one more modification of the sufficient conditions for instability of the solution $x = 0$ of the system (4.1.1). We denote by $N(t, c)$ the set of points x from the domain D. Let there exist a sequence $t_k \to +\infty$ such that $\lim_{t_k \to +\infty} V(t + t_k, x) = c$, such that where $c = \text{const} > 0$.

Corollary 2 to Theorem 4.3.1 (Andreev [1])   Suppose that

(1) there exists a function $V \in C(R_+ \times R^n, R_+)$, with $V(t, x)$ locally Lipschitzian in x, taking positive values for some $t \geq t_0 \geq 0$ in an arbitrary small neighbourhood of $x = 0$, and bounded in the domain $V(t, x) \geq 0$;
(2) for $(t, x) \in R_+ \times D$ the inequality

$$D^+ V(t, x) \geq W(t, x) \geq 0$$

holds;
(3) there exists a sequence $t_n \to +\infty$, for which the limiting set $N(t, c)$ and the limiting pair (G, W^0) are such that, given $c > 0$, the set $N(t, c) \cap \{W^0(t, x) = 0\}$ does not contain the solutions of the system

$$\frac{dx}{dt} = G(t, x).$$

Then the zero solution of the system (4.1.1) is unstable.

4.4   Asymptotic Stability and Uniform Asymptotic Stability with Respect to Some Variables

We shall consider the system (4.1.1) under the following assumptions. The vector x is presented as $x = (y^T, z^T)^T$, $y \in R^m$, $z \in R^p$, $m + p = n$, $m > 0$, $p \geq 0$, and vector function $f(t, x)$ is defined in the domain $R_+ \times R^m \times D$, where $D = \{z \in R^p : \|z\| < +\infty\}$. Assume that in this domain the vector function $f(t, x)$ satisfies the conditions of z-continuity of solutions, the existence and uniqueness conditions for the solutions of system (4.1.1), the existence conditions for limiting functions $G(t, x)$ of $f(t, x)$ and the conditions of initial continuity of solutions to the initial system (4.1.1)

$$\frac{dx}{dt} = f(t, x), x(t_0) = x_0$$

and solutions to the limit systems (4.1.5)

$$\frac{dx}{dt} = G(t, x).$$

Definition 4.4.1   The solution $(y^T, z^T) = 0$ of the system (4.1.1) is

(a) *y-stable* iff for any $t_0 \in R_+$ and $\varepsilon > 0$ there exists a $\beta(t_0, \varepsilon) > 0$ such that, provided $\|x_0\| < \beta$, the inequality $\|y(t; t_0, x_0)\| \leq \varepsilon$ holds for all $t \geq t_0$;
(b) *uniformly y-stable* iff the condition in (a) is satisfied and β does not depend on t_0;
(c) *y-attractive* iff for any $t_0 \in R_+$ an $\eta = \eta(t_0)$ can be found and for any $\varepsilon > 0$ and $\|x_0\| < \eta$ a $\sigma = \sigma(t_0, \varepsilon, x_0) > 0$ can be found such that $\|y(t; t_0, x_0)\| < \varepsilon$ for all $t \geq t_0 + \sigma, t_0 \in R_+$;
(d) *uniformly y-attractive* iff for some $\eta > 0$ and any $\varepsilon > 0$ a $\sigma = \sigma(\varepsilon) > 0$ can be found such that $\|y(t; t_0, x_0)\| < \varepsilon$, provided that $\|x_0\| < \eta, t_0 \in R_+$ and $t \geq t_0 + \sigma$;

(e) *asymptotically y-stable* iff it is *y*-stable and *y*-attractive;
(f) *uniformly asymptotically y-stable* iff it is uniformly *y*-stable and uniformly *y*-attractive.

Definition 4.4.2   Every solution $(y^T(t), z^T(t))$ of the system (4.1) is *z-bounded* iff for any $\alpha > 0$ and $t_0 \in R_+$ there exists a positive function $\beta = \beta(t_0, \alpha)$ continuous in t_0 for every value of α and such that the condition $\|x_0\| \leqslant \alpha$ implies that $\|z(t, t_0, x_0)\| < \beta(t_0, \alpha)$ for $t \geqslant t_0$.

Theorem 4.4.1   Suppose that

(1) the solution $x = 0$ of the system (4.1.1) is *z*-bounded;
(2) there exists a function $V \in C(R_+ \times R^m \times R^p, R_+)$, with $V(t, x)$ locally Lipschitzian in x and such that

$$V(t, x) \geqslant a(\|y\|) \quad \forall(t, x) \in R_+ \times R^m \times R^p;$$

(3) for all $(t, x) \in R_+ \times R^m \times R^p$ the inequality

$$D^+ V(t, x) \leqslant g(t, V(t, x), x)$$

is satisfied, where $g \in C(R_+ \times R_+ \times R^n, R)$, with $g(t, w, x)$ a quasimonotonic function nondecreasing in w for $\|x\| < +\infty$;
(4) the functions $g(t, w, x)$ and $g^0(t, w, x)$ satisfy all conditions of Lemma 4.1.1;
(5) for every limiting pair (G, g^0) the set

$$M^+(G, g^0) \subset \{x : y = 0\}.$$

Then the zero solution of the system (4.1.1) is asymptotically *y*-stable.

Proof   Condition (4) implies that the solution $w = 0$ of the comparison equation

$$\frac{dw}{dt} = g(t, w, x), \quad w(t_0) = w_0 \geqslant 0$$

is stable. By conditions (1)–(3) and Theorem 3.1.5 by Lakshmikantham, Leela and Martynyuk [1] on the comparison principle, the solution $x = 0$ of the system (4.1.1) is *y*-stable. We define the neighbourhood N of the point $x = 0$ by the formula

$$\sup \{V(t, x); \ \forall x \in N \subset D\} \leqslant a(H_2), \quad H_2 < H.$$

The solution $x(t) = x(t; t_0, x_0)$ of the system (4.1.1) is bounded for all $t \geqslant t_0$ by virtue of conditions (1)–(4). According to Lemma 4.1.2, the set of limit points $\Omega^+(x(t; t_0, x_0)) \subset M^+_*(G, g^0)$. If condition (5) of the theorem is satisfied, we have $M^+_*(G, g^0) \subset \{x : y = 0\}$. Therefore $\Omega^+(x(t; t_0, x_0)) \subset \{x : y = 0\}$ and $\lim_{t \to +\infty} y(t; t_0, x_0) = 0$. Thus the theorem is proved.

Definition 4.4.3   The set $N(t, c), c \geqslant 0, t \geqslant 0$ will be referred to as the *limit set of points from the product $R_+ \times D$* if for a sequence $t_n \to \infty$ there exists a sequence $x_n \to x \in D$ such that $V(t + t_n, x_n) \to c$ as $t_n \to +\infty$ and $x_n \to x$.

Theorem 4.4.2 Suppose that

(1) conditions (1)–(4) of Theorem 4.4.1 are satisfied;
(2) for a sequence $t_n \to +\infty$ the limiting pair (G, g^0) and the set $N(t, c)$ are such that, given any $c_0 > 0$, the set $N \cap \{g^0 = 0\}$ does not contain the solutions of the system

$$\frac{dx}{dt} = G(t, x), \quad x(t_0) = x_0.$$

Then the zero solution of the system (4.1.1) is asymptotically *y*-stable uniformly relative to x_0.

Proof This is similar to that of Theorem 4.4.1. ∎

Further, we assume that the function $V(t, x)$ satisfies the Lipschitz condition in t and x on every compact set $K = [t_0, t_0 + \tau] \times D_1$, $D_1 \subset D$, $t_0 \geqslant 0$.
Suppose that the limits (4.2.2)–(4.2.4) hold for the same sequence $t_n \to +\infty$. Therefore (ρ, G, g^0) is a coordinated (with respect to convergence) total limit for the function $V(t, x)$ of the system (4.1.1) and equation (4.1.2).

Theorem 4.4.3 Suppose that

(1) conditions (1), (3) and (4) of Theorem 4.4.1 are satisfied;
(2) the function $V(t, x)$ described in condition (2) of Theorem 4.4.1 satisfies

$$a(\| x \|) \leqslant V(t, x) \leqslant b(\| x \|)$$

where $a, b \in \mathcal{K}$;
(3) for the coordinated total limit (ρ, G, g^0) the set $\{\rho(t, x) = c > 0\} \cap \{g^0(t, w, x) = 0\}$ does not contain the solutions of the system

$$\frac{dx}{dt} = G(t, x), \quad x(t_0) = x_0.$$

Then the zero solution of the system (4.1.1) is uniformly asymptotically *y*-stable.

Proof According to conditions (1) and (2) of the theorem, the zero solution of the system (4.1.1) is uniformly *y*-stable. The solutions $x(t; t_0, x_0)$ starting in the domain $D_0 = b^{-1}(a(H_1)) \subset D_1(H_1 < H)$ are bounded. By Theorem 4.4.2, the function $V(t, x(t; t_0, x_0))$ is decreasing on the solutions $x(t; t_0, x_0)$ for $x_0 \in D_0$ and tends to zero as $t \to +\infty$. If it can be shown that $V(t, x(t; t_0, x_0)) \to 0$ uniformly in t_0 and x_0, the theorem will be proved.
We shall first demonstrate the following property of the function $\rho_0(t, \psi(t; t_0, x_0))$. Let (ρ_0, G_0, g^*) be an arbitrary coordinated total limit of functions. It can easily be shown that $\rho(t, \psi(t, t_0, x_0))$ decreases and vanishes uniformly along every solution $\psi(t; t_0, x_0)$ of the system

$$\frac{dx}{dt} = G_0(t, x), \quad x(t_0) = x_0 \in D_0.$$

Following the arguments from the proof of Theorem 4.2.2, we can easily show that $\rho_0(t, \psi(t, t_0, x_0))$ decreases and vanishes.

Suppose now that $V(t, x(t; t_0, x_0)) \to 0$ as $t \to +\infty$ nonuniformly in $t_0 \in R_+$ and $x_0 \in D_0$, i.e. there exists an $\varepsilon > 0$ such that for the sequence $\tau_n \to +\infty$ a sequence $t_n, x_n, \tau_n \geqslant 0, x_n \in D$, can be found such that $V(t_n + \tau_n, x(t_n + \tau_n), t_n, x_n) \geqslant \varepsilon_0$. Clearly, for $t, t_n \leqslant t \leqslant t_n + \tau_n$, we have

$$V(t, x(t; t_n, x_n)) \geqslant \varepsilon_0. \tag{4.4.1}$$

Since D_0 is precompact, there exists a subsequence $\{x_k\}$ such that $x_k \to x_0^* \in D_0$. The sequence $\{t_k\}$ is unbounded, otherwise the condition $V(t, x(t; t_0, x_0^*)) \to 0$ and the continuity conditions for solutions of the system (4.1.1) and the function $V(t, x)$ are violated. We take from $t_k \to +\infty$ a subsequence $\{t_j\} \subset \{t_k\}$ such that

$$f(t + t_j, x) \to G_0(t, x),$$
$$V(t + t_j, x) \to \rho_0(t, x),$$
$$g(t + t_j, w, x) \to g^*(t, w, x).$$

Moreover the sequence of solutions of the system (4.1.1) $x_j(t) = x(t + t_j, t_j, x_j)$ will converge uniformly in $t \in [0, \tau], \tau > 0$, to the solution $x = \psi(t, 0, x^*)$ of the system

$$\frac{dx}{dt} = G_0(t, x), \quad x(t_0) = x_0 \in D_0.$$

According to (4.4.1), we have the estimate $V(t + t_j, x_j(t)) \geqslant \varepsilon_0$ for $t \in [0, \tau_j]$. Then, taking the limit as $t_j \to +\infty$, we find $\rho(t, \psi(t; 0, x^*)) \geqslant \varepsilon_0 > 0$ for $t \geqslant 0$, which contradicts the previously established fact that $\rho_0(t, \psi(t, 0, x_0^*))$ decreases and vanishes. This proves Theorem 4.4.3. ∎

4.5 Instability with Respect to Some Variables

The sufficient conditions for y-instability of the solution $x = 0$ of the system (4.1) are derived in a form similar to Theorem 4.3.1.

Theorem 4.5.1 Suppose that

(1) the solutions of the system (4.1.1) are z-bounded in the neighborhood of $x = 0$;
(2) there exists a function $V \in C(R_+ \times R^n, R_+)$, with $V(t, x)$ locally Lipschitzian in x, bounded in the domain $V(t, x) \geqslant 0$ and with $V(t, x) \geqslant a(\|y\|)$;
(3) for $(t, x) \in R_+ \times D$ the inequality

$$D^+ V(t, x) \geqslant F(t, V(t, x), x)$$

holds, where $F \in C(R_+ \times R_+ \times R^n, R)$, with $F(t, w, x)$ a function quasimonotonically nondecreasing in w;
(4) the function F is precompact on $R_+ \times W \times W_1$, and the solution $w = 0$ of the equation

$$\frac{dw}{dt} = F^0(t, w, x), \quad w(t_0) = w_0 \geqslant 0, \tag{4.5.1}$$

is unstable;

(5) for some sequence $t_n \to +\infty$ there exists a limiting pair of functions (G, F^0) and a set $N(t, c)$ such that, given any $c_0 > 0$, the set $N(t, c) \cap \{F^0(t, w, x) = 0\}$ does not contain the solutions of the system (4.5.1).

Then the zero solution of the system (4.1.1) is y-unstable.

Proof   This is similar to that of Theorem 4.3.1.

Corollary to Theorem 4.5.1 (Andreev [2])   Suppose that

(1) the solution $x = 0$ of the system (4.1.1) is z-bounded;
(2) there exists a function $V \in C(R_+ \times R^n, R_+)$, with $V(t, x)$ locally Lipschitzian in x, for which condition (2) of Theorem 4.5.1 is satisfied;
(3) for $(t, x) \in R_+ \times D$ the inequality

$$D^+ V(t, x) \geqslant W(t, x) \geqslant 0$$

is satisfied, where $W(t, x)$ is the function described in the corollary to Theorem 4.2.1;
(4) for some sequence $t_n \to +\infty$ there exists a limiting pair of functions (G, W^0) associated with the functions $f(t, x)$ and $W(t, x)$ and a limit set $N(t, c)$ such that, given any $c_0 > 0$, $N(t, c_0) \cap \{W^0(t, x) = 0\}$ does not contain the solutions of the system

$$\frac{dx}{dt} = G(t, x), \qquad x(t_0) = x_0.$$

Then the zero solution of the system (4.1.1) is y-unstable.

4.6   Some Models of Lagrangian Mechanics

We shall consider a mechanical system with time-dependent behaviour described by the Lagrange equations

$$\frac{d}{dt}\left(\frac{\partial L}{\partial \dot{q}}\right) - \frac{\partial L}{\partial q} = Q, \tag{4.6.1}$$

where

$$q = [q_1 \quad q_2 \quad \cdots \quad q_n]^T, \qquad L = L_2 + L_1 + L_0,$$

$$2L_2 = (\dot{q})^T A(t, q)\dot{q}, \qquad L_1 = B^T(q)\dot{q}, \qquad L_0 = L_0(t, q),$$

$Q = Q(t, q, \dot{q})$ is the resultant of the generalized gyroscopic and dissipative forces, and $\partial L/\partial q \equiv 0$, $Q \equiv 0$ for $\dot{q} = q = 0$, so that the point

$$\dot{q} = q = 0 \tag{4.6.2}$$

is the equilibrium state for the system (4.6.1).

Assumption 4.6.1   Assume that for the system (4.6.1)

(1) $L_0(t, 0) = 0 \quad \forall t \in R_+$;
(2) $\partial L/\partial t \geqslant 0 \quad \forall t \in R_+$;

(3) the matrices $A(t, q)$, $\partial A/\partial t$, $\partial A/\partial q$ and $\partial B/\partial q$, the vector Q and the function $\partial L/\partial q$ are bounded on R_+ and satisfy the Lipschitz condition with respect to the variables $(t, q) \in R_+ \times R^n$;

(4) the matrices A_0, C_0 and D_0 and the vectors F_0 and Q_0 are limits of the corresponding matrices and vectors from equation (4.6.1); in particular,

$$F_0(t, q) = \lim_{t_n \to +\infty} \left\{ \frac{\partial L_0}{\partial q}(t + t_n, q) \right\}.$$

Then there exists a limiting system of equations

$$A_0^T \ddot{q} + \dot{q}^T C_0 \dot{q} + D^T \dot{q} + F_0 = Q_0. \tag{4.6.3}$$

Let $M_*^+ \subset \{\dot{q}_1 = \cdots = \dot{q}_n = 0\}$ be invariant with respect to the solutions of the limiting system (4.6.3).

Theorem 4.6.1   Assume that

(1) conditions (1)–(4) of Assumption 4.6.1 are satisfied;
(2) the function $L_0(t, q) \leqslant 0$ for all $t \in R_+$;
(3) there exists a function $g \in C(R_+ \times R_+ \times R^n \times R^n, R)$, with $g(t, 0, 0, 0) = 0$ for all $t \in R_+$, such that

$$Q^T(t, q, \dot{q})\dot{q} \leqslant g(t, L_2 - L_0, q, \dot{q});$$

(4) there exists a local L_1 function $\alpha(t)$ such that

$$|g(t, w, q, \dot{q}) - g^0(t, w, q, \dot{q})| \leqslant \alpha(t),$$

$$\int_{t_0}^{t} \alpha(s)\, ds \leqslant k < +\infty,$$

where

$$g^0(t, w, q, \dot{q}) = \lim_{t_n \to +\infty} \left\{ g(t + t_n, w, q, \dot{q}) \right\};$$

(5) the inequality

$$g^0(t, w, q, \dot{q}) \leqslant 0$$

holds for all $(t, w, q, \dot{q}) \in R_+ \times R_+ \times R^n \times R^n$.

Then every bounded motion of the system (4.6.1) unboundedly approaches the connected subset of the set of equilibrium states of the system (4.6.3), i.e. the set of points defined by the equalities

$$\lim_{t_n \to +\infty} \left\{ \frac{\partial L}{\partial q}(t + t_n, q) \right\} \equiv 0 \quad \forall t \in R_+.$$

Proof   When conditions (3)–(5) of the theorem are satisfied, the solution $w = 0$ of the comparison equation

$$\frac{dw}{dt} = g(t, w, q, \dot{q}), \qquad w(t_0) = w_0 \geqslant 0, \tag{4.6.4}$$

is stable. Since from condition (2) of the theorem, the function $L_2 - L_0$ is nonnegative and from condition (3) its total derivative by vitrue of the system (4.6.1) has the estimate

$$(L_2 - L_0)^{\cdot} = -\frac{\partial L}{\partial t} + Q^T \dot{q} \leqslant g(t, L_2 - L_0, q, \dot{q}), \tag{4.6.5}$$

using Lemma 4.1.2, the theorem is proved. ∎

Corollary to Theorem 4.6.1 (Andreev [2])   Assume that

(1) conditions (1)–(4) of Assumption 4.6.1 are satisfied;
(2) the function $V = -L_0(t, q)$ is positive-definite;
(3) the equilibrium state

$$q = \dot{q} = 0 \tag{4.6.6}$$

of the system (4.6.1) is nondegenerate and isolated, i.e.

$$\left\| \left(\frac{\partial L_0}{\partial q} \right) \right\| \geqslant c(\| \dot{q} \|),$$

with $C(a) = 0$ for $a = 0$;
(4) the dissipative forces Q are forces of complete dissipation, i.e.

$$Q^T \dot{q} \leqslant -a(\| \dot{q} \|), \quad a(0) = 0, \quad a(r) > 0 \quad \text{for } r > 0.$$

Then the equilibrium state (4.6.6) of the system (4.6.1) is uniformly asymptotically stable.

Proof   Since the functions $A(t, q)$ and $\partial L_0 / \partial q$ are bounded under condition (2) of the corollary, the function $L_2 - L_0$ is positive-definite and decreasing relative to $\dot{q}$ and q. Therefore

$$(L_2 - L_0)^{\cdot} \leqslant -a(\| \dot{q} \|) \leqslant 0. \tag{4.6.7}$$

Analysis of the structure of the system (4.6.3) shows that its only solution lying in the set $\{a(\| \dot{q} \|) = 0\} = \{\dot{q}_1 = \cdots = \dot{q}_n = 0\}$ is the solution $q = \text{const}$, defined by the equations

$$F_0(t, q) \equiv 0, \quad \text{for } q \equiv 0,$$

i.e. the solution y of system (4.6.1) can only be zero. Applying the corollary to Theorem 4.2.1, we find that the corollary to Theorem 4.6.1 is proved. ∎

4.7   Comments and References

4.1   The results of this section (Lemma 4.1.1 and 4.1.2 and Theorem 4.1.1) are new. They come from the development of the comparison principle and Lyapunov's direct method (see Theorem 3.1.5 by Lakshmikantham, Leela and Martynyuk [1]) in view of results by Andreev [1].

4.2   Theorems 4.2.1 and 4.2.2 are new.

4.3   Theorem 4.3.1 is new.

4.4   Theorems 4.4.1 and 4.4.3 are new.

4.5   Theorem 4.5.1 is new.

4.6   Equations (4.6.1) and the system of limiting equations (4.6.3) are from Andreev [1], where they were investigated in terms of the Corollaries to Theorems 4.2.1 and 4.5.1.

5 Stability of Integro-Differential Systems

5.0 Introduction

In this chapter we establish sufficient conditions for μ-stability of systems of integro-differential equations with small parameters and delays. Some theorems are applied investigate population dynamics models and to analyse the μ-stability of motion of an "ageing" viscoelastic cylinder interacting with an elastic shell.

5.1 Statement of the Problem

We consider the system of integro-differential equations

$$\frac{dx}{dt} = X(t, x) + \mu R\left(t, x, \int_0^t f(t, s, x(s))\, ds \right),\tag{5.1.1}$$

where μ is a small parameter, $x \in R^n$, and the vector functions $X(t, x)$, $R(t, x, y)$ and $f(t, s, x)$ are defined and continuous in the domains

$$D = R_+ \times \Omega, \quad D \times \Omega_1, \quad R_+ \times R_+ \times \Omega,$$

with

$$\Omega = \{x \in R^n : \|x\| < H = \text{const} > 0\},$$
$$\Omega_1 = \{y \in R^m : \|y\| < H_1 = \text{const} > 0\},$$

and satisfy the existence and uniqueness conditions of the solution of the Cauchy problem for the system (5.1.1).

Assumption 5.1.1   We assume the following.

(1) For $\mu = 0$ in the system (5.1.1) the generating system

$$\frac{dx}{dt} = X(t, x), \quad x(t_0) = x_0,\tag{5.1.2}$$

has an equilibrium solution $x = 0$ that is nonasymptotically stable, and its Lyapunov function $V \in C^{1,1}(R_+ \times \Omega, R_+)$ is such that in the domain D

$$\frac{dV}{dt} = \frac{\partial V}{\partial t} + (\nabla V)^T X(t, x) \leqslant V^*(x), \qquad (5.1.3)$$

where $V^* \in C(\Omega, R_-)$ is a nonpositive function.
(2) For the limiting system

$$\frac{dx}{dt} = g(t, x) \qquad (5.1.4)$$

associated with the system (5.1.2) the general solution $x(t; g) = x(t; t_0, x_0; g)$ is known and condition (5) of Assumption 3.1.1 is satisfied.

We introduce the notation

$$\mathscr{E}(V^* = 0) = \{x \in \Omega : V^*(x) = 0\}, \qquad \psi(t, x, y) = (\nabla V)^T R(t, x, y),$$

$$\psi^0(\theta, t_0, x_0) = \int_{t_0}^{t_0 + \theta} \psi \left(t, x(t; g), \int_0^t f(t, s, \tilde{x}(s)) \, ds \right) dt,$$

where $\tilde{x}(s) = x_0$ for $0 \leqslant s \leqslant t_0$ and $\tilde{x}(s) = x(s; t_0, x_0; g)$ is a solution of the limiting system (5.1.4) for $s \geqslant t_0$.

Definition 5.1.1 The mean $\psi^0(\theta, t_0, x_0) < 0$ in the set $\mathscr{E}(V^* = 0)$ is said to be *negative-definite* if for any numbers α and β, $0 < \alpha < \beta < H$, numbers $r = r(\alpha, \beta) > 0$, $\delta = \delta(\alpha, \beta) > 0$ and $l_1 = l_1(\alpha, \beta) > 0$ can be found such that $\psi^0(\theta, t_0, x_0) < -\delta\theta$ for $\theta > l_1$, $\alpha \leqslant \|x_0\| \leqslant \beta$ and $\rho(x_0, \mathscr{E}(V^* = 0)) < r$ for all $t_0 \geqslant 0$.

Assumption 5.1.2 Investigation of the μ-stability of the solution $x = 0$ to the system (5.1.1) is done under the following conditions.

(1) There exist summable functions $M(t)$ and $N(t)$ and positive constants M_0 and N_0 such that

$$\psi(t, x, y) \leqslant M(t), \qquad \int_{t_1}^{t_2} M(t) \, dt \leqslant M_0(t_2 - t_1)$$

for $(t, x, y) \in R_+ \times (\Omega \backslash \mathscr{E}(V^* = 0)) \times \Omega_1$, where $[t_1, t_2] \subset R_+$ is a finite segment.
(2) The function $\psi(t, x, y)$ satisfies a Lipschitz condition in (x, y) and

$$\|R(t, x, y)\| \leqslant N(t), \qquad \int_{t_1}^{t_2} N(t) \, dt \leqslant N_0(t_2 - t_1).$$

We shall investigate the μ-stability of the system (5.1.1) using the method of Lyapunov functions and the solution of the limiting system (5.1.4).

If there exists no limiting system (5.1.4) for the system (5.1.2), investigation of the stability of the system (5.1.1) involves the general solution of the system (5.1.2).

5.2 Stability Theorems

We shall need the following definition.

Definition 5.2.1 The system (5.1.1) is said to be μ-*stable* if for every $\varepsilon > 0$ there exist $\eta(\varepsilon) > 0$ and $\mu_0(\varepsilon) > 0$ such that for $0 < \mu < \mu_0$ each solution $x(t, \mu)$ of the system (5.1.1) for which $\| x(0, \mu) \| < \eta(\varepsilon)$ satisfies the inequality $\| x(t, \mu) \| < \varepsilon$ for all $t \geqslant 0$.

Theorem 5.2.1 Assume that

(1) there exist a function $V \in C^{1,1}(R_+ \times \Omega, R_+)$ and functions $a, b \in \mathcal{K}$ such that

$$a(\| x \|) \leqslant V(t, x) \leqslant b(\| x \|)$$

and the condition (5.1.3) is satisfied;
(2) condition (5) of Assumption 3.1.1 is satisfied;
(3) conditions (1) and (2) of both Assumptions 5.1.1 and 5.1.2 are satisfied;
(4) the mean $\psi^0(\theta, t_0, x_0) < 0$ in the set $\mathscr{E}(V^* = 0)$;
(5) the vector function $f(t, s, x)$ satisfies a Lipschitz condition in x, and there exist functions $f_1(t, s)$ and $p_0(t)$ that tend to zero as $t \to \infty$ such that

$$\| f(t, s, x) \| \leqslant f_1(t, s),$$

$$p(t, \theta_1) = \int_0^{\theta_1} f_1(t, s) \, ds < H_1 \quad \text{for} \quad t \geqslant \theta_1 \geqslant 0$$

$$\int_{\theta_1}^{\theta_1 + l_2} p(t, \theta_1) \, dt \leqslant p_0(l_2) l_2.$$

Then the system (5.1.1) is μ-stable.

Proof Let $\varepsilon \in \,]0, H[$ be given. By condition (1) of the theorem, every point of the moving surface $V(t, x) = a(\tfrac{1}{2}\varepsilon)$ lies in the annular domain $b^{-1}(a(\tfrac{1}{2}\varepsilon)) \leqslant \| x \| \leqslant \tfrac{1}{2}\varepsilon$ for $t \in R_+$. For the numbers $b^{-1}(a(\tfrac{1}{2}\varepsilon))$ and $\tfrac{1}{2}\varepsilon$ positive numbers δ, l_1 and r can be chosen that satisfy condition (4) of the theorem. According to Lemma 3.1.2, there exists a time τ_0 for

$$\lambda = (2l)^{-1} \exp(-2Ll) \min \{\delta/32L^2 l, \delta/16Ll, \tfrac{1}{2}\varepsilon\}$$

such that when $\tau_1 > \tau_0$ and $t > \tau_1$, for the solutions $\bar{x}(t; \tau_1, x(\tau_1))$ and $x(t; \tau_1, x(\tau_1); g)$ of the systems (5.1.2) and (5.1.4) the inequality

$$\| \bar{x}(t; \tau_1, x(\tau_1)) - x(t; \tau_1, x(\tau_1); g) \| \leqslant \lambda(t - \tau_1) \exp[L(t - \tau_1)]$$

holds; the number l will be chosen below. We now consider an arbitrary solution $x(t, \mu)$ of the system (5.1.1) for every $t_0 \in [0, \tau_1]$ and $\| x_0 \| < \eta = \tfrac{1}{2} b^{-1}(a(\tfrac{1}{2}\varepsilon)) \exp(-L\tau_1)$. By the Lipschitz condition on $X(t, x)$ and condition (3) of the theorem, we can obtain from the system of equations (5.1.1) for every $t_0 \in [0, \tau_1]$ and $t \in [t_0, \tau_1]$

$$\| x(t, \mu) \| \leqslant b^{-1}(a(\tfrac{1}{2}\varepsilon)) \quad \text{for} \quad \| x_0 \| < \eta,$$

$$\mu < \mu_1 = \frac{b^{-1}(a(\tfrac{1}{2}\varepsilon))}{2N_0 \tau_1 \exp(L\tau_1)}. \tag{5.2.1}$$

Therefore, in order to prove the theorem, it is sufficient to show that $\| x(t; \tau_1, x(\tau_1)) \| < \varepsilon$ for all $t > \tau_1$. Suppose that the solution $x(t, \mu)$ left the domain $\| x \| < b^{-1}(a(\frac{1}{2}\varepsilon))$ and $V(t, x(t, \mu)) = a(\frac{1}{2}\varepsilon)$ for $t = \tau_2$. By the conditions of the theorem, for the solutions $x(t, \mu)$ and $\bar{x}(t)$ of the systems (5.1.1) and (5.1.2) the estimate

$$\| x(t, \mu) - \bar{x}(t) \| \leqslant 2\mu N_0 l \exp(2Ll) \qquad (5.2.2)$$

holds on $[\tau_2, \tau_2 + 2l]$.

In view of the stability condition $\| \bar{x}(t) \| < \frac{1}{2}\varepsilon$, we get for the solution $x = 0$ of the system (5.1.2) $\| x(t, \mu) \| < \varepsilon$ for $t \in [\tau_2, \tau_2 + 2l]$ and

$$\mu < \mu_2 = \frac{\varepsilon}{4 N_0 \exp(2Ll)}.$$

Consider the behaviour of the function $V(t, x)$ along the solution $x(t, \mu)$ of the system (5.1.1). Integrating the expression for the total derivative of $V(t, x)$ in view of the system (1.1) and condition (1) of the theorem, we obtain for $t \geqslant \tau_2$

$$V(t, x(t, \mu)) \leqslant V(\tau_2, x(\tau_2, \mu))$$

$$+ \int_{\tau_2}^{t} V^*(x(s)) \, ds$$

$$+ \mu \int_{\tau_2}^{t} \psi\left(\tau, x(\tau), \int_0^{\tau} f(\tau, s, x(s)) \, ds\right) d\tau. \qquad (5.2.3)$$

We deal with two possible cases.

Case 1   Let $\rho(x(\tau_2), \mathscr{E}(V^* = 0)) < r$. In order to estimate the last integral on the right-hand side of the inequality (5.2.3), we represent it in the form

$$\int_{\tau_2}^{t} \psi\left(\tau, x(\tau), \int_0^{\tau} f(\tau, s, x(s)) \, ds\right) d\tau = J_1 + J_2 + J_3 + J_4 + J_5, \qquad (5.2.4)$$

where

$$J_1 = \int_{\tau_2}^{t} \left[\psi\left(\tau, x(\tau), \int_0^{\tau} f(\tau, s, x(s)) \, ds\right) - \psi\left(\tau, \bar{x}(\tau), \int_0^{\tau} f(\tau, s, x(s)) \, ds\right) \right] d\tau,$$

$$J_2 = \int_{\tau_2}^{t} \left[\psi\left(\tau, \bar{x}(\tau), \int_0^{\tau} f(\tau, s, x(s)) \, ds\right) - \psi\left(\tau, \bar{x}(\tau), \int_0^{\tau} f(\tau, s, \bar{x}(s)) \, ds\right) \right] d\tau,$$

$$J_3 = \int_{\tau}^{t} \left[\psi\left(\tau, \bar{x}(\tau), \int_0^{\tau} f(\tau, s, \bar{x}(s)) \, ds\right) - \psi\left(\tau, \bar{x}(\tau), \int_0^{\tau} f(\tau, s, \bar{x}(s)) \, ds\right) \right] d\tau,$$

$$J_4 = \int_{\tau}^{t} \left[\psi\left(\tau, \bar{x}(\tau), \int_0^{\tau} f(\tau, s, \bar{x}(s)) \, ds\right) - \psi\left(\tau, x(\tau; g), \int_0^{\tau} f(\tau, s, \bar{x}(s)) \, ds\right) \right] d\tau,$$

$$J_5 = \int_{\tau}^{t} \psi\left(\tau, x(\tau; g), \int_0^{\tau} f(\tau, s, \bar{x}(s)) \, ds\right) d\tau.$$

In view of the inequality (5.2.1), owing to the choice of λ and to the stability of the system (5.1.2), we can show that the solutions $x(t; g) = x(t; \tau_2, x(\tau_2); g)$ of the system (5.1.4) do not leave the domain Ω, and therefore conditions (3) and (5) of the theorem can be used to estimate the integrals J_2 and J_4.

We estimate

$$J_2 \leqslant L \int_{\tau_2}^t \left\| \int_0^\tau f(\tau, s, x(s)) \, ds - \int_0^\tau f(\tau, s, \bar{x}(s)) \, ds \right\| d\tau$$

$$\leqslant \int_{\tau_2}^t \left\| \int_0^{\tau_2} [f(\tau, s, x(s)) - f(\tau, s, \bar{x}(s))] \, ds \right\| d\tau$$

$$+ L \int_{\tau_2}^t \left\| \int_{\tau_2}^\tau [f(\tau, s, x(s)) - f(\tau, s, \bar{x}(s))] \, ds \right\| d\tau$$

$$\leqslant 2L \int_{\tau_2}^t \int_0^{\tau_2} f_1(\tau, s) \, ds \, d\tau + L^2 \int_{\tau_2}^t \int_{\tau_2}^\tau \| x(s) - \bar{x}(s) \| \, ds \, dt$$

$$\leqslant 2L \int_{\tau_2}^t p(\tau, \tau_2) \, d\tau + 2L^2 l \mu N_0 \exp(2Ll)(t - \tau_2)$$

$$\leqslant 2L p_0(t - \tau_2)(t - \tau_2) + 2L^2 l \mu N_0 \exp(2Ll)(t - \tau_2).$$

As a consequence of condition (5) of the theorem, a number l_2 can be found for the number $\delta/32L$ such that for all $t \geqslant \tau_2 + l_2$ the inequality

$$p_0(t - \tau_2) \leqslant \delta/32L \tag{5.2.5}$$

holds. We take

$$\mu_3 = \frac{\delta}{32L^2 l N_0 \exp(2Ll)}.$$

Then we have for $\mu < \mu_3$ and $t \in [\tau_2 + l, \tau_2 + 2l]$, where $l = \max(l_1, l_2)$,

$$J_2 \leqslant \tfrac{1}{8} \delta(t - \tau_2). \tag{5.2.6}$$

Similarly, we can obtain for J_3 the estimate

$$J_3 \leqslant 2L p_0(t - \tau_2) + 4L^2 l^2 \lambda \exp(2Ll)(t - \tau_2).$$

Using the expression for λ and in view of the inequality (5.2.5), we have for $t \in [\tau_2 + l, \tau_2 + 2l]$

$$J_3 \leqslant \tfrac{1}{8} \delta(t - \tau_2). \tag{5.2.7}$$

Taking into account the inequalities (5.2.1) and (5.2.2) and the values of λ, and by virtue of condition (3) of the theorem, for $t \in [\tau_2, \tau_2 + 2l]$ when

$$\mu < \mu_4 = \frac{\delta}{16N_0 L \exp(2Ll)}$$

we find

$$J_4 \leqslant \tfrac{1}{8} \delta(t - \tau_2). \tag{5.2.8}$$

Since $J_5 = \psi(t - \tau_2, \tau_2, x(\tau_2))$, according to condition (4) of the theorem 2.1 the inequality

$$J_5 \leqslant -\delta(t - \tau_2) \tag{5.2.9}$$

holds for $t \in [\tau_2 + l, \tau_2 + 2l]$.

Combining the estimates (5.2.6)–(5.2.9) and taking into account the fact that $V^*(x) \leqslant 0$ and $V(\tau_2, x(\tau_2)) = a(\tfrac{1}{2}\varepsilon)$, we obtain from (5.2.3) for $t \in [\tau_2 + l, \tau_2 + 2l]$ and $\mu < \mu_5 = \min\{\mu_1, \mu_2, \mu_3, \mu_4\}$

$$V(t, x(t, \mu)) < a(\tfrac{1}{2}\varepsilon) - \tfrac{1}{2}\mu\delta(t - \tau_2). \tag{5.2.10}$$

Thus we have shown that even if the solution $x(t; \tau_2, x(\tau_2))$ of the system (5.1.1) leaves the surface $V(t, x) = a(\tfrac{1}{2}\varepsilon)$, it will remain in the domain B_ε for $t \in [\tau_2, \tau_2 + 2l]$ because of the choice of μ. The inequality (5.2.10) shows that at some time from $]\tau_2, \tau_2 + l[$ the solution $x(t, \mu)$ will return to the domain bounded by the surface $V(t, x) = a(\tfrac{1}{2}\varepsilon)$.

Case 2   Assume that $\rho(x(\tau_2), \mathscr{E}(V^* = 0)) \geqslant r$. Then, on some interval $[\tau_2, t_2]$, where $\|x\| \geqslant b^{-1}(a(\tfrac{1}{2}\varepsilon)), \rho(x(\tau_2)$ and $\mathscr{E}(V^* = 0)) > r_1 (0 < r_1 < r)$ for the solution $x(t; \tau_2; x(\tau_2))$, a $\mu_0(\varepsilon)$ can be found such that $\|x(t; \tau_2, x(\tau_2))\| < \tfrac{1}{2}\varepsilon$. Actually, under condition (1) of the theorem, there exists a $\delta_1 > 0$ such that

$$\sup_{x \in Q} V^*(x) \leqslant -\delta_1,$$

where

$$Q = \{x: \rho(x, \mathscr{E}(V^* = 0)) \geqslant r_1, \eta \leqslant \|x\| \leqslant \tfrac{1}{2}\varepsilon\}.$$

The inequality (5.2.3) and condition (3) of the theorem imply

$$V(t, x(t; \tau_2, x(\tau_2))) \leqslant a(\tfrac{1}{2}\varepsilon) + (-\delta_1 + \mu M_0)(t - \tau_2).$$

Taking $\mu_0 = \min\{\mu_5, \delta_1/2M_0\}$, we get for $\mu < \mu_0$ and $t \in [\tau_2, \tau_2 + 2l]$

$$V(t, x(t; \tau_2, x(\tau_2))) \leqslant a(\tfrac{1}{2}\varepsilon) - \tfrac{1}{2}\delta_1(t - \tau_2).$$

This shows that the function $V(t, x)$ does not increase along the solution $x(t, \mu)$, which does not leave the domain $\|x\| < \tfrac{1}{2}\varepsilon$.

The estimates obtained in cases 1 and 2 are uniform relative to τ_2 and $x(\tau_2)$, and hence cases 1 and 2 will occur on every subsequent time interval $[t_i, t_{i+1}]$, where $t_i = \tau_2 + 2il$, and therefore the solution $x(t, \mu)$ of the system (5.1.1) will remain in the domain $\|x\| < \varepsilon$ for all $t \geqslant t_0$, when $\|x_0\| < \eta$ and $\mu < \mu_0$. Thus Theorem 5.2.1 is proved. ∎

We shall consider the case when there exists no limiting system (5.1.4) for the system (5.1.2), and the system (5.1.2) can be integrated.

We introduce the notation

$$\psi_1(t_0, x_0) = \lim_{\theta \to +\infty} \left\{ \left[\int_{t_0}^{t_0+\theta} \psi\left(\tau, \bar{x}(\tau), \int_0^r f(\tau, s, \tilde{\bar{x}}(s))\, ds \right) dr \right] \theta^{-1} \right\} \tag{5.2.11}$$

where $x(t)$ is a solution of the system (5.1.1), $x(t_0) = x_0$, $\bar{x}(t)$ is a solution of the system (5.1.2), $\bar{x}(t_0) = x(t_0) = x_0$, $\tilde{\bar{x}}(t) = x(t)$ for $0 \leqslant t \leqslant t_0$ and $\tilde{\bar{x}}(t) = \bar{x}(t)$ for $t_0 \leqslant t < +\infty$.

Similarly, we define

$$\psi_2(t_0, x_0) = \lim_{\theta \to +\infty} \left\{ \left[\int_{t_0}^{t_0+\theta} \psi\left(\tau, \bar{x}(\tau), \int_0^r f(\tau, s, \bar{x}) \, ds \right) dr \right] \theta^{-1} \right\} \quad (5.2.12)$$

where $\bar{x}(t) = x_0$ for $0 \leqslant t \leqslant t_0$ and $\bar{x}(t) = \bar{x}(t)$ for $t_0 \leqslant t < \infty$.

Theorem 5.2.2 Assume that in the domain $R_+ \times R_+ \times \Omega \times \Omega_1$ the following conditions we satisfied.

(1) There exists a function $V \in C^{1,1}(D, R_+)$ and functions $a, b \in \mathcal{K}$ such that

$$a(\|x\|) \leqslant V(t, x) \leqslant b(\|x\|),$$

and moreover V satisfies a Lipschitz condition in x with constant L_1.

(2) There exists a constant $M > 0$ such that

$$\|R(t, x, y)\| \leqslant M,$$

the vector function $R(t, x, y)$ satisfies a Lipschitz condition in x and y with constant L_2, and the vector function $X(t, x)$ satisfies a Lipschitz condition in x with constant L_3.

(3) There exists a mean $\psi_1(t_0, x_0)$ uniformly in t_0 and x_0, and for every $\eta > 0 (\eta < H)$ and $t_0 \in R_+$ a $\delta = \delta(\eta) > 0$ can be found such that $\psi_1(t_0, x_0) < -\delta$ for $\|x_0\| > \eta$.

(4) The vector function $f(t, s, x)$ satisfies a Lipschitz condition in x with constant L_4, and there exists a function $f_1(t, s)$ integrable on R_+ such that

$$\|f(t, s, x)\| \leqslant f_1(t, s), \quad \int_0^t f_1(t, s) \, ds \leqslant H_1.$$

Then the system (5.1.1) is μ-stable.

Proof This is similar to that of Theorem 5.2.1.

Theorem 5.2.3 Let the system (5.1.1) be such that in the domain $R_+ \times R_+ \times \Omega \times \Omega_1$ the following conditions are satisfied.

(1) Conditions (1) and (2) of Theorem 5.2.2 are satisfied.

(2) There exists a mean $\psi_2(t_0, x_0)$ uniformly in t_0 and x_0 that is negative-definite relative to variable x_0.

(3) The vector function $f(t, s, x)$ satisfies a Lipschitz condition in x with constant L_4, and there exist functions $f_1(t, s)$ and $p(t, \theta)$ such that

$$\|f(t, s, x)\| \leqslant f_1(t, s),$$

$$p(t, \theta) = \int_0^\theta f_1(\tau, s) \, ds \leqslant H_1.$$

(4) There exists a function $p_0(l) \to 0$ as $l \to +\infty$ such that

$$\int_\theta^{\theta+l} p(t, \theta) \, dt \leqslant p_0(l) l.$$

Then the system (5.1.1) is μ-stable.

Proof   This is similar to that of Theorem 5.2.1.

5.3   Almost-Periodic Solutions

We shall now consider the integro-differential system

$$\frac{dx}{dt} = f(t, x) + \int_{-\infty}^{0} F(t, s, x(t + s), x(t)) \, ds, \tag{5.3.1}$$

where $f: R \times R^n \to R^n$ is continuous and is almost-periodic in t uniformly for $x \in R^n$, and $F(t, s, x, y)$ is continuous on $R \times]-\infty, 0] \times R^n \times R^n$ and is almost-periodic in t uniformly for $(s, x, y) \in R_- =]-\infty, 0] \times R^n \times R^n$. For $\varphi, \psi \in BC$, let

$$\rho(\varphi, \psi) = \sum_{m=1}^{\infty} \frac{\rho_m(\varphi, \psi)}{2^m[1 + \rho_m(\varphi, \psi)]},$$

where $\rho_m(\varphi, \psi) = \sup_{-m \leqslant s \leqslant 0} |\varphi(s) - \psi(s)|$. Then $\rho(\varphi^k, \varphi) \to 0$ as $k \to \infty$ iff $\varphi^k(s) \to \varphi(s)$ uniformly on any compact interval in $]-\infty, 0]$.

Assumption 5.3.1   For the system (5.3.1)

(1) for any $\varepsilon > 0$ and any $r > 0$ there exists an $s = s(\varepsilon, r) > 0$ such that $\int_{-\infty}^{-s} |F(t, s, x(t + s), x(t))| \, ds \leqslant \varepsilon$ for all $t \in R$ when $x(\sigma)$ is continuous and $|x(\sigma)| \leqslant r$ for all $\sigma \leqslant t$;
(2) the system (5.3.1) has a bounded solution $u(t)$ defined on $[0, \infty[$ that passes through $(0, \varphi^0)$, $\varphi^0 \in BC$.

Remark 5.3.1   It follows from (1) that for any $r > 0$ there exists an $L(r) > 0$ such that

$$\int_{-\infty}^{0} |F(t, s, x(t + s), x(t))| \, ds \leqslant L(r) \quad \text{for } t \in R$$

when $x(\sigma)$ is continuous and $|x(\sigma)| \leqslant r$ for all $\sigma \leqslant t$. Moreover, $\int_{-\infty}^{0} F(t, s, x(t + s), x(t)) \, ds$ is continuous in t when $x(\sigma)$ is continuous and bounded for $\sigma \leqslant t$. Thus, under condition (1), for any $t_0 \in R$ and any $\varphi^0 \in BC$ there exists a solution of (5.3.1) that passes through (t_0, φ^0). Moreover, a solution can be continued up to $t = \infty$ if it remains in a compact set in R^n.

Let K be a compact set in R^n such that $u(t) \in K$ for all $t \in R$, where $u(t) = \varphi^0(t)$ for $t < 0$.

Definition 5.3.1   The bounded solution $u(t)$ is said to be *relatively eventually totally* (K, ρ)-*stable* if for any $\varepsilon > 0$ there exist $\delta(\varepsilon) > 0$ and $\alpha(\varepsilon) \geqslant 0$ such that if $t_0 \geqslant \alpha(\varepsilon)$, $\rho(u_{t_0}, x_{t_0}) < \delta(\varepsilon)$ and $p(t)$ is any continuous function satisfying $|p(t)| < \delta(\varepsilon)$ for $t \geqslant t_0$ then $\rho(u_t, x_t) < \varepsilon$ for all $t \geqslant t_0$. Here x is a solution through (t_0, x_{t_0}) of

$$\frac{dx}{dt} = f(t, x) + \int_{-\infty}^{0} F(t, s, x(t + s), x(t)) \, ds + p(t)$$

such that $x_{t_0}(s) \in K$, $s \leqslant 0$, and $x(t) \in K$ for $t \geqslant t_0$.

Moreover, if we can choose $\alpha(\varepsilon) \equiv 0$, $u(t)$ is said to be *relatively totally (K, ρ)-stable*. In the case where $p(t) \equiv 0$, this gives the definition of the *relative uniform (K, ρ)-stability* of $u(t)$.

Hamaya [1] obtained the following result under a slightly stronger condition on F, but the proof is essentially the same as in their paper (see Hamaya, Yoshizawa [1, Theorem 3]).

Theorem 5.3.1 If a bounded solution $u(t)$ of (5.3.1) is relatively eventually totally (K, ρ)-stable then $u(t)$ is asymptotically almost-periodic in t, and consequently the system (5.3.1) has an almost-periodic solution in K.

Denote by $\Omega(f, F)$ the set of all limiting functions (g, G) such that for some sequence $\{t_k\}$, $t_k \to \infty$ as $k \to \infty$, $f(t + t_k, x) \to g(t, x)$ uniformly on $R \times S$ for any compact subset S in R^n $F(t + t_k, s, x, y) \to G(t, s, x, y)$ uniformly on $R \times S^*$ for any compact subset S^* in R_- as $k \to \infty$. Moreover, let $(v, g, G) \in \Omega(u, f, F)$ denote that, for the same sequence $\{t_k\}$, $u(t + t_k) \to v(t)$ uniformly on any compact subset in R as $k \to \infty$. Clearly, if $(g, G) \in \Omega(f, F)$ then $g(t, x)$ is almost-periodic in t uniformly for $x \in R^n$ and $G(t, s, x, y)$ is almost-periodic in t uniformly for $(s, x, y) \in R_-$.

The following lemma can be easily proved.

Lemma 5.3.1 Assume that $F(t, s, x, y)$ satisfies condition (1) of Assumption 5.3.1 and $G \in \Omega(F)$ satisfies condition (1) for the same $s = s(\varepsilon, r) > 0$ as for F. Then the estimate

$$\int_{-\infty}^{s} |G(t, s, x(t + s), x(t))| \, ds \leqslant \varepsilon \quad \forall t \in R$$

holds when $x(\sigma)$ is continuous and $|x(\sigma)| \leqslant r$ for all $\sigma \leqslant t$.

Lemma 5.3.2 When $(v, g, G) \in \Omega(u, f, F)$, $v(t)$ is a solution defined on R of

$$\frac{dx}{dt} = g(t, x) + \int_{-\infty}^{0} G(t, s, x(t + s), x(t)) \, ds \tag{5.3.2}$$

and $v(t) \in K$ for all $t \in R$.

Proof Since $(v, g, G) \in \Omega(u, f, F)$, there exists a sequence $\{t_k\}$, $t_k \to \infty$ as $k \to \infty$, such that $f(t + t_k, x) \to g(t, x)$ uniformly on $R \times K$, $F(t + t_k, s, x, y) \to G(t, s, x, y)$ uniformly on $R \times T \times K \times K$ for any compact subset T in $]-\infty, 0]$, and $u(t + t_k) \to v(t)$ uniformly on any compact subset in R as $k \to \infty$.

Set $u^k(t) = u(t + t_k)$. Then $u^k(t)$ is a solution defined for $t \geqslant -t_k$ of

$$\frac{dx}{dt} = f(t + t_k, x) + \int_{-\infty}^{0} F(t + t_k, s, x(t + s), x(t)) \, ds.$$

Therefore, if k is sufficiently large, for $a \in R$ and $t \geqslant a$,

$$u^k(t) = u^k(a) + \int_{a}^{t} \left[f(w + t_k, u^k(w)) + \int_{-\infty}^{0} F(w + t_k, s, u^k(w + s), u^k(w)) \, ds \right] dw. \tag{5.3.3}$$

There exists a $c > 0$ such that $|u^k(t)| \leqslant c$ for all $t \in R$, and hence for any $\varepsilon > 0$ there is an $s = s(\varepsilon, c) > 0$ such that

$$\int_{-\infty}^{-s} |F(w + t_k, s, u(w + t_k + s), u(w + t_k))| \, ds \leqslant \varepsilon,$$

$$\int_{-\infty}^{-s} |G(w, s, v(w + s), v(w))| \, ds \leqslant \varepsilon$$

by condition (1) of Assumption 5.3.1 and Lemma 5.3.1. Then we have

$$\left| \int_{-\infty}^{0} F(w + t_k, s, u^k(w + s), u^k(w)) \, ds - \int_{-\infty}^{0} G(w, s, v(w + s), v(w)) \, ds \right|$$

$$\leqslant 2\varepsilon + \int_{-s}^{0} |F(w + t_k, s, u^k(w + s), u^k(w)) - G(w, s, v(w + s), v(w))| \, ds.$$

From this, we have

$$\int_{-\infty}^{0} F(w + t_k, s, u^k(w + s), u^k(w)) \, ds \to \int_{-\infty}^{0} G(w, s, v(w + s), v(w)) \, ds$$

as $k \to \infty$, because $u^k(t) \to v(t)$ uniformly on any compact set in R, and $G \in \Omega(F)$. Then, by the Lebesgue convergence theorem,

$$\int_{a}^{t} \left[\int_{-\infty}^{0} F(w + t_k, s, u^k(w + s), u^k(w)) \, ds \right] dw$$

$$\to \int_{a}^{t} \left[\int_{-\infty}^{0} G(w, s, v(w + s), v(w)) \, ds \right] dw.$$

Thus, letting $k \to \infty$ in (5.3.3), we have

$$v(t) = v(a) + \int_{a}^{t} g(w, v(w)) \, dw + \int_{a}^{t} \left[\int_{-\infty}^{0} G(w, s, v(w + s), v(w)) \, ds \right] dw.$$

From this, we can see that $v(t)$ is a solution of (5.3.2) defined on $]-\infty, \infty[$, and $v(t) \in K$ for all $t \in R$. ∎

We shall now discuss the relationship between total stability and the separation condition, which also implies the existence of an almost-periodic solution. This kind of problem has also been discussed for almost-periodic functional differential equations on phase space B by Hino [1]. If $x(t)$ is a solution defined on $]-\infty, \infty[$ such that $x(t) \in K$ for all $t \in R$, we say that x *is in* K and denote this by $x \in K$.

Definition 5.3.2   We say that the system (5.3.1) satisfies a ρ-*separation condition in* K if for any $(g, G) \in \Omega(f, F)$, there exists a $\lambda(g, G) > 0$ such that if x and y are distinct solutions of (5.3.2) in K then $\rho(x_t, y_t) \geqslant \lambda(g, G)$ for all $t \in R$.

Lemma 5.3.3   If the system (5.3.1) satisfies a ρ-separation condition in K then for each $(g, G) \in \Omega(f, F)$ the number of solutions in K is finite.

Proof If for some $(g, G) \in \Omega(f, F)$ the system (5.3.2) has an infinite number of solutions $x^k(t)$ in K then there exists a subsequence $\{x^{k_j}(t)\}$ of $\{x^k(t)\}$ that tends to a continuous function $x(t)$ defined on $]-\infty, \infty[$ uniformly on any compact set in R. Therefore, in the same way as in Lemma 5.3.2, we can see that $x(t)$ is a solution of (5.3.2) in K. But, for any $a \in R$, $\rho(x_a^{k_j}, x_a) \to 0$ as $j \to \infty$, and hence we cannot have a constant $\lambda(g, G) > 0$. ∎

The following lemma can be proved by the same argument as in the proof for ordinary differential equations (see Yoshizawa [3, pp. 198–200].

Lemma 5.3.4 Assume that the system (5.3.1) satisfies a ρ-separation condition in K. Then we can choose a positive constant λ_0 independent of (g, G) for which $\rho(x_t, y_t) \geq \lambda_0$ for all $t \in R$. We shall call λ_0 the ρ-separation constant in K.

Theorem 5.3.2 Suppose that the system (5.3.1) satisfies a ρ-separation condition in K. Then for any $(g, G) \in \Omega(f, F)$ any solution x of (5.3.2) in K is relatively totally (K, ρ)-stable. Moreover, we can choose the number $\delta(\cdot)$ in Definition 5.3.1 such that $\delta(\varepsilon)$ depends only on ε and is independent of (g, G) and the solutions.

Proof We shall show that for any $\varepsilon > 0$ there exists a $\delta(\varepsilon) > 0$ such that for any $(g, G) \in \Omega(f, F)$ and any solution x of (5.3.2) in K, $\rho(x_t, y_t) < \varepsilon$ for all $t \geq t_0$, whenever y is a solution of

$$\frac{dx}{dt} = g(t, x) + \int_{-\infty}^{0} G(t, s, x(t+s), x(t)) \, ds + p(t),$$

such that $y_{t_0}(s) \in K, s \leq 0, y(t) \in K$ for $t \geq t_0$ and $\rho(x_{t_0}, y_{t_0}) < \delta(\varepsilon)$ at some $t_0 \in R$. Here $p(t)$ is any continuous function satisfying $|p(t)| < \delta(\varepsilon)$ on $[t_0, \infty[$. Suppose the converse. Then there exist an $\varepsilon > 0$ and sequences $\{(g_k, G_k)\}, (g_k, G_k) \in \Omega(f, F), \{x^k(t)\}, \{y^k(t)\}, \{p_k(t)\}, \{t_k\}$ and $\{\tau_k\}$ such that $\tau_k > t_k, |p_k(t)| < 1/k$ on $[t_k, \infty[, \rho(x_{t_k}^k, y_{t_k}^k) < 1/k, \rho(x_{\tau_k}^k, y_{\tau_k}^k) = \varepsilon$ and $\rho(x_t^k, y_t^k) < \varepsilon$ on $[t_k, \tau_k[$. Here $x^k(t)$ is a solution in K of

$$\frac{dx}{dt} = g_k(t, x(t)) + \int_{-\infty}^{0} G_k(t, s, x(t+s), x(t)) \, ds,$$

and $y^k(t)$ is a solution of

$$\frac{dx}{dt} = g_k(t, x(t)) + \int_{-\infty}^{0} G_k(t, s, x(t+s), x(t)) \, ds + p_k(t),$$

such that $y_{t_k}^k(s) \in K, s \leq 0$, and $y^k(t) \in K$ for $t \geq t_k$. We can assume that $\varepsilon \leq \frac{1}{2}\lambda_0$, where λ_0 is the ρ-separation constant in K.

Set $u^k(t) = x^k(t + \tau_k)$ and $v^k(t) = y^k(t + \tau_k)$. Then $u^k(t)$ is a solution in K of

$$\frac{dx}{dt} = g_k(t + \tau_k, x(t)) + \int_{-\infty}^{0} G_k(t + \tau_k, s, x(t+s), x(t)) \, ds,$$

and $v^k(t)$ is defined for $t \geqslant t_k - \tau_k$ and is a solution of

$$\frac{dx}{dt} = g_k(t + \tau_k, x(t)) + \int_{-\infty}^{0} G_k(t + \tau_k, s, x(t+s), x(t)) \, ds + p_k(t + \tau_k),$$

such that $v_{t_k - \tau_k}^k(s) = y_{t_k}^k(s) \in K$, $s \leqslant 0$, and $v^k(t) \in K$ for $t \geqslant t_k - \tau_k$ $(t_k - \tau_k < 0)$. Since $(g_k(t + \tau_k, x), G_k(t + \tau_k, s, x, y)) \in \Omega(f, F)$, taking a subsequence if necessary, $g_k(t + \tau_k, x) \to h(t, x)$ uniformly on $R \times K$, and $G_k(t + \tau_k, s, x, y) \to H(t, s, x, y)$ uniformly on $R \times T \times K \times K$ for any compact set T in $]-\infty, 0]$ as $k \to \infty$, and $(h, H) \in \Omega(f, F)$.

Set $r_k = t_k - \tau_k$. Then we can assume that $r_k \to r$ as $k \to \infty$, where $r = -\infty$ or $r > -\infty$. Since $\{u^k(t)\}$ is uniformly bounded and equicontinuous, we can assume that $u^k(t) \to \xi(t)$ uniformly on any compact set in $]-\infty, \infty[$. Thus, by the same argument as in the proof of Lemma 5.3.2, we can see that $\xi(t)$ is a solution in K of

$$\frac{dx}{dt} = h(t, x(t)) + \int_{-\infty}^{0} H(t, s, x(t+s), x(t)) \, ds. \tag{5.3.4}$$

In the case where $r > -\infty$, since for some constants c and L, $|v^k(t)| \leqslant c$ for $t \geqslant t_k - \tau_k = r_k$ and $|dv^k(t)/dt| \leqslant L$ for $t \geqslant r_k$, we can assume that there exists a continuous function $\eta(t)$ defined on $[r, \infty[$ and $v^k(t) \to \eta(t)$ uniformly on any compact set in $]r, \infty[$. Thus we have $\eta(t) \in K$ for all $t \geqslant r$ and $v^k(r_k) \to \eta(r)$ as $k \to \infty$. Now we shall see that $y_{t_k}^k \to \xi_r$ as $k \to \infty$. Since $u^k(t) \to \xi(t)$ uniformly on any compact set in R, we have $\rho(u_r^k, \xi_r) \to 0$ as $k \to \infty$. Moreover, $|du^k/dt| \leqslant L$ for all $t \in R$ and some constant L, and hence $\rho(u_{r_k}^k, u_r^k) \to 0$ as $k \to \infty$. Therefore the inequality

$$\rho(y_{t_k}^k, \xi_r) \leqslant \rho(y_{t_k}^k, u_{r_k}^k) + \rho(u_{r_k}^k, \xi_r)$$
$$\leqslant \rho(y_{t_k}^k, x_{t_k}^k) + \rho(u_{r_k}^k, u_r^k) + \rho(u_r^k, \xi_r)$$

implies that $\rho(y_{t_k}^k, \xi_r) \to 0$ as $k \to \infty$.

Thus $\xi(r) = \eta(r)$, because $y^k(t_k) = v^k(r_k) \to \eta(r)$. Then we have a solution $z(t)$ in K of (5.3.4), where

$$z(t) = \begin{cases} \eta(t) & \text{for } t \geqslant r, \\ \xi(t) & \text{for } t < r. \end{cases}$$

Note here that $\xi(t)$ is a solution in K of (5.3.4). In the case where $r = -\infty$, $z(t) = \eta(t)$ is a solution in K of (5.3.4).

Thus we have two solutions in K of (5.3.4). But $\rho(\xi_0, z_0) = \lim_{k \to \infty} \rho(u_0^k, v_0^k) = \varepsilon > 0$, which shows that $\xi(t)$ and $z(t)$ are distinct solutions in K. Therefore $\rho(\xi_t, z_t) \geqslant \lambda_0$ for all $t \in R$. However, $\rho(\xi_0, z_0) = \varepsilon < \frac{1}{2}\lambda_0$. This contradiction proves Theorem 5.3.2. ∎

Corollary to Theorem 5.3.2   Suppose that the system (5.3.1) satisfies a ρ-separation condition in K. Then it has an almost-periodic solution in K.

Proof   This follows immediately from Theorem 5.3.1 and 5.3.2. ∎

5.4 Population Models

Among the mathematical models applied in population dynamics, the delay equations describe the dynamics of an n species population with regard to its previous history. We shall consider some problems of population dynamics modelled by integro-differential delay equations.

5.4.1 Investigation of periodic (almost-periodic) systems

We consider the system

$$\frac{dx_i(t)}{dt} = x_i(t)\left[b_i(t) - a_{ii}(t)x_i(t) - \sum_{\substack{j=1, \\ j \neq i}}^{n} a_{ij}(t) \int_{-\infty}^{t} K_{ij}(t-u)x_j(u)\,du \right],$$

$$i = 1, 2, \ldots, n, \quad t \geqslant t_0, \quad t_0 \in R, \tag{5.4.1}$$

where $b_i(t)$ and $a_{ij}(t)$, $i, j = 1, 2, \ldots, n$, are continuous positive periodic functions with general period ω, and $K_{ij}: R_+ \to R_+$, $(i \neq j) \in [1, n]$, are delay kernels. The system (5.4.1) models the dynamics of an n-species population where every individual competes with all the others for the common general resources, and interspecies competition causes a delay extended over the whole past of the system. The latter is described by the kernels K_{ij} in (5.4.1). The assumption of periodicity of the parameters $b_i(t)$ and $a_{ij}(t)$, $i, j = 1, \ldots, n$, reflects the periodicity of environmental phenomena (for example, seasons changes is weather and food reserves).

We shall present easily verified sufficient conditions for the global asymptotic stability of strictly positive (component wisely) periodic solutions of the system (5.4.1).

We shall need the following auxiliary assertions.

Lemma 5.4.1 Assume that the delay kernels K_{ij} $(i \neq j) \in [1, n]$ are piecewise-continuous and the series $\sum_{r=0}^{\infty} K_{ij}(u + r\omega)$ are uniformly convergent relative to u on $[0, \omega]$.

Then an ω-periodic solution of the system (5.4.1) is also an ω-periodic solution of the system

$$\frac{dx_i(t)}{dt} = x_i(t)\left[b_i(t) - a_{ii}(t)x_i(t) - \sum_{\substack{j=1, \\ j \neq i}}^{n} a_{ij}(t) \int_{t-\omega}^{t} H_{ij}(t-u)x_j(u)\,du \right], \tag{5.4.2}$$

$i = 1, 2, \ldots, n$, where

$$H_{ij}(u) = \sum_{r=0}^{\infty} K_{ij}(u + r\omega), \quad i, j = 1, 2, \ldots, n, \quad i \neq j. \tag{5.4.3}$$

Conversely, an ω-periodic solution of the system (5.4.2), (5.4.3) is an ω-periodic solution of the system (5.4.1).

Proof This follows immediately from the fact that for an ω-periodic solution $[x_1(t) \cdots x_n(t)]$ of system (5.4.1) the relation

$$\int_{-\infty}^{t} K_{ij}(t-s)x_j(s)\,ds = \sum_{r=0}^{\infty} \int_{t-(r+1)\omega}^{t-r\omega} K_{ij}(t-s)x_j(s)\,ds$$

$$= \sum_{r=0}^{\infty} \int_{t-\omega}^{t} K_{ij}(t-s+r\omega) x_j(s-r\omega)\, ds$$

$$= \int_{t-\omega}^{t} H_{ij}(t-s) x_j(s)\, ds$$

is satisfied.   ∎

We consider a cone R_{+}^{n} in R^{n} with the natural order. We define constants b_i, B_i, a_{ij} and A_{ij}, $(i,j)\in[1,n]$, as follows:

$$\inf_{t\in R} b_i(t) = \min_{t\in[0,\omega]} b_i(t) = b_i,$$

$$\inf_{t\in R} a_{ij}(t) = \min_{t\in[0,\omega]} a_{ij}(t) = a_{ij},$$

$$\sup_{t\in R} b_i(t) = \max_{t\in[0,\omega]} b_i(t) = B_i,$$

$$\sup_{t\in R} a_{ij}(t) = \max_{t\in[0,\omega]} a_{ij}(t) = A_{ij},$$

Assumption 5.4.1   For the system (5.4.1)

(1) the delay kernels K_{ij} are normed and

$$\int_{0}^{\infty} K_{ii}(s)\, ds = 1, \qquad \int_{0}^{\infty} s K_{ij}(s)\, ds < \infty, \qquad (i \neq j)\in[1,n]; \tag{5.4.4}$$

(2) $b_i > 0, \quad a_{ii} > 0, \quad i = 1, 2, \dots, n;$ \hfill (5.4.5)

(3) $b_i > \displaystyle\sum_{\substack{j=1, \\ i\neq j}}^{n} A_{ij}\frac{B_i}{a_{jj}}, \quad i = 1, 2, \dots, n.$ \hfill (5.4.6)

For each i we define

$$\alpha_i = \frac{B_i}{a_{ii}}, \qquad \beta_i = \left(b_i - \sum_{\substack{j=1, \\ j\neq i}}^{n} A_{ij}\frac{B_j}{a_{jj}} \right) \Big/ A_{ii}.$$

The solutions of the system (5.4.1) correspond to initial conditions in the form

$$x_i(s) = \varphi_i(s) \geq 0, \qquad \sup \varphi_i(s) < \infty, \qquad \varphi_i(0) > 0, \tag{5.4.7}$$

where φ_i are piecewise-continuous on $]-\infty, 0]$.

Lemma 5.4.2   Let $x(t; t_0, \varphi) = [x_1(t, \cdot) \ \dots \ x_n(t, \cdot)]^T$ be a solution of the system (5.4.2), (5.4.3) with initial conditions

$$x_i(s; t_0, \varphi) = \varphi_i(s), \qquad s\in[t_0 - \omega, t_0], \qquad t_0\in R, \qquad \varphi = [\varphi_1 \ \dots \ \varphi_n]^T.$$

If

$$0 < \beta^* = \max_{1\leq i\leq n} \beta_i \leq \varphi_i(s) \leq \alpha^* = \min_{1\leq i\leq n} \alpha_i, \tag{5.4.8}$$

$$s\in[t_0 - \omega, t_0], \qquad i = 1, 2, \dots, n, \qquad t_0\in R,$$

then

$$\beta^* \leqslant x_i(t; t_0, \varphi) \leqslant \alpha^*, \qquad t \geqslant t_0, \qquad t_0 \in R, \tag{5.4.9}$$

$$i = 1, 2, \ldots, n.$$

Lemma 5.4.3 (Halanay [1]) Every bounded and uniformly stable solution of the system (5.4.2) converges a symptotically (as $t \to \infty$) to an almost-periodic function.

Lemma 5.4.2 and 5.4.3 are applied to prove the following assertion.

Theorem 5.4.1 Let

(1) conditions (1)–(3) of Assumption 5.4.1 be satisfied;
(2) there exist a positive constant m such that

$$\min_{t \in [0, \omega]} a_{ij}(t) > \sum_{\substack{i=1, \\ i \neq j}}^{n} \left[\max_{t \in [0, \omega]} a_{ij}(t) \right] + m, \qquad j = 1, 2, \ldots, n.$$

Then the system (5.4.2) has a periodic solution $x(t) = [x_1(t) \; \ldots \; x_n(t)]^T$ with period ω such that

$$\beta^* \leqslant x_i(t) \leqslant \alpha^*, \qquad i = 1, 2, \ldots, n, \qquad t \in [0, \omega].$$

Proof See Gopalsamy [1]. ∎

Let $p(t) = [p_1(t) \; \ldots \; p_n(t)]^T$ be a strictly positive (componentwise) periodic solution of the system (5.4.1)–(5.4.3) such that

$$\beta^* \leqslant p_i(t) \leqslant \alpha^*, \qquad i = 1, 2, \ldots, n, \qquad t \in [0, \omega]. \tag{5.4.10}$$

A periodic solution of the system (5.4.1) is globally asymptotically stable, if any other solution $x(t) = [x_1(t) \; \ldots \; x_n(t)]^T$ of is such that under the condition

$$x_i(s) = \varphi_i(s) \geqslant 0, \qquad s \in]-\infty, t_0], \qquad \varphi_i(t_0) > 0, \qquad \sup_{s \leqslant t_0} \varphi_i(s) < \infty \tag{5.4.11}$$

the limit

$$\lim_{t \to \infty} \sum_{i=1}^{n} |x_i(t) - p_i(t)| = 0 \tag{5.4.12}$$

is satisfied. Then it follows that global asymptotic stability of the solution $p(t)$ implies its uniqueness.

Theorem 5.4.2 Let the conditions of Theorem 5.4.1 be satisfied.
Then every periodic solution $p(t)$ of the system (5.4.1) with strictly positive components is globally asymptotically stable.

Proof Let $x(t) = [x_1(t) \; \ldots \; x_n(t)]^T$ be a solution of the system (5.4.1) with initial conditions (5.4.11), and let $p(t) = [p_1(t) \; \ldots \; p_n(t)]^T$ be a periodic solution of (5.4.1) with strictly positive components. Consider the Lyapunov functional

$$v(t) = V(t, x(t), p(t)) = \sum_{i=1}^{n} \left\{ \left| \ln\left[\frac{x_i(t)}{p_i(t)}\right] \right| \right.$$

$$\left. + \sum_{\substack{j=1, \\ j \neq i}}^{n} \int_0^\infty K_{ij}(s) \int_{t-s}^t a_{ii}(s+u)|x_j(u) - p_j(u)|\, du\, ds \right\}, \quad t > t_0$$

$$(5.4.13)$$

for all $t_0 \in R$. Since the solutions $x(t)$ and $p(t)$ are bounded and strictly positive for $t > t_0$,

$$v(t_0) \leqslant \sum_{i=1}^{n} \left\{ \left| \ln\left[\frac{x_i(t_0)}{p_i(t_0)}\right] \right| + \sum_{\substack{j=1, \\ j \neq i}}^{n} A_{ij}\left[\sup_{u \leqslant t_0} |x_j(u) - p_j(u)| \right] \right\} < \infty \qquad (5.4.14)$$

for $t_0 \in R$. Moreover,

$$v(t) \geqslant \sum_{i=1}^{n} \left| \ln\left[\frac{x_i(t)}{p_i(t)}\right] \right|, \qquad s > t_0. \qquad (5.4.15)$$

For the Dini derivative of the functional (5.4.13) we have the estimate

$$D^+ v(t) \leqslant -m \sum_{j=1}^{n} |x_j(t) - p_j(t)| < \infty, \qquad (5.4.16)$$

provided that $\sum_{j=1}^{n} |x_j(t) - p_j(t)| > 0$ for all $t > t_0$. We shall demonstrate that the condition (5.4.16) implies the limit (5.4.12). Suppose that this is not true. Then the exists a sequence $\{t_s\}$, $s = 0, 1, 2, \ldots$, such that $\{t_s\} \to \infty$ as $s \to \infty$ and for some positive ε

$$\sum_{j=1}^{n} |x_j(t_s) - p_j(t_s)| > \varepsilon, \qquad s = 0, 1, 2, \ldots;$$

i.e.

$$D^+ v(t_s) < -m\varepsilon, \qquad s = 0, 1, 2, \ldots \quad . \qquad (5.4.17)$$

Since $x_i(t)$ and $p_i(t)$ are bounded for $t > t_0$, with bounded derivatives, $v(t)$ is uniformly continuous on $[t_0, \infty[$. If we take ε sufficiently small then

$$D^+ v(u) < -m(\tfrac{1}{2}\varepsilon) \quad \text{for } u \in [t_s - \varepsilon, t_s[, \qquad s = 0, 1, 2, \ldots, \qquad (5.4.18)$$

and consequently

$$v(t_s) - v(t_s - \varepsilon) \leqslant \int_{t_s - \varepsilon}^{t} D^+ v(u)\, du \leqslant -m(\tfrac{1}{2}\varepsilon^2). \qquad (5.4.19)$$

We get from (5.4.19)

$$v(t_s) \leqslant v(t_s - \varepsilon) - m(\tfrac{1}{2}\varepsilon^2) \leqslant v(t_{s-1}) - m(\tfrac{1}{2}\varepsilon^2)$$

$$\leqslant v(t_{s-2}) - m2(\tfrac{1}{2}\varepsilon^2) \leqslant v(t_0) - ms(\tfrac{1}{2}\varepsilon^2) \to -\infty$$

as $s \to \infty$. This contradicts the nonnegativeness of $v(t)$. Thus Theorem 5.4.2 is proved. ∎

Remark 5.4.1   For the case where $a_{ij}(t)$ and $b_i(t)$ in the system (5.4.1) are almost-periodic functions, Murakami [2] showed that (5.4.1) has almost periodic solutions under the same conditions.

5.4.2   Existence of a strictly positive (componentwise) almost-periodic solution

We shall now consider the system of integro-differential equations

$$\frac{dx_i(t)}{dt} = x_i(t)\left[b_i(t) - a_{ii}(t)x_i(t) \right.$$

$$\left. - \sum_{\substack{j=1,\\j\neq i}}^{n} a_{ij}(t) \int_{-\infty}^{t} K_{ij}(t-u)x_j(u)\,du \right]. \tag{5.4.20}$$

Assumption 5.4.2   For the system (5.4.2)

(1) $a_{ij}(t)$ and $b_i(t)$ are real-valued continuous almost-periodic functions on R, and

$$a_{ij} = \inf_{t\in R} a_{ij}(t) \geq 0, \qquad b_i = \inf_{t\in R} b_i(t) > 0, \qquad a_{ii} > 0;$$

(2)
$$K_{ij}(s) \geq 0, \qquad \int_0^\infty K_{ij}(s)\,ds = 1, \qquad \int_0^\infty sK_{ij}(s)\,ds < \infty;$$

(3)
$$b_i > \sum_{\substack{j=1,\\j\neq i}}^{n} \frac{A_{ij}B_j}{a_{jj}},$$

where $A_{ij} = \sup_{t\in R} a_{ij}$ and $B_i = \sup_{t\in R} b_i(t)$.

For each i we define

$$\alpha_i = \frac{B_i}{a_{ii}}, \qquad \beta_i = \left(b_i - \sum_{\substack{j=1,\\j\neq i}}^{n} \frac{A_{ij}B_j}{a_{jj}} \right)\bigg/ A_{ii}.$$

Then $0 < \beta_i \leq \alpha_i$. Let K be the following closed bounded set in R^n:

$$K = \{(x_1, x_2, \ldots, x_n) \in R^n; \beta_i \leq x_i \leq \alpha_i \quad \text{for each } i\}.$$

Then we can see that for any $t_0 \in R$ and any φ such that $\varphi(s) \in K$, $s \leq 0$, every solution of (5.4.20) through (t_0, φ) remains in K for all $t \geq t_0$. In addition to (1)–(3), under the condition

(4)
$$a_{ii} > \sum_{\substack{j=1,\\j\neq i}}^{n} A_{ji} \quad \forall i = 1, 2, \ldots, n,$$

there exists a strictly positive almost-periodic solution for ordinary differential equations and for the periodic case in (5.4.20) (Gopalsamy [1]) and for the system (5.4.20) (Murakami [2]).

We shall now see that the existence of a strictly positive almost-periodic solution of (5.4.20) can be obtained under conditions (1)–(3) of Assumption 5.4.2 (without (4)).

Condition (3) implies that

$$-b_i + \sum_{\substack{j=1,\\ j\neq i}}^{n} b_j \frac{A_{ij}}{a_{jj}} < 0 \quad \text{for} \quad i = 1, 2, \ldots, n. \tag{5.4.21}$$

Let C^n and D^n be the following $n \times n$ matrices:

$$C^n = \begin{bmatrix} -a_{11} & A_{12} & \cdots & A_{1n} \\ A_{21} & -a_{22} & \cdots & A_{2n} \\ \vdots & \vdots & & \vdots \\ A_{n1} & A_{n2} & \cdots & -a_{nn} \end{bmatrix},$$

$$D^n = \begin{bmatrix} d_{11} & d_{12} & \cdots & d_{1n} \\ d_{21} & -d_{22} & \cdots & d_{2n} \\ \vdots & \vdots & & \vdots \\ d_{n1} & d_{n2} & \cdots & -d_{nn} \end{bmatrix},$$

where $d_{ii} = b_i$ and $d_{ij} = b_j A_{ij}/a_{jj}$, and denote by C^n_{ij} and D^n_{ij} the cofactors of the elements in C^n and D^n respectively. Clearly

$$C^n = \det D^n \prod_{k=1}^{n} \frac{a_{kk}}{b_k}, \quad C^n_{ij} = D^n_{ij} \prod_{\substack{k=1,\\ k\neq j}}^{n} \frac{a_{kk}}{b_k}.$$

Lemma 5.4.4 Under condition (3) of Assumption 5.4.2, $(-1)^n C^n_{ii} < 0$ and $(-1)^n C^n_{ij} \leqslant 0$ for $i, j = 1, 2, \ldots, n$, and $(-1)^n \det C^n > 0$.

Proof It is sufficient to show that $(-1)^n D^n_{ii} < 0$, $(-1)^n D^n_{ij} \leqslant 0$ and $(-1)^n \det D^n > 0$. Consider the matrix

$$A^m = \begin{bmatrix} -c_{11} & c_{12} & \cdots & c_{1m} \\ c_{21} & -c_{22} & \cdots & c_{2m} \\ \vdots & \vdots & & \vdots \\ c_{m1} & c_{m2} & \cdots & -c_{mm} \end{bmatrix}.$$

Now assume that $c_{ii} > 0$, $c_{ij} \geqslant 0$ and $-c_{ii} + \sum_{j=1, j\neq i}^{m} c_{ij} < 0$ for $i, j = 1, 2, \ldots, m$, which imply that $0 \leqslant c_{ij} \leqslant c_{ii}$ $(i \neq j)$. Since we have

$$\det A^m = \det \begin{bmatrix} -c_{11} & c_{12} & \cdots & c_{1m} \\ 0 & -c_{22} + \dfrac{c_{12}c_{21}}{c_{11}} & \cdots & c_{2m} + \dfrac{c_{1m}c_{21}}{c_{11}} \\ \vdots & \vdots & & \vdots \\ 0 & c_{m2} + \dfrac{c_{12}c_{m1}}{c_{11}} & \cdots & -c_{mm} + \dfrac{c_{1m}c_{m1}}{c_{11}} \end{bmatrix},$$

and the sum of the elements of the ith row $(i \geqslant 2)$ is

$$-c_{ii} + \sum_{\substack{j=2,\\ j\neq i}}^{m} c_{ij} + \frac{c_{i1}}{c_{11}} \sum_{j=2}^{m} c_{ij} < 0,$$

we can show that $(-1)^m \det A^m > 0$ by induction. In the case where $c_{mm} < 0$ and $-c_{ii} + \sum_{j=1, j \neq i}^m c_{ij} < 0$ for $i = 1, 2, \ldots, m-1$ we can show that $(-1)^{m-1} \det A^m \geqslant 0$ in the same way as above.

Therefore we can show that $(-1)^n \det D^n > 0$, since we have the condition (5.4.21). To simplify notation, we denote the $k \times k$ matrix

$$M^k = \begin{bmatrix} -d_{i_1 i_1} & d_{i_1 i_2} & d_{i_1 i_3} & \cdots & d_{i_1 i_k} \\ d_{i_2 i_1} & -d_{i_2 i_2} & d_{i_2 i_3} & \cdots & d_{i_2 i_k} \\ \vdots & \vdots & \vdots & & \vdots \\ d_{i_k i_1} & d_{i_k i_2} & d_{i_k i_3} & \cdots & -d_{i_k i_k} \end{bmatrix}$$

for $i_1 < i_2 < \cdots < i_k$. Then, for the same reason as above, we have $(-1)^{n-1} \det M^{n-1} > 0$ for $i_1, i_2, \ldots, i_{n-1} = 1, 2, \ldots, n$ and $i_1 < i_2 < \cdots < i_{n-1}$. Then it follows from this that $(-1)^n D_{ii}^n < 0$ for $i = 1, 2, \ldots, n$.

To show that $(-1)^n D_{ij}^n \leqslant 0$, consider

$$\det \begin{bmatrix} M^{n-2} & d^T \\ c & e \end{bmatrix} \tag{5.4.22}$$

for $i_1, i_2, \ldots, i_{n-2} = 1, 2, \ldots, n$ and $i_1 < i_2 < \cdots < i_{n-2}$, where e, c_k and d_k $(k = 1, \ldots, n-2)$ are nonnegative constants, $c = [c_1 \ c_2 \ \cdots \ c_{n-2}]$ and $d = [d_1 \ d_2 \ \cdots \ d_{n-2}]$, and the sum of the elements of the ith row $(i = 1, 2, \ldots, n-2)$ is negative. Then we can see that

$$(-1)^{n-2} \det \begin{bmatrix} M^{n-2} & d^T \\ c & e \end{bmatrix} \geqslant 0.$$

Since $D_{ij}^n = (-1) \times$ (a determinant of the type (5.4.22)), we have $(-1)^n D_{ij}^n \leqslant 0$. This proves Lemma 5.4.4. ∎

For the system (5.4.20) we now consider the functionals

$$V_i(t, x(\cdot), y(\cdot)) = |\log x_i(t) - \log y_i(t)|$$

$$+ \sum_{\substack{j=1, \\ j \neq i}}^{n} \int_0^\infty K_{ij}(s) \left[\int_{t-s}^t a_{ij}(s+u) |x_j(u) - y_j(u)| \, du \right] ds,$$

$i = 1, 2, \ldots, n$, where $x(\cdot)$ and $y(\cdot)$ are solutions of (5.4.20) that remain in K. Then we have

$$D^+ V_i(t, x(\cdot), y(\cdot)) \leqslant -a_{ii} |x_i(t) - y_i(t)| + \sum_{\substack{j=1, \\ j \neq i}}^{n} A_{ij} |x_j(t) - y_j(t)|.$$

Putting $V = [V_1 \ V_2 \ \cdots \ V_n]^T$ and $z(t) = [|x_1(t) - y_1(t)| \ \cdots \ |x_n(t) - y_n(t)|]^T$, we have $D^+ V(t, x(\cdot), y(\cdot)) \leqslant C^n z(t)$. We denote the $n \times n$ matrix

$$\bar{C}^n = \begin{bmatrix} |c_{11}^n| & |c_{21}^n| & \cdots & |c_{n1}^n| \\ |c_{12}^n| & |c_{22}^n| & \cdots & |c_{n2}^n| \\ \vdots & \vdots & & \vdots \\ |c_{1n}^n| & |c_{2n}^n| & \cdots & |c_{nn}^n| \end{bmatrix}$$

and define the functional $W = [W_1 \ W_2 \ \ldots \ W_n]^T$ by $W(t, x(\cdot), y(\cdot)) = \bar{C}^n V(t, x(\cdot), y(\cdot))$. Then we have

$$D^+ W(t, x(\cdot), y(\cdot)) \leqslant \bar{C}^n C^n z(t) = - |\det C^n| z(t).$$

Now consider the Lyapunov functional

$$U(t, x(\cdot), y(\cdot)) = \sum_{i=1}^{n} W_i(t, x(\cdot), y(\cdot)).$$

Then we have

$$D^+ U(t, x(\cdot), y(\cdot)) \leqslant - |\det C^n| \sum_{i=1}^{n} |x_i(t) - y_i(t)|.$$

Using this Lyapunov functional, we can see that a solution $u(t)$ of (5.4.20) in K is relatively weakly uniformly asymptotically (K, ρ)-stable in Ω (Murakami [2]). Then it follows from Hamaya and Yoshizawa [1, Proposition 1, Remark 3] that $u(t)$ is relatively eventually totally (K, ρ)-stable, and therefore it follows from Theorem 5.3.2 that the system (5.4.20) has an almost-periodioc solution $p(t) = [p_1(t) \ \ldots \ p_n(t)]^T$ such that $0 < \beta_i \leqslant p_i(t) \leqslant \alpha_i$ for all $t \in R$.

5.5 Stability of Motion of a Viscoelastic Cylinder

We shall consider the oscillations of an ageing viscoelastic cylinder of inner and outer radii a and b covered by an elastic shall of thickness h. We shall consider radial oscillations of the cylinder, assuming the incompressibility of its material. A cross-section of the cylinder by a plane orthogonal to its axis of symmetry is a plane ring of inner and outer radii a and b. We take the origin of a polar coordinate system (r, φ) at the centre of the ring. It is clear that the radial motion u of the ring points does not depend on the coordinate φ, i.e. $u = u(r, t)$. The boundary conditions are

$$\sigma_{rr}|_{r=a} = -p(t), \qquad \sigma_{rr}|_{t=b} = -q(t).$$

The motion is described by an integro-differential equation

$$\frac{d^2 v}{dt^2} + \lambda^2 \theta(t) v = F(t) + \mu \lambda^2 \int_0^t \theta(\tau) \tilde{\Gamma}(t - \tau) v(\tau) \, d\tau, \qquad (5.5.1)$$

where

$$v(t) = u(a, t) = \frac{c(t)}{a},$$

$$\lambda^2 = \frac{2G_0(b^2 - a^2)/a^2 b^2 + E_0 h/b^3}{\rho \ln(b/a) + \rho_0 \ln(h/b)},$$

$$\mu = \left[2G_0 + \frac{E_0 h a^2}{b(b^2 - a^2)} \right]^{-1},$$

$$F(t) = \frac{p(t)}{a(\rho \ln(b/a) + \rho_0 \ln(h/b)},$$

$E = E_0 Q(t)$ and $G = G_0 \theta(t)$ are the elastic and shift moduli respectively, and $\theta(t)$ is a nondecreasing bounded function with $\lim_{t \to \infty} \theta(t) = \theta_0 > 0$.

Let $v = v_0(t)$ be a particular solution of equation (5.5.1). Introducing a new variable $x_1 = v - v_0$, we get

$$\frac{d^2 x_1}{dt^2} + \lambda^2 \theta(t) x_1 = \mu \lambda^2 \int_0^t \theta(\tau) \tilde{F}(t - \tau) x_1(\tau) d\tau. \tag{5.5.2}$$

This equation is equivalent to the system

$$\frac{dx_1}{dt} = x_2,$$
$$\frac{dx_2}{dt} = -\lambda^2 \theta(t) x_1 + \mu \lambda^2 \int_0^t \theta(\tau) \tilde{F}(t - \tau) x_1(\tau) d\tau. \tag{5.5.3}$$

We investigate the stability of the equilibrium state $x_1 = x_2 = 0$ of the system (5.5.3). It should be noted that equation (5.5.2) for $\theta(t) \equiv 1$ describes radial oscillations of a viscoelastic cylinder without ageing. For the unperturbed system

$$\frac{dx_1}{dt} = x_2,$$
$$\frac{dx_2}{dt} = -\lambda^2 \theta(t) x_1 \tag{5.5.4}$$

the Lyapunov function is taken in the form

$$V = \frac{1}{2} \left[x_1^2 + \frac{x_2^2}{\lambda^2 \theta(t)} \right].$$

The derivative of this function in view of the system (5.5.4) is nonpositive:

$$\frac{dV}{dt} = -\frac{\dot{\theta}(t) x_2^2}{2\lambda^2 \theta^2(t)} \leqslant 0.$$

The limiting system corresponding to (5.5.4),

$$\frac{dx_1}{dt} = x_2,$$
$$\frac{dx_2}{dt} = -\lambda_0^2 x_1,$$

where $\lambda_0^2 = \lambda^2 \theta_0$, has the general solution

$$x_{1g}(t) = A \sin(\lambda_0(t - t_0) + \alpha),$$
$$x_{2g}(t) = A\lambda_0 \cos(\lambda_0(t - t_0) + \alpha).$$

Where

$$A = x_{10}^2 + x_{20}^2, \quad \sin \alpha = \frac{x_{10}}{A}, \quad \cos \alpha = \frac{x_{20}}{A\lambda_0}.$$

If

$$\tilde{F}(t - \tau) = \sum_k c_k \exp[-\lambda_k(t - \tau)], \quad \sum_k \frac{c_k}{\lambda_0^2 + \lambda_k^2} > 0,$$

then there exists a negative-definite mean relative to (x_{10}, x_{20}) of the function

$$\psi\left(t, x_{1g}(t), x_{2g}(t), \int_0^t \tilde{F}(t-\tau)x_{1g}(\tau)\,d\tau \right)$$

in the form

$$\bar{\psi}\left(t, x_g(t), \int_0^t \theta(\tau)\tilde{F}(t-\tau)x_{1g}(\tau)\,d\tau \right) \leqslant - A^2\lambda_0^2 \sum_k \frac{c_k(\lambda_0^2 + \lambda_k^2)}{2\theta_1} < 0;$$

i.e. condition (4) of Theorem 5.2.1 is satisfied. The other conditions of Theorem 5.2.1 are easily verified with reference to the system (5.5.3). Therefore the solution $v = v_0(t)$ of equation (5.5.1) is μ-stable.

5.6   Comments and References

5.1   The statement of the problem on μ-stability of the system (5.1.1) generalizes the similar notion from Chapter 3.

5.2   Theorem 5.2.1 was proved by Karimzhanov [2] as a part of a general investigations supervised by A.A. Martynyuk (see Karimzhanov [3]). Theorems 5.2.2 and 5.2.3 are due to Hapaev and Falin [1].

5.3   The results of this section correspond to those of Hamaya and Yoshizawa [1].

5.4   This section is based on results of Gopalsamy [1] and Hamaya and Yoshizawa [1].

5.5   Integro-differential equations are applicable to a number of viscoelasticity problems (see Filatov [1] and references therein). The investigation of oscillations of a viscoelastic cylinder described here was carried out together with A. Karimzhanov.

6 Optimal Stabilization of Controlled Motion and Limiting Equations

6.0 Introduction

The use of limiting equations in the analysis of qualitative properties of motion developed in Chapter 4 is applied here to the investigation of the optimal stabilization of mechanical systems with lumped parameters. The theorems obtained modify well-known results by weakening the conditions imposed on the derivative of the optimal Lyapunov function. Moreover, an optimal stabilization problem is solved with regard to the equilibrium state of a mechanical system under control forces explicitly independent of time. Sufficient conditions for optimal stabilization are established for controlled systems with neutral (uncontrolled) parts. The possibility of decentralized control is shown by an example of an integro-differential controlled system.

6.1 Problems of Controlled System Stabilization

We consider a controlled dynamical system

$$\frac{dy}{dt} = Y(t, y, u), \quad y(t_0) = y_0, \tag{6.1.1}$$

where $Y \in C(R_+ \times R^n \times R^m, R^n)$ and u is a control function. We analyse the partial motion of the system (6.1.1) generated by the control forces $V_j = p_j(t), j = 1, 2, \ldots, m$. This motion corresponds to a partial solution $y_s = f_s(t), s = 1, 2, \ldots, n$, and is considered to be unperturbed. Alongside the unperturbed motion $\{f_s(t)\}$, we shall examine the perturbed motion $\{y_s(t)\}$, which is assumed to be described by the same equations (6.1.1) but for values of $V_j(t)$ different from those of $p_j(t)$.

The deviations $\Delta V_j = V_j - p_j(t)$ of the variables V_j from $p_j(t)$ are such that the motion $y_s = f_s(t)$ is stable. By means of the substitutions

$$x_s = y_s - f_s(t), \quad u_j = V_j - p_j(t), \quad s = 1, 2, \ldots, n, \quad j = 1, 2, \ldots, m, \qquad (6.1.2)$$

where x_s are perturbations of motion and u_j are the deviations of the control forces from the values $p_j(t)$, the system (6.1.1) is reduced to the form

$$\frac{dx}{dt} = f(t, x, u), \quad x(t_0) = x_0, \qquad (6.1.3)$$

where $f \in C(R_+ \times R^n \times R^m, R^n)$ and $f(t, x, u) = Y(t, x + f, u + p) - Y(t, f, p)$.

The system (6.1.3) will be taken as the system of perturbed equations of motion.

Following Krasovsky [2], we adopt the following formulations of the stabilization problem

Problem A (on stabilization)   Control forces $u_1(t, x_1, \ldots, x_n), \ldots, u_m(t, x_1, \ldots, x_n)$ must be found that ensure the asymptotic stability of the unperturbed motion $x = 0$ according to equation (6.1.3) for $u_j = u_j^0(t, x_1, \ldots, x_n)$, $j = 1, 2, \ldots, m$.

Many engineering problems require the transition process to be of the highest possible quality. Such a requirement can frequently be expressed in terms of a minimization condition on some integral

$$I = \int_{t_0}^{\infty} \omega(t, x[t], u[t]) \, dt, \qquad (6.1.4)$$

where $\omega \in C(R_+ \times D \times R^m, R_+)$ and $D \subset R^n$.

It should be emphasized that the particular function $\omega(t, \cdot)$ depends on the problem under investigations but, the following conditions are common.

(1) The minimization condition on (6.1.4) must ensure sufficiently rapid damping of the motion $x[t]$ (whose $x[t]$ indicates that the motion is generated by a fixed control, $u = u^*(t, x)$).
(2) The value of integral I must give a good estimate of the resources spent on generating the control forces $u[t]$.
(3) The function $\omega(t, \cdot)$ must allow solution of the problem.

In particular, in many cases the function

$$\omega(t, x, u) = x^T A x + u^T B u,$$

where A is an $n \times n$ constant matrix and B as $m \times m$ constant matrix, satisfies conditions (1)–(3).

Problem B (on optimal stabilization)   Let the quality criterion for the process $x(t)$ be formulated in terms of the integral (6.1.4). A control $u(t, x)$ must be found such that ensures asymptotic stability of the solution $x = 0$ of the system (6.1.3) (for $u = u^0(t, x)$) and, moreover,

$$\int_{t_0}^{\infty} \omega(t, x^0[t], u^0[t]) \, dt \leqslant \int_{t_0}^{\infty} \omega(t, x^*[t], u^*[t]) \, dt \qquad (6.1.5)$$

for any other control $u = u^*(t, x)$ under all initial conditions

$$t_0 \geqslant 0 \quad \text{and} \quad \|x(t_0)\| \leqslant \delta.$$

The value of δ is either previously defined or has the same meaning as in the definition of stability of the solution $x = 0$ of the system (6.1.3).

Functions $u^0(t, x)$ satisfying the conditions of Problem B are called the *optimal control*.

6.2 Stabilization Theorems

Our aim is to present a new approach based on a combination of the method of Lyapunov functions, the comparison technique and the method of limiting equations employed in solving the problems of controlled motion stabilization.

This allows the requirements on the Lyapunov function and its total derivative due to the perturbed equations of motion to be weakened when solving stabilization problems. To this end, we consider an auxiliary function $V \in C(R_+ \times R^n, R_+)$, where $V(t, x)$ is locally Lipschitzian in x, a function $g(t, w, v)$, where $g \in C(R_+^3, R)$, and the scalar comparison equation

$$\frac{dw}{dt} = g(t, w, v), \qquad w(t_0) = w_0 \geqslant 0. \tag{6.2.1}$$

We assume that a maximal solution $r(t) = r(t, t_0, w_0)$ of equation (6.2.1) exists on the interval $[t_0, \infty[$.

Let E denote the admissible set of controls:

$$E = \{u \in R^m : U(t, u) \leqslant r(t), t \geqslant t_0\}, \tag{6.2.2}$$

where $U \in C(R_+ \times R^m, R)$.

For a fixed control vector $u^0(t, x) \in E$ we consider the system of perturbed equations of motion

$$\frac{dx}{dt} = f^0(t, x, u^0(t, x)) \tag{6.2.3}$$

and the limiting system

$$\frac{dx}{dt} = G(t, x, u^0(t, x)), \tag{6.2.4}$$

where $G \in \Omega(f^0)$. Here $\Omega(f^0)$ is defined as in Chapter 1. Namely,

$$\Omega(f^0) = \{f^0 \in C(R_+ \times W \times R^m, R^n) : \{t_n\} \subset R_+, t_n \to +\infty, f_n^0(t, x, u^0) \to G(t, x, u^0)$$

in the compact open topology$\}$,

where

$$f_n^0(t, x, u^0) = f^0(t + t_n, x, u^0).$$

For the comparison equation (6.2.1) we shall consider the limiting comparison equation

$$\frac{dw}{dt} = g_0(t, w, v), \qquad w(t_0) = w_0 \geqslant 0, \tag{6.2.5}$$

where $g_0 \in \Omega(g)$.

It is clear that the vector function $f^0(t, x, u^0(t, x))$ and the function $g(t, w, v)$ are assumed to satisfy the precompactness conditions (see Section 1.1).

We now formulate the following result.

Theorem 6.2.1.   For a fixed control $u^0 \in U(t, u)$ let the vector function f in the system (6.1.3) be continuous on $R_+ \times N$ and precompact. Let there exist

(1) a time-invariant neighbourhood N of the point $x = 0$;
(2) a function $V \in C(R_+ \times N, R_+)$, with $V(t, x)$ locally Lipschitzian in x and

$$\alpha(\|x\|) \leqslant V(t, x) \leqslant b(\|x\|) \quad \forall (t, x) \in R_+ \times N,$$

where $a, b \in \mathcal{K}$;
(3) a function $g \in C(R_+^3, R)$, with $g(t, w, v)$ is continuous in v for every (t, w), and $U \in C(R_+ \times R^m, R)$ such that

$$D^+ V(t, x) = \lim_{\theta \to \infty +} \sup \left\{ [V(t + \theta, x + \theta f(t, x, u)) - V(t, x)] \theta^{-1} \right\} \leqslant g(t, V(t, x), U(t, u(t)));$$

(4) a limiting function $g_0(t, w, v) \in \Omega(g)$ such that $g_0(t, w, v) \leqslant 0$ for all $t \geqslant t_0$;
(5) a local L_1 function $m(t)$ such that $|g(t, w, v) - g_0(t, w, v)| \leqslant m(t)$ for all $t \in R_+, w \in W$ and $u \in E$, and

$$\int_{t_0}^t m(s) \, ds \leqslant k < +\infty, \qquad t \geqslant t_0;$$

(6) at least one limiting pair (G, g_0) corresponding to (f^0, g) and such that the set $\{g_0(t, w, v) = 0\}$ does not contain any solution of the system (6.2.4) except for $x = 0$.

Then the control $u^0 \in U(t, x)$ ensures asymptotic stability of the unperturbed motion $x = 0$ of (6.1.3).

Proof   First we shall prove that the unperturbed motion $x = 0$ of (6.1.3) is stable in the Lyapunov sense. We find from conditions (1)–(5) that the solution $x = 0$ of the comparison equation

$$\frac{dw}{dt} = g(t, w, v), \qquad w(t_0) = w_0 \geqslant 0, \tag{6.2.6}$$

is stable. This is provided by Lemma 4.1.1 applied to equations (6.2.1) and (6.2.5). Assume that the stability of the solution $w = 0$ of equation (6.2.6) means that the condition $w_0 < b(\delta)$ implies

$$w(t, t_0, w_0) < a(\varepsilon), \qquad t \geqslant t_0, \tag{6.2.7}$$

where $w(t, t_0, w_0)$ is a solution of equation (6.2.6) on $[t_0, \infty[, \varepsilon \in]0, H[, \delta = \delta(t_0, \varepsilon), t_0 \geqslant 0$. We take for any $\varepsilon < \varepsilon_1 \in]0, H[$ a value $\delta = \delta(t_0, \varepsilon)$ such that

$$b(\delta) < a(\varepsilon). \tag{6.2.8}$$

Let $\|x_0\| < \delta$. We claim that $\|x(t, \cdot)\| < \varepsilon$ for all $t \geqslant t_0$, where $x(t, \cdot) = x(t; t_0, x_0, u^0)$ is a solution of the system (6.1.3) corresponding to the control $u^0 \in E$. If this is not true then there must exist another control $u^* \in E$ and the corresponding solution $x(t) = x(t; t_0, x_0, u^*)$ of (6.1.3) such that $\|x(t_1)\| = \varepsilon$ and $\|x(t)\| < \varepsilon$ for $t_0 \leqslant t \leqslant t_1$. Conditions (1) and (2) imply

$$a(\varepsilon) \leqslant V(t_1, x(t_1)). \tag{6.2.9}$$

We take $w_0 = V(t_0, x_0)$ and find in terms of condition (3) that $D^+ n(t) \leqslant g(t, n(t), U(t, u(t)))$, $t_0 \leqslant t \leqslant t_1$, where $n(t) = V(t, x(t))$. Since $u^* \in E$ and the function $g(t, w, v)$ is nondecreasing in v, we have

$$D^+ n(t) \leqslant g(t, n(t), r(t)), \qquad t_0 \leqslant t \leqslant t_1, \tag{6.2.10}$$

where $r(t) = r(t, t_0, w_0)$ is the maximal solution of equation (6.2.6). According to the comparison theorem (Theorem 3.1.5 by Lakshmikantham, Leela and Martynyuk [1]), we have the estimate

$$n(t) \leqslant r(t), \qquad t_0 \leqslant t \leqslant t_1, \tag{6.2.11}$$

where $r(t)$ is the maximal solution of the equation

$$\frac{dw}{dt} = g(t, w, r(t)), \qquad w(t_0) = w_0 \geqslant 0. \tag{6.2.12}$$

Taking together the estimates (6.2.7), (6.2.9) and (6.2.11), and in view of the inequality (6.2.8), we obtain

$$a(\varepsilon) \leqslant V(t_1, x(t_1)) \leqslant r(t_1) < a(\varepsilon).$$

The contradiction thus obtained proves the stability of the unperturbed motion $x = 0$ under the control $u^0 \in E$.

Further, by condition (6), we get for the system (6.2.4) and equation (6.2.5)

$$M^+(G, g_0) \equiv \{x = 0\},$$

and therefore $M_*^+(G, g_0) \equiv \{x = 0\}$.

In addition, $M_*^+(G, g_0)$ is a combination of the subsets $M^+(G, g_0) \subset \{g_0(t, w, v) = 0\}$ with respect to all pairs (G, g_0). The former are invariant relative to the solutions of the system (6.2.4).

Hence it follows that

$$\Omega^+(x(t)) \equiv \{x = 0\},$$

where $\Omega^+(x(t))$ is the set of limit points of the solutions of system (6.2.3) under the control $u^0 \in E$ and $\lim_{t \to +\infty} \|x(t)\| = 0$. This proves Theorem 6.2.1.

Next we shall present a result with the choice of conditions for the controls $u^0(t, x)$ solving Problem B (on optimal stabilization).

Theorem 6.2.2   Let the vector function f in the system (6.1.3) be continuous on $R_+ \times N$, $N \subseteq R^n$, and precompact under the control $u^0 \in E_1 \subset E$. Assume that

(1) there exist
   (i) a time-invariant neighbourhood N of the point $x = 0$;
   (ii) a function $V \in C^1(R_+ \times N, R_+)$ and

$$a(\|x\|) \leqslant V(t, x) \leqslant b(\|x\|) \quad \forall (t, x) \in R_+ \times N,$$

   where $a, b \in \mathcal{K}$;
   (iii) at least one limiting pair (G, g_0) corresponding to (f, g) and such that the set $\{g_0(t, w, x, u) = 0\}$ does not contain any solution of the limiting system

$$\frac{dx}{dt} = G(t, x, u^0(t, x))$$

   except for $x = 0$;
(2) the following conditions are satisfied:
   (i) $\mathscr{B}[V; t, x, u^0, g] \equiv V_t(t, x) + [V_x(t, x)]^T f(t, x, u^0(t, x)) + g(t, V(t, x), x, u^0(t, x)) \leqslant 0$;
   (ii) for any $u \in E_1$, $\mathscr{B}[V; t, x, u, g] \geqslant 0$;
   (iii) the solution $w = 0$ of the comparison equation

$$\frac{dw}{dt} = g(t, w, x, u^0), \quad w(t_0) = w_0 \geqslant 0,$$

is stable.

Then the controlled system (6.1.3) is asymptotically stable, and, moreover, the inequality

$$\int_{t_0}^{\infty} g(s, V(s, x^0(s)), x^0(s), u^0(s)) \, ds = \min_U \int_{t_0}^{\infty} g(s, V(s, x(s)), x(s), u(s)) \, ds$$

$$\leqslant V(t_0, x_0). \tag{6.2.13}$$

holds.

Proof   It is necessary to establish two facts. First the control $u^0(t, x) \in E_1$ must be shown to ensure the asymptotic stability of the unperturbed motion $x = 0$ of (6.1.3). Secondly the conditon (6.2.13) must be shown to be satisfied under this control. Under conditions (1)(i)–(iii) and (2)(ii), for $u = u^0(t, x)$ the estimate of the total derivative of the function $V(t, x)$ due to the system (6.1.3) has the form

$$\frac{dV}{dt} \leqslant -g(t, V(t, x), x, u^0(t, x)), \tag{6.2.14}$$

and the scalar comparison equation is

$$\frac{dw}{dt} \leqslant -g(t, w(t, x), x, u^0(t, x)), \quad w(t_0) = w_0 \geqslant 0. \tag{6.2.15}$$

By the precompactness conditions for the system

$$\frac{dx}{dt} = f(t, x, u^0(t, x))$$

and equation (6.2.15), there exists a sequence $t_n \to +\infty$ as $n \to +\infty$ (common to the vector function f and the function g) such that

$$f_n(t, x, u^0(t, x)) \to G(t, x, u^0(t, x)),$$

$$g_n(t, w, x, u^0(t, x)) \to g_0(t, w, x, u^0(t, x)), \tag{6.2.16}$$

where $f_n(t, \cdot) = f(t + t_n, \cdot)$ and $g_n(t, \cdot) = g(t + t_n, \cdot)$.

According to condition (2)(ii) and the comparison principles (see Theorem 3.1.5 by Lakshmikantham, Leela and Martynyuk [1]) the solution $x = 0$ of the system (6.1.3) is stable. Conditions (1)(ii) and (iii) ensure, in view of the stability of the solution $x = 0$ of (6.1.3), its asymptotic stability, since $M_*^+(G, g_0) \equiv \{x = 0\}$ and $\Omega^+(x^0[t]) \equiv \{x = 0\}$.

Moreover, the initial condition x_0 of the solution $x^0[t, t_0, x_0]$ are taken from the domain

$$\|x_0\| < \eta, \tag{6.2.17}$$

where the number η defining the domain (6.2.17), can be found from the inequality (see Krasovsky [2, p. 486])

$$\sup_x \{V(t, x) \text{ for } \|x\| \leqslant \eta\} < \inf_x \{V(t, x) \text{ for } \|x\| = h\}, \tag{6.2.18}$$

where $0 < h < H < \infty$ determines the time-invariant neighbourhood N of the point $x = 0$.

We shall now prove (6.2.13). Under the conditions (6.2.17) and (6.2.18), the motion $x^0[t]$ satisfies $\|x^0[t]\| \leqslant h < H$. Therefore, condition (2)(i) or the inequality (6.2.14) are satisfied along this solution. Besides, because of the asymptotic stability of the solution $x = 0$ of (6.1.3), the limit

$$\lim_{t \to \infty} V(t, x^0[t]) = 0 \tag{6.2.19}$$

holds. Integrating the inequality (6.2.14) along the motion $x^0[t]$ in from $t = t_0$ to $t = +\infty$, and in view of (6.2.15), we arrive at

$$\int_{t^0}^{\infty} g(s, V(s, x^0[s]), x^0[s], u^0[s]) \, ds \leqslant V(t_0, x_0). \tag{6.2.20}$$

Suppose some other control $u^*(t, x) \in E_1$ also yields asymptotic stability of the solution $x = 0$ (6.1.3) and the relation (6.2.13), i.e $\|x^*[t]\| < \varepsilon$ for all $t \geqslant t_0$ when $\|x_0\| < \eta$ and $\lim_{t \to \infty} x^*[t] = 0$. This means that

$$\lim_{t \to \infty} V(t, x^*[t]) = 0.$$

According to condition (2)(ii), we have

$$\frac{dV}{dt} \geqslant -g(t, V(t, x^*[t]), x^*[s], u^*[t])$$

and

$$V(t_0, x_0) \leqslant \int_{t_0}^{\infty} g(s, V(s, x^*[s]), x^*[s], u^*[s]) \, ds. \tag{2.21}$$

The inequality (6.2.21) together with (6.2.20) proves the relation (6.2.13).   ■

Corollary to Theorem 6.2.1 (Andreev [3])   Let

(1) conditions (1)(i) and (ii) of Theorem 6.2.1 be satisfied;
(2) there exist a function $W(t, x, u) \geqslant 0$ satisfying the precompactness conditions for $u = u^0(t, x) \in E_1$;
(3) $\mathscr{B}[V; t, x, u^*, w] = V_t(t, x) + (V_x(t, x))^T f(t, x, u^0(t, x)) + W(t, x, u^0(t, x)) = 0$;
(4) $\mathscr{B}[V; t, x, u, w] \geqslant 0$ for any $u \neq u^0 \in E_1$;
(5) there exist at least one limiting pair (G, λ_0) corresponding to (f, W) such that the set $\{\lambda_0(t, x) = 0\}$ does not contain the solutions of the system

$$\frac{dx}{dt} = G(t, x, u^0(t, x))$$

except for $x = 0$.

Then the control $u = u^0(t, x)$ ensures asymptotic stability of the solution $x = 0$ of (6.1.3), and, moreover, the inequality

$$\int_{t_0}^{\infty} W(s, x^0[s], u^0[s]) \, ds = \min_U \int_{t_0}^{\infty} W(s, x(s), u(s)) \, ds \leqslant V(t_0, x_0)$$

is satisfied.

Corollary to Theorem 6.2.2 (Krasovsky [2])   Let

(1) conditions (1)(i) and (ii) of Theorem 6.2.2 be satisfied;
(2) the function $g(t, w, x, u) \equiv w(t, x) = \omega(t, x, u^0(t, x))$ be positive-definite;
(3) $\mathscr{B}[V; t, x, u^0] = 0$;
(4) for every $u \in E_1$ the inequality

$$\mathscr{B}[V; t, x, u] \geqslant 0$$

be valid.

Then the control $u = u^0(t, x)$ ensures asymptotic stability of the solution $x = 0$ of (6.1.3), and, moreover, the inequality

$$\int_{t_0}^{\infty} w(t, x[t], u^0[t]) \, dt = \min_u \int_{t_0}^{\infty} \omega(t, x[t], u[t]) \, dt = V(t_0, x_0).$$

is satisfied.

In optimal control theory for a given system it is important to define the type of integrand in the quality criterion and the class of control forces so that the known Lyapunov function for the system without control can be applied to the system under control forces. We shall consider a similar problem for a Lyapunov function with a derivative of constant sign for a nonautonomous system.

Let the right-hand side of the system (6.1.3) satisfy for $u(t, x) = 0$ the condition $f(t, 0) = 0$ and the precompactness condition, and there exist a nonnegative function

$V \in C^1(R_+ \times N, R)$, with $V(t, 0) = 0$ and a derivative $V'(t, x) \leqslant -S(t, x) \leqslant 0$ in view of the system

$$\frac{dx}{dt} = f(t, x, 0). \tag{6.2.22}$$

Let the system (6.2.22) be under the control forces

$$X(t, x, u) = m(t, x)u, \qquad u^T = [u_1 \quad u_2 \quad \cdots \quad u_r], \tag{6.2.23}$$

where $m(t, x)$ is a $p \times r$ matrix.

For the controlled system

$$\frac{dx}{dt} = f(t, x) + X(t, x, u) \tag{6.2.24}$$

the functional

$$J = \int_{t_0}^{\infty} W(t, x(t), u(t)) \, dt, \qquad W \geqslant 0,$$

with integrand

$$W(t, x, u) = F(t, x) + u^T R(t, x)u, \tag{6.2.25}$$

is the used as the optimality criterion, where $F(t, x)$ is a function to be defined and $R(t, x)$ is a positive-definite $r \times r$ matrix.

It is easy to find that the minimum of $\mathscr{B}[V; t, x, u]$ for the system (6.2.24) with a given function $V = V(t, x)$ and quality criterion J is gives by the control force

$$u = u^0(t, x) = -\tfrac{1}{2}(R^{-1}(t, x)m^T(t, x) \operatorname{grad} V(t, x). \tag{6.2.26}$$

The function $F(t, x)$ is determined from the relation

$$\mathscr{B}[V; t, x, u^0(t, x)] \equiv 0$$

as

$$F = \tfrac{1}{2}(\operatorname{grad} V)^T m R^{-1} m^T \frac{\partial V}{\partial x} + S. \tag{6.2.27}$$

Thus if the Lyapunov function $V = V(t, x)$ for the system (6.2.22) is taken to be optimal then the control force (6.2.26) can be found from the class of additional forces (6.2.23) that are optimal relative to the functional

$$J = \int_{t_0}^{\infty} [S + \tfrac{1}{4}(\operatorname{grad} V)^T m R^{-1} m^T \operatorname{grad} V + u^T Ru] \, dt.$$

Under this force, the derivative of the function $V(t, x)$ has the form

$$V'(t, x) = -W^0(t, x) = -[S + \tfrac{1}{2}(\operatorname{grad} V)^T m R^{-1} m^T \operatorname{grad} V].$$

Let the function $S(t, x)$ satisfy the precompactness condition and let the partial derivatives $\partial V/\partial x$ and the coefficients of the matrices $R(t, x)$ and $m(t, x)$ be bounded. Then the right-hand side of (6.2.24) with u given by (6.2.26) and with the function $W(t, x)$ will satisfy the precompactness and existence conditions for limiting functions. In addition, the function $V(t, x)$ admits an infinitesimally small upper bound.

Applying the corollary to Theorem 6.2.1 to the problem in question, we find that, in order that the function $V(t,x)$ constructed for the system (6.2.22) be applicable in Problem B for the system (6.2.24) with quality criterion

$$J = \int_{t_0}^{\infty} W(t, x(t), u(t))\, dt$$

and a function W defined by (6.2.25) under the control (6.2.26), it is sufficient that the following conditions be satisfied:

(1) the function $V \in C^1(R_+ \times N, R_+)$ and

$$a(\|x\|) \leqslant V(t,x) \leqslant b(\|x\|) \quad \forall (t,x) \in R_+ \times N;$$

(2) there exists at least one limiting pair (G, λ) corresponding to $f(t,x) + m(t,x)u^0(t,x),\ W)$ such that the set $\{\lambda(t,x) = 0\}$ does not contain the solutions of the limiting system

$$\frac{dx}{dt} = G(t,x),$$

except for $x = 0$.

6.3 Optimal Stabilization in the Neutral Case

We proceed with examining the system (6.2.24) under additional assumptions on the properties of the solution $x = 0$ of the system without control, i.e. the system

$$\frac{dx}{dt} = f(t,x). \tag{6.3.1}$$

For (6.3.1) let a function $V(t,x)$ be known satisfy the conditions

$$a(\|x\|) \leqslant V(t,x) \leqslant b(\|x\|) \quad \forall (t,x) \in R_+ \times N, \tag{6.3.2}$$

$$\frac{dV}{dt} = \frac{\partial V}{\partial t} + (\operatorname{grad} V)^T f(t,x) \leqslant 0 \quad \forall (t,x) \in R_+ \times N, \tag{6.3.3}$$

where $a, b \in \mathcal{K}$.
Let

$$\varphi(t,x,u) = (\operatorname{grad} V)^T X(t,x,u),$$

where $u = u(t,x) \in U$. Consider the expression

$$\mathscr{B}[V; t, x, u] = \frac{dV}{dt} + \varphi(t,x,u) + \omega(t,x,u), \tag{6.3.4}$$

where $\omega(t,x,u) \geqslant 0$.
    We suppose that for $(t,x,u) \in R_+ \times N \times U$ the inequality

$$\|X(t,x,u)\| \leqslant N_1 \|x\|^{r_1}, \tag{6.3.5}$$

is satisfied, where $r_1 > 1$, N_1 and r_1 are constant, and the function $\varphi(t, x, u)$ is differentiable in x and satisfies the condition

$$\| \operatorname{grad} \varphi \| \leqslant N_2 \| x \|^{d-1}, \qquad (6.3.6)$$

where $d \geqslant r_1$ and $N_2 = \text{const} > 0$.

If the above assumptions are satisfied, the following result on optimal stabilization of the nonlinear nonstationary system (6.2.24) holds.

Theorem 6.3.1 Let

(1) there exist a function $V \in C^1(R_+ \times N, R_+)$ and functions a, $b \in \mathcal{K}$ such that the conditions (6.3.2) and (6.3.3) are satisfied;
(2) $\min_{u \in U} \mathcal{B}[V; t, x, u] = \mathcal{B}[V; t, x_0[t], u_0[t]] = 0$;
(3) there exist positive numbers δ and l and a function $c \in K$ such that for any $(t', x') \in G$ one of the conditions
  (i) $\omega(t', x', u_0[t']) \geqslant c(\| x' \|)$,
  or
  (ii) $\int_{t'}^{t'+T_1} \varphi(t, x^0(t, t', x'), u_0[t]) \, dt \leqslant -\delta T_1 \| x' \|^d$ for $T_1 > l$.
  is satisfied.

Then the control vector $u = u_0(t, x) \in E$ solves the problem on optimal stabilization.

Proof The proof of asymptotic stability of the unperturbed motion $x = 0$ of the system (6.2.24) for $u = u_0(t, x)$ is similar to the proof of Theorem 6.2.1.

The optimality of the control force $u = u_0(t, x) \subset E$ is proved in the same way as in Theorem 6.2.2. ∎

We now consider the nonlinear system with linear control

$$\frac{dx}{dt} = X(t, x) + R(t, x) + u(t, x)^T M(t, x), \qquad (6.3.7)$$

where $M(t, x)$ is a matrix defined in the domain $R_+ \times N$.

We suppose that the right-hand side of the system (6.3.7) satisfies the existence and uniqueness conditions for the solution of the Cauchy problem for this system. We define the type of function $\omega(t, x, u)$ and control $u = u_0(t, x)$ for which the Lyapunov function $V(t, x)$, known for the free system corresponding to (6.3.7) and ensuring stability of the unperturbed motion $x = 0$ of the system, can serve as an optimal function for (6.3.7).

The function $\omega(t, x, u)$ in (6.3.4) is taken in the form

$$\omega(t, x, u) = F(t, x) + u^T A u. \qquad (6.3.8)$$

Here $F(t, x)$ is a nonnegative function and $u^T A u$ is a given positive-definite quadratic form, where A is a symmetric matrix such that

$$(M \operatorname{grad} V)^T A^{-1}(M \operatorname{grad} V) - 8\varphi_1 \geqslant 0, \qquad (6.3.9)$$

with $\varphi_1 = (\text{grad } V)^T R$. We form the expression

$$\mathscr{B}_1[V; t, x, u] = V' + u^T M \text{ grad } V + \varphi_1 + F(t, x) + u^T A u, \qquad (6.3.10)$$

which reaches a minimum relative to $u \in U$ for $u = u_0$, where

$$u_0 = -\tfrac{1}{2}(M \text{ grad } V)^T A^{-1}. \qquad (6.3.11)$$

In fact, in view of (6.3.11) and the symmetry of the matrix A, the terms in (6.3.10) that depend on u can be represented in the form

$$u^T M \text{ grad } V + u^T A u = (u - u_0)^T A(u - u_0) - u_0^T A u_0.$$

Hence it is clear that the minimum of (6.3.10) relative to $u \in E$ is attainable for $u = u_0$.

For $(t', x') \in D$ it can be shown from $\mathscr{B}_1[V_1; t', x', u_0[t']] = 0$ for a function ω of the form (6.3.8) that there exists a function $c \in K$ satisfying condition (3)(i) of Theorem 6.3.8, provided that M grad $V \neq 0$ when $x \neq 0$.

Thus we arrive at the following conclusion. If the function $V(t, x)$ ensures stability of unperturbed motion $x = 0$ of the system (6.3.1), it is be the optimal Lyapunov function for the system (6.3.7), optimized by the control vector u_0 defined by (6.3.11) with respect to the functional

$$J_1 = \int_{t_0}^{\infty} [F(t, x) + u^T A u] \, dt, \qquad (6.3.12)$$

provided that $M\nabla V \neq 0$ when $x \neq 0$ and for any $(t', x') \in R_+ \times N$ there exist positive numbers δ and l such that

$$\int_{t'}^{t' + T_1} \varphi_1(t, x^0(t, t', x(t'))) \, dt \leqslant -\delta T_1 \|x'\|^d$$

for $T_1 > l$.

6.4   Decentralized Control Systems

This section gives an algorithm for solution of a large-scale system with stabilization of viscoelastic elements. This algorithm is based on the method of averaging from non-linear mechanics together with the method of Lyapunov functions. We behave that one of the positive aspects of this approach is that it allows us to consider problems of stabilization in cases when the noncontrolled nonlinear approach to the system gives only a neutrally stable solution or the solution becomes such as a result of local decentralized controls, i.e. we actually consider optimization of solutions in the critical case. The effective application of this approach requires the solution of two problems: alongside construction of the Lyapunov function for a contracted system, it is necessary to construct the general solution.

We shall employ the following rotation here, in addition to that real elsewhere: $d(x, M) = \inf_{y \in M} \|x - y\|$ is the distance from the point $x \in R^n$ to the set $M \subset R^n$, $T = [0, +\infty[$, $T_0 = [t_0, +\infty[$, (x, y) is the scalar product of vectors x and y, and U is the set of continuous functions $u(t, x)$ in the domain $\Omega \subset T \times R^n$.

6.4.1 Statement of the problem

We consider a controlled viscoelastic system, whose perturbed state is described by

$$\frac{dx}{dt} = f\left(t, x, u, \mu X\left(t, x, \int_0^t R(t, s, x(s))\, ds \right) \right), \tag{6.4.1}$$

where $x \in R^n$, $u \in R^m$, $t \in T_0$, $f: T_0 \times R^n \times R^m \times \Lambda \times R^n \to R^n$, and $\mu \in \Lambda$ is a small parameter. Suppose that the system (6.4.1) can be decomposed into m subsystems

$$\frac{dx_s}{dt} = f_s(t, x_s) + m_s(t, x_s)u_s^l + \mu X_s\left(t, x, \int_0^t R(t, s, x(s))\, ds \right), \quad s = 1, 2, \ldots, m. \tag{6.4.2}$$

Here $x_s \in R^{n_s}$ is a state of the sth subsystem, so that

$$R^n = R^{n_1} \times R^{n_2} \times \cdots \times R^{n_m}; \quad f_s: T_0 \times R^{n_s} \to R^{n_s},$$

$m_s(t, x_s)$ is a matrix, $u_s^l \in R^{r_s}$ is a vector of the decentralized local control of the sth subsystem, and $X_s: T_0 \times R^n \times R^n \to R^{n_s}$. The functions f_s, m_s, X_s and R are defined and continuous in the corresponding domains $D_1 \subset T_0 \times R^{n_s}$, $D_2 \subset T_0 \times R^{n_s} \times R^{r_s}$, $D_3 \subset T_0 \times R^n \times R^d$ and $D_4 \subset T_0 \times T \times R^n$, and they vanish for $x_s = x = 0$ for all $s \in [1, m]$.

The quality of the process of transition for any $s \in [1, m]$ can be characterized by the functional

$$J_s = \int_0^\infty [\psi_s(t, x[t], \mu) + (B_s u_s^l, u_s^l)]\, dt. \tag{6.4.3}$$

In (6.4.3) the nonnegative functions ψ_s must be defined, $(B_s u_s^l, u_s^l)$ are positive-definite quadratic forms with symmetric matrices $B_s, s \in [1, m]$, as given before.

For the algorithmic determination of stable solutions, it is necessary to determine controls $u \in U$ so that

(1) the zero solution of the system (6.4.1) will be uniformly μ-stable;
(2) the functional

$$J = \sum_{s=1}^m J_s[t_0, x_{s0}, u_s^l, \mu]$$

on the trajectories of the system (6.4.1), starting from the point $(t_0, x_0) \in D_1^0 \subset T_0 \times R^n$, will take its minimum value for $\mu < \mu_0, \mu_0 \in \Lambda$.

6.4.2 Auxiliary transformations

We suppose for the system (6.4.1) that the Lyapunov functions $V_s(t, x_s)$ are found for $\mu = 0$, and are positive, continuous and differentiable for $(t, x_s) \in D_1$, and for which

$$\frac{\partial V_s}{\partial t} + (\mathrm{grad}\, V_s, f_s(t, x_s)) \equiv \omega_s(t, x_s) \leqslant 0 \quad \forall s \in [1, m]. \tag{6.4.4}$$

From the minimum condition for the expressions

$$\mathscr{B}_s[V_s; t, x_s, u_s^l] = \omega_s(t, x_s) + (\text{grad } V_s, m_s(t, x_s)u_s^l) + (\text{grad } V_s, X_s(t, x, y))$$
$$+ \psi_s(t, x_s, \mu) + (B_s u_s^l, u_s^l), \quad s \in [1, m], \tag{6.4.5}$$

we determine

$$u_s^l(t, x_s) = -\tfrac{1}{2} B_s^{-1} m_s^T(t, x_s) \text{grad } V_s, \quad s \in [1, m]. \tag{6.4.6}$$

Taking into account the fact that

$$(\text{grad } V_s, m_s(t, x_s)u_s^l) + (B_s u_s^l, u_s^l) = (B_s(u_s^l - \bar{u}_s^l), u_s^l - \bar{u}_s^l) + (B_s \bar{u}_s^l, \bar{u}_s^l), \quad s \in [1, m],$$

the functions ψ_s in the quality criterion (6.4.3) can be determined from the expressions

$$\psi_s(t, x, \mu) = -\omega_s(t, x_s) - \mu(\text{grad } V_s, X(t, x, y)) + (B_s \bar{u}_s^l, \bar{u}_s^l), \quad s = 1, 2, \ldots, m. \tag{6.4.7}$$

In view of (6.4.6), the quality criterion (6.4.3) has a specific meaning. Using (6.5.6), we transform the system (6.4.2) to the form

$$\frac{dx}{dt} = g(t, x) + \mu X\left(t, x, \int_0^t R(t, s, x(s)) \, ds \right), \tag{6.4.8}$$

where

$$g(t, x) = [g_1(t, x_1) \quad \cdots \quad g_m(t, x_m)]^T,$$
$$g_s(t, x_s) = f_s(t, x_s) - \tfrac{1}{2} m_s B_s^{-1} m_s^T \text{grad } V_s, \quad s \in [1, m].$$

With the system

$$\frac{dx}{dt} = g(t, x), \quad x(t_0) = x_0, \quad t_0 \geqslant 0, \tag{6.4.9}$$

which is obtained from (6.52) for $\mu = 0$, we shall associate the Lyapunov function

$$V = \sum_{s=1}^m V_s(t, x_s), \tag{6.4.10}$$

about which corresponding assumptions are made in the theorem.

We assume that the solution $x(t) = x(t; t_0, x_0)$ of the system (6.4.8) is unknown, but that we know the solution $\bar{x}(t) = \bar{x}(t; t_0, x_0)$ of the system (6.4.9), and also $\bar{x}(t_0) = x(t_0) = x_0$, $x(t) = \tilde{x}(t)$ for $0 \leqslant t \leqslant t_0$, and $\tilde{x}(t) = \bar{x}(t)$ for $t \in T_0$.

In constructing the algorithm, the mean

$$\varphi_0(t_0, x_0) = \lim_{T \to \infty} \frac{1}{T} \int_{t_0}^{t_0 + T} \varphi\left(\tau, \bar{x}(\tau), \int_0^\tau R(\tau, s, \tilde{x}(s)) \, ds \right) d\tau, \quad t_0 \geqslant 0, \tag{6.4.11}$$

of the function

$$\varphi(t, x, y) = \left(\text{grad } V, X\left(t, x, \int_0^t R(t, s, x(s)) \, ds \right) \right),$$

calculated using the solutions of the system (6.4.9) is important.

Let $V^*(x)$ be a nonpositive scalar function, defined and continuous on $D_0 \subset R^n$ and let $\mathscr{E}(V^* = 0)$ in D_0 be the set of points x for which $V^*(x) = 0$.

In the set $\mathscr{E}(V^* = 0)$ the mean $\varphi_0(t_0, x_0)$ is definitely less than zero if for any numbers α and ε $(0 < \alpha < \varepsilon < H)$ we can find $r(\alpha, \varepsilon)$ and $\delta(\alpha, \varepsilon)$ $(r > 0,\ \delta > 0)$ such that $\varphi_0(t_0, x_0) < -\delta(\alpha, \varepsilon)$ for $\alpha < \|x_0\| < \varepsilon$ and $\rho(x_0, \mathscr{E}(V^* = 0)) < r(\alpha, \varepsilon)$ for all $t_0 \geqslant 0$.

6.4.3   The main result

Definition 6.4.1   The zero solution of the system (6.4.8) is

(a) *μ-stable with respect to T_i* if for all $t_0 \in T_i$ and any $\varepsilon > 0$ there exist $\delta(t_0, \varepsilon) > 0$ and $\mu^*(\varepsilon) < \bar{\mu}$ such that $\|x(t; t_0, x_0)\| < \varepsilon$ for all $t \in T_0$ follows from $\|x_0\| < \delta$ and $\mu < \mu^*$;

(b) *uniformly μ-stable with respect to T_i* if (a) is satisfied and there is a maximum δ_M satisfying (a) and such that

$$\inf \{\delta_M(t_0, \varepsilon) : t_0 \in T_i\} > 0$$

corresponding to any $\varepsilon > 0$.

In (a) and (b) we can omit the expression "with respect to T_i" iff $T_i = R$.

Theorem 6.4.1   Let the following conditions be satisfied.

(1) There exists a positive-definite differential function $V(t, x)$ admitting as infinitesimally small upper bound.

(2) There exists a function $V^*(x)$ such that the total derivative of the function $V(t, x)$ in view of the system (6.4.9) satisfies the condition

$$\frac{\partial V}{\partial t} + (\operatorname{grad} V, g(t, x)) \leqslant V^*(x) \leqslant 0$$

in the domain D_1^0.

(3) The partial derivatives $\partial V / \partial x_i$, $i \in [1, n]$, are continuous along $(t, x) \in D_1^0$ and bounded.

(4) There exist constants $M_1, M_2, L_1, \ldots, L_4$ such that
   (i) $g(t, x) \in \operatorname{Lip}_x(L_1)$, $g(t, 0) = 0$ $\forall t \in T_0$;
   (ii) $\|X(t, x, y)\| \leqslant M_1$ $\forall (t, x, y) \leqslant D_3$, $X(t, 0, 0) = 0$ $\forall t \in T_0$;
   (iii) $\varphi(t, x, y) \in \operatorname{Lip}_x(L_2)$;
   (iv) $\varphi(t, x, y) \in \operatorname{Lip}_y(L_3)$;
   (v) $R(t, s, x) \in \operatorname{Lip}_x(L_4)$;
   (vi) $|\varphi(t, x, y)| \leqslant M_2$ $\forall x \in D_0 \backslash \mathscr{E}(V^* = 0)$ $\forall t \in T_0$.

(5) The mean (6.4.1) exist uniformly with respect to $(t_0, x_0) \in D_1^0$.

(6) $\varphi(t_0, x_0) < 0$ in the set $\mathscr{E}(V^* = 0)$.

Then the zero solution of the system (6.4.8) is uniformly μ-stable.
If we add one more condition to the above, namely

(7) $\min_{u \in \mathscr{U}} \sum_{s=1}^{m} \mathscr{B}_s[V_s; t, x_s, u] = 0$,

then the functional

$$J = \sum_{s=1}^{m} \int_0^{\tau_s} [\psi_s(t, x[t], \mu) + (B_s u_s^l, u_s^l)] \, dt, \qquad 0 < \tau_s < +\infty,$$

takes a minimum value on the solutions of the system (6.4.1).

Proof In order to prove the uniform μ-stability of the zero solution of the system (6.4.8) we should set $0 < \varepsilon < H$ and construct $\delta(\varepsilon)$ and $\mu^*(\varepsilon)$ for which the conditions of Definition 6.4.1(b) are satisfied. For a function $V(t, x)$ satisfying condition (1) of the theorem we can take constants $c(\varepsilon_1) > 0$ $(0 < \varepsilon_1 < \varepsilon)$ and $0 < \delta(\varepsilon_1) < \varepsilon_1$ such that the moving surface $V(t, x) = c$ will be located in the domain $\|x\| < \varepsilon_1$ when the domain $\|x\| < \delta$ is contained inside the surface $V(t, x) = c$ for all $t \in T$. Let the solution $x(t; t_0, x_0)$ of the system (6.4.8) with initial conditions $t_0 \geqslant 0$ and $\|x_0\| < \delta$ leave the domain $\|x\| < \delta$ and intersect the surface $V(t, x) = c$ at the point x_0' at time $t = t_0'$. We consider the behaviour of the function $V(t, x)$ along the solution $x(t; t_0, x_0)$ of (6.4.8). In view of condition (2) of the theorem , we obtain the estimate

$$V(t, x(t)) \leqslant V(t_0', x_0') + \int_{t_0'}^t V^*(x(\tau)) \, d\tau + \mu \int_{t_0'}^t \varphi(\tau, x(\tau), y(\tau)) \, d\tau. \qquad (6.4.12)$$

It is necessary to estimate the third term in this inequality in two cases:

(1) when $\rho(x_0', \mathscr{E}(V^* = 0)) < r(\delta, \varepsilon)$;
(2) when $\rho(x_0', \mathscr{E}(V^* = 0)) \geqslant r(\delta, \varepsilon)$.

Case 1 Here we illustrate the method of choosing $\mu^*(\varepsilon)$ and $l > 0$ such that the solution $x(t; t_0, x_0')$, having left the surface $V(t, x) = c$ for $t \in [t_0', t_0' + 2l]$, will remain in the domain $\|x\| < \varepsilon$ and at some moment $t_1 \in [t_0', t_0' + 2l]$ will return back inside the surface $V(t, x) = c$. With this aim in view, we represent the integral

$$I(t) = \int_{t_0'}^t \varphi(\tau, x(\tau), y(\tau)) \, d\tau \qquad (6.4.13)$$

in the form

$$I(t) = \int_{t_0'}^t \varphi\left(\tau, \bar{x}(\tau), \int_0^\tau R(\tau, s, \tilde{\bar{x}}(s)) \, ds\right) d\tau$$

$$+ \int_{t_0'}^t \left[\varphi\left(\tau, x(\tau), \int_0^\tau R(\tau, s, x(s)) \, ds\right) - \varphi\left(\tau, \bar{x}(\tau), \int_0^\tau R(\tau, s, x(s)) \, ds\right)\right] d\tau$$

$$+ \int_{t_0'}^t \left[\varphi\left(\tau, \bar{x}(\tau), \int_0^\tau R(\tau, s, x(s)) \, ds\right) - \varphi\left(\tau, \bar{x}(\tau), \int_0^\tau R(\tau, s, \tilde{\bar{x}}(s)) \, ds\right)\right] d\tau$$

$$= I_1(t) + I_2(t) + I_3(t).$$

From condition (6) of the theorem,

$$I_1(t) \leqslant -\tfrac{3}{4}(t - t_0')\delta \quad \text{for} \quad t > t_0' + l, \qquad (6.4.14)$$

where $\delta = \delta(\varepsilon_1)$ and $l = l(\delta)$. Note that, by virtue of conditions (4)(i) and (ii), for the norm of the deviation, $\| x(t) - \bar{x}(x) \|$ the estimate

$$\| x(t) - \bar{x}(t) \| \leqslant \mu k$$

holds for $t_0' \leqslant t \leqslant t_0' + 2l$, where $k = M_1 2l \exp(2L_1 l)$. By virtue of condition (4)(iii), having chosen $\mu_0 = \frac{1}{4}\delta L_2 k$, we obtain

$$I_2(t) \leqslant \int_{t_0'}^{t} \| \varphi(t, x, y) - \varphi(t, \bar{x}, y) \| \, dt \leqslant \tfrac{1}{4}(t - t_0')\delta, \ \ t_0' \leqslant t \leqslant t_0' + 2l. \qquad (6.4.15)$$

From condition (4)(iv), having chosen $\mu_0' = \frac{1}{4}\delta L_3 L_4 k$, we obtain

$$I_3(t) \leqslant \int_{t_0'}^{t} \| \varphi(t, x, y) - \varphi(t, x, \bar{y}) \| \, dt \leqslant \tfrac{1}{4}(t - t_0')\delta, \ \ t_0' \leqslant t \leqslant t_0' + 2l. \qquad (6.4.16)$$

Taking the estimates (6.4.14)–(6.4.16) together, we find that

$$V(t, x(t)) \leqslant V(t_0', x_0') - \tfrac{1}{4}\delta(t - t_0'), \quad t_0' \leqslant t \leqslant t_0' + 2l. \qquad (6.4.17)$$

Case 2 Here, on any interval $[t_1, t_2] \subset T_0$ on which, for a solution $x(t; t_0, x_0)$, the condition $\| x \| \geqslant \delta, \rho(x, \mathscr{E}(V^* = 0)) > r_1 (0 < r_1 < r(\delta, \varepsilon))$ is satisfied, we can choose $\mu_0''(\varepsilon)$ such that the solution $x(t; t_0', x_0')$ will not leave the domain $\| x \| < \varepsilon_1$. Indeed, according to condition (2), there exists a $\sigma > 0$ such that $\sup\{V^*(x) \text{ for } x \in Q\} \leqslant -\sigma$, where

$$Q = \{x : x \in R^n, \rho(x, \mathscr{E}(V^* = 0)) \geqslant r_1, \ \delta \leqslant \| x \| < \varepsilon\}.$$

We determine

$$V(t, x(t)) \leqslant V(t_0', x_0') - \tfrac{1}{2}\sigma(t - t_0') \qquad (6.4.18)$$

from the estimate (6.4.12), according to condition (4)(vi) for $\mu < \mu_0''$, where $\mu_0'' = \frac{1}{2}\sigma M_2$.

It follows from the last estimate that the function $V(t, x(t))$ does not increase along the solution of the system (6.4.8) in either of the two possible cases. Hence the solution $x(t; t_0, x_0)$ will neither leave the level $V(t, x) = c$ nor the domain $\| x \| < \varepsilon_1$. Choosing $\mu^*(\varepsilon) = \min(\mu_0, \mu_0', \mu_0'')$, we determine that for $\mu < \mu^*(\varepsilon)$ at every subsequent interval of time either Case 1 or Case 2 will hold, and the solution $x(t; t_0, x_0)$ will not leave the domain $\| x \| < \varepsilon$. All estimates are uniform along t_0' and x_0', and the values of δ and μ^* are chosen independently of t_0. By this, we prove the uniform μ-stability in the sense of Definition 6.4.1(b).

The proof of optimality is similar to that given in section 6.2. ∎

6.5   A Model from Analytical Mechanics

We consider a holonomic mechanical system with n-degree of freedom, subjected to gyroscopic and control forces:

$$\frac{d}{dt}\left(\frac{\partial L}{\partial q'}\right) - \frac{\partial L}{\partial q} = Gq' + U, \qquad (6.5.1)$$

$L(q, q') = L_2 + L_1 + L_0.$

Here $q = R^n$ are generalized coordinates, $G = G(t, q)$ is a skew-symmetric matrix (i.e. $G^T = -G$) whose components are assumed to be bounded uniformly continuous functions $(t, q) \in R_+ \times N$, $N \subseteq R^n$, and $U = [u_1 \ u_2 \ \cdots \ u_r]$ is the desired control force. We take $L_0(q) = 0$ and $\partial L_0 / \partial q = 0$ for $q = 0$; i.e. for $U = 0$ the system (6.5.1) has the equilibrium state

$$q' = q = 0. \tag{6.5.2}$$

Let the control force be of the form

$$Q = M(t, q)U, \tag{6.5.3}$$

where $M(t, q)$ is a given $n \times m$ matrix whose components are bounded uniformly continuous functions $(t, q) \in R_+ \times N$, and the functional to be minimizes is

$$J = \int_{t_0}^{\infty} [F(t, q, q') + u^T \beta(t, q)u] \, dt. \tag{6.5.4}$$

Here $\beta(t, q)$ is a positive definite $r \times r$ matrix of the same class as M, and $F(t, q, q')$ is a function to be defined.

We shall establish the conditions under which the function $L_2 - L_0$ is optimal for optimal stabilization of the equilibrium state (6.5.2) under the forces (6.5.3) with quality estimation of the transition process using (6.5.4). Following the arguments from Section 6.3, we find from the minimization condition for the expression

$$\mathcal{B}[V; t, q, q', u] \quad \text{for} \quad u = u^0(t, q, q')$$

that

$$u = u^0(t, q, q') = -\tfrac{1}{2} \beta^{-1}(t, q) M^T(t, q) q'. \tag{6.5.5}$$

The value of $F(t, q, q')$ is determined from the condition $\mathcal{B}[V; t, q, q', u(t, q, q')] \equiv 0$ as

$$F(t, q, q') = \tfrac{1}{4}(q')^T M(t, q) \beta^{-1}(t, q) M^T(t, q) q'. \tag{6.5.6}$$

The system (6.5.1) subjected to the control force (6.5.5) is precompact and the limiting system corresponding to it is of the form

$$\frac{d}{dt}\left(\frac{\partial L}{\partial q'}\right) - \frac{\partial L}{\partial q} = G^* q - \tfrac{1}{2}(\beta^*)^{-1} M^{*T} q',$$

where G^*, β^* and M^* are the limiting matrices corresponding to G, β and M respectively. Under the force (6.5.5) the derivative of the function $L_2 - L_0$ in view of the system (6.5.1) is

$$(L_2 - L_0)' = -W(t, q, q') = -2F(t, q, q').$$

The function $\lambda = \lambda(t, q, q')$, defined for $W = W(t, q, q')$, is

$$\lambda(t, q, q') = \tfrac{1}{2}(q')^T M^*(\beta^*)^{-1} M^{*T} q'.$$

For a sequence of intervals $[t_n, t_n + t_0]$ $(t_n \to +\infty, t_0 > 0)$, let the matrix $M\beta^{-1}M^T$ be positive-definite and such that $x^T M\beta^{-1}M^T x \geqslant \alpha_0 x^T x$ for some $\alpha_0 > 0$. Then the corresponding limiting matrix $M^*(\beta^*)^{-1}(M^*)^T$ relative to the sequence $t_n \to +\infty$ is also positive-definite for $t \in [0, t_0]$.

The equality $(q')^T M^* (\beta^*)^{-1} (M^*)^T q' = 0$ for $t \in [0, t_0]$ holds iff $q' = 0$.

Theorem 6.5.1   Suppose that

(1) the function $V = -L_0(q)$ is positive-definite;
(2) the matrix $M\beta^{-1}M^T$ is positive-definite;
(3) the equilibrium state (6.5.2) is isolated.

Then the control force (6.5.5) is a solution of the problem of optimal stabilization with respect to (q', q) of the equilibrium state (6.5.2) of the systems (6.5.1) and (6.5.3) relatively to the functional (6.5.4).

Theorem 6.5.2   Suppose that

(1) the function $V = -L_0(q)$ is positive-definite relative to $q_1, q_2, \ldots, q_m$;
(2) the matrix $M\beta^{-1}M^T$ is positive-definite;
(3) there is no equilibrium state on the set $\{L_0(q) < 0\}$;
(4) under any control force (6.5.5), the motions of the system (6.5.1) are bounded relative to $q_{m+1}, q_{m+2}, \ldots, q_n$.

Then the control force (6.5.5) is a solution of the problem of optimal stabilization with respect to $q'_1, q_1, q'_2, \ldots, q_m$ of the equilibrium state (6.5.2) of the systems (6.5.1) and (6.5.3) relative to the functional (6.5.4) and (6.5.6).

Proofs   Both of these theorems results allow immediately from the theorems is Sections 6.1 and 6.2.

6.6   Comments and References

6.1   The consideration of the stabilization of controlled motions follows the approach of Krasovsky [2].

6.2   Theorems 6.2.1 and 6.2.2. are both new.

6.3   Theorem 6.3.1 is due to Martynyuk and Karimzhanov [2].

6.4   The results of this section are due to Martynyuk [7].

6.5   For the mechanical models of this section see Andreev [3].

7 Stability of Abstract Compact and Uniform Dynamical Processes

7.0 Introduction

The conceptual similarity of results on stability properties in terms of limiting equations for various classes of nonautonomous equations in finite-dimensional Banach spaces prompted the development of a theory of stability of motion for abstract compact and uniform dynamical processes in terms of limiting processes.

In this chapter we deal with stability theory for dynamical processes where motions are continuous relative to the coordinates only (for fixed t); the solutions of partial differential equations in general are processes possessing these properties.

Dafermos has established that any nonautonomous dynamical process on a metric space M is equivalent to an autonomous dynamical process on some other phase space. As a result, the properties of positive limiting sets in compact dynamical processes can be derived from the properties of positive limiting sets in the generating autonomous dynamical processes. This approach has the advantage that the algebraic structure of an autonomous dynamical process is simpler than that of a nonautonomous one. Although the topological structure of the phase space of an autonomous process is more complicated than that of the phase space of a nonautonomous one, the basic properties of the generating autonomous process do not depend upon the specific character of the topological structure of the phase space. This approach allows the so-called limiting Lyapunov functionals to be used in investigation of the stability of motions modelled by abstract dynamical processes.

7.1 The Generalized Direct Lyapunov Method for Abstract Weak $\mathscr{D}^+$ processes on M

7.1.1 Description of system and definitions

Definition 7.1.1 A *Weak $\mathscr{D}^+$ process on a metric space M* is a two-parameter family of mappings such that

(1) $S(0, \tau, p) = p$ for all $(\tau, p) \in R \times M$;

(2) $S(s + t, \tau, p) = S(t, \tau + s, S(s, \tau, p))$ for all $(\tau, p) \in R \times M$ and all $t, s \in R$;

(3) for every $t \in R_+$ the one-parameter family $S(t, \tau, \cdot): M \to M$ is equicontinuous in τ.

Definition 7.1.2   The set

$$H[S] = \overline{\{S^\sigma : \sigma \in R_+\}}$$

is called the *shell* of the $\mathscr{D}^+$ process $S(t, \tau)$ on M, where S^σ is the σ shift with respect to the second argument of the $\mathscr{D}^+$ process:

$$S^\sigma(t, \tau) = S(t, \tau + \sigma).$$

A random $\mathscr{D}^+$ process $S(t, \tau): M \to M$ is called *autonomous* or a *weak $\mathscr{D}^+$ system* on M if $S^\sigma = S$ for all $\sigma \in R$.

Definition 7.1.3 The set

$$\Omega(S, \tau, p) = \bigcap_{\sigma \geqslant 0} \bigcup_{t \geqslant 0} S(t, \tau, p)$$

is called the *positive limiting set of motion through the point* $(\tau, p) \in R \times M$.

Definition 7.1.4   The mappings defined by the expressions

$$\hat{T}(t, \bar{S}) = \bar{S}^t, \quad \bar{S} \in H[S], \quad t \in R_+, \tag{7.1.1}$$

$$T(t, \bar{S})x = (\bar{S}^t, \bar{S}(t, 0, x)), \quad \bar{S} \in H[S], \quad t \in R_+, \quad x \in M, \tag{7.1.2}$$

are called *weak $\mathscr{D}^+$ systems*. The weak $\mathscr{D}^+$ system (7.1.1) is called the *system of shifts*, while the weak $\mathscr{D}^+$ system (7.1.2) is called the *generated $\mathscr{D}^+$ system*.

Definition 7.1.5   A $\mathscr{D}^+$ process $S(t, \tau)$ on M is called *compact* if the trajectory of motion through the point $\bar{S}$ of the $\mathscr{D}^+$ system of shifts $\hat{T}(t, S)$ is sequentially precompact in $H[S]$.

The positive limiting set of motions of a compact $\mathscr{D}^+$ process is denoted by $L[S]$ and referred to as its *asymptotic hull*.

Every $\mathscr{D}^+$ process $\bar{S}$ on the asymptotic hull $L[S]$ is called a *limiting $\mathscr{D}^+$ process*.

7.1.2   Localization of the limiting set of a $\mathscr{D}^+$ process on M

The following result is known.

Theorem 7.1.1   Let

(1) $S(t, \tau)$ be a compact $\mathscr{D}^+$ process on $M \to M$;

(2) $q = \lim_{n \to \infty} S(t_n, 0)p \in \Omega(S, 0, p)$, $\bar{S} = \lim_{n \to \infty} S^{t_n} \in L[S]$.

Then

(a) $\bar{S}(t, 0)q \in \Omega(S, 0, p)$ for all $\tau \in R_+$.

If, moreover, an additional condition holds, namely

(3) if the trajectory of motion of the $\mathscr{D}^+$ process through the point $(0, p)$ is precompact in M
then there exists a $g(q, \cdot): R \to \Omega(S, 0, p)$ such that
(b) $g(q, 0) = q$;
(c) $g(q, s + \tau) = \bar{S}(\tau, s) g(q, \tau)$ for all $(s, \tau) \in R \times R_+$.

We now consider the $\mathscr{D}^+$ system $T(t)$ on $H[S] \times M$, generated by the $\mathscr{D}^+$ process $S(t, \tau): M \to M$. For the $\mathscr{D}^+$ system $T(t)$ the invariance of its limiting set is expressed by Theorem 7.1.1. This property is referred to as its *unbiasedness* (*generalized invariance*).

Definition 7.1.6 Let $S(t, \tau)$ be a compact $\mathscr{D}^+$ process on M. A functional $V: R \times M \to R$ is called an *L functional for* $S(t, \tau)$ *of Dafermos type* if

(1) $V(t + \tau, S(t, \tau), p) \leqslant V(\tau, p)$ for all $(t, \tau, p) \in R_+ \times R \times M$;
(2) for a sequence σ_n in R such that S^{σ_n} converges, the sequence $V(t + \sigma_n, p)$ also converges for all $p \in M$.
Let V be an L functional of Dafermos type for the $\mathscr{D}^+$ process $S(t, \tau)$ and $\bar{S} \in H[S]$. Then $\bar{S} = \lim_{n \to \infty} S^{t_n}$.

We set

$$W(\bar{S}, p) = \lim_{n \to \infty} V(t_n, p). \tag{7.1.3}$$

The inequality (7.1.3) implies

$$W(T^\tau(S, p)) = W(\bar{S}^t, \bar{S}(t, 0)p) = \lim V(t + t_n, \bar{S}(0, \tau)p) \leqslant \lim_{n \to \infty} V(t_n, p) = W(\bar{S}, p).$$

Therefore $W(\bar{S}, p)$ is an L functional with respect to the $\mathscr{D}^+$ system $T(t)$ on $H[S] \times M$ generated by the $\mathscr{D}^+$ process $S(t, \tau): M \to M$.

Theorem 7.1.2 Let

(1) V be an L functional of Dafermos type relative to the compact $\mathscr{D}^+$ process $S(t, \tau)$ on M such that the family of mappings $V(t, \cdot): M \to R$, $t \in R$, is equicontinuous;
(2) for some $p \in M$

$$q = \lim_{n \to \infty} S(t_n, 0)p \in \Omega(S, 0, p), \quad \bar{S} = \lim_{n \to \infty} \bar{S}^{t_n} \in L[S].$$

Then

$$W(\bar{S}^t, \bar{S}(t, 0)q) = W(\bar{s}, q) \quad \forall t \in R_+. \tag{7.1.4}$$

7.1.3   Localization of the limiting set for a uniform $\mathscr{D}^+$ process on M

Definition 7.1.7 A compact $\mathscr{D}^+$ process $S(t, \tau)$ on M is called *uniform* if the family of mappings

$$\{S'(t, \tau)p: M \to M, \ t \in R, \tau \in R_+\}$$

is equicontinuous.

Theorem 7.1.3   Let

(1) $S(t, \tau)$ be a uniform $\mathscr{D}^+$ process on M;
(2) $q = \lim_{n \to \infty} S(t_n, 0) \in \Omega(S, 0, p)$, $\bar{S} = \lim_{n \to \infty} S^{t_n} \in L[s]$.

Then

$$\Omega(S, 0, p) = \Omega(\bar{S}, 0, q). \tag{7.1.5}$$

Moreover, let

(3) the trajectory through $(0, p)$ be precompact in M.

Then for every $\varepsilon > 0$ the set of t values for which

$$d(g(q, s + t), g(q, s)) < \varepsilon \quad \forall s \in R, \tag{7.1.6}$$

is relatively dense. Here $g(q, \cdot): R \to \Omega(s, \cdot, p)$ is the mapping introduced in Theorem 7.1.1.

If the motion through the point (t, p) is continuous, the set of all $\bar{v}$ defined on $\Omega(\bar{S}, 0, p)$ is equi-almost-periodic (in the classical sense).

Proof   Applying the theorem 7.1.2 on localization to the $\mathscr{D}^+$ system $T(t)$ generated by the $\mathscr{D}^+$ process, we arrive at (7.1.5).

Theorem 7.1.4   Let

(1) $S(t, \tau)$ be a uniform $\mathscr{D}^+$ process on M;
(2) V be an L functional of Dafermos type such that the family of mappings $V(t, \cdot): M \to R \ (t \in R_+)$ is equisemicontinuous from below;
(3) there exist $p \in M$ such that $q = \lim_{n \to \infty} S(t_n, 0)p \in \Omega(S, 0, p)$, and $\bar{S} = \lim_{n \to \infty} S^{t_n} \in L[s]$.

Then

$$W(\bar{S}^t, \bar{S}(t, 0)q) = W(\bar{S}, q) \quad \forall t \in R_+. \tag{7.1.7}$$

Theorems 7.1.2–7.1.4 constitute the generalized Lyapunov direct method for the investigation of stability-like properties of uniform and compact $\mathscr{D}^+$ processes $S(t, \tau)$ on M. We shall utilize this method in the subsequent sections to examine the asymptotic stability of some classes of partial differential equations and integro-differential equations.

7.2   The Continuous L Functional for a Linear Wave Equation

In this section we consider the problem

$$u''(s) + A(s)u'(s) + Bu(s) = f(s) \quad \forall s \in [t, \infty), \tag{7.2.1}$$

$$u(t) = u_0, \quad u(t) = u_1, \tag{7.2.2}$$

which generates an asymptotically autonomous $\mathscr{D}^+$ process.

The asymptotic properties of the solutions of the problem (7.2.1), (7.2.2) are investigated using the generalized Lyapunov direct method, employing a continuous L functional in the Dafermos sense and a limiting L functional.

Many boundary-value problems for hyperbolic partial differential equations can be reduced to the problem (7.2.1), (7.2.2).

Assumption 7.2.1 The following conditions are satisfied.

(1) B is a positive selfconjugate operator, dense in a separable Hilbert space H_0 and such that B^{-1} is a compact operator.

(2) $f(s) \in L_1([t, \infty), H_0) \quad \forall t \in R;$ (7.2.3)

(3) $A(s) \xrightarrow{\;L(H_0)\;} A, \quad s \to \infty, \quad A(s) \in L_1^{loc}(R, L(H))$ (7.2.4)

We denote the norm in H_0 by $\|\cdot\|_0$. The domains of definition of the operators B and $B^{1/2}$ with the corresponding norms $\|v\|_1 = \|B^{1/2}v\|_0 = \langle Bv, v \rangle^{1/2}$ and $\|v\|_2 = \|Bv\|_0$ are Hilbert spaces H_1 and H_2 respectively. The Hilbert space conjugate to H_1 is denoted by H_{-1}. It is clear that H_i is dense in H_{i-1} for $i = 0, 1, 2$ and the inclusions of H_i into H_{i-1} are compact mappings.

We present some assertions about the existence and properties of the solutions of the problem (7.2.1), (7.2.2).

Assertion 7.2.1 Let $u_0 \in H_1$ and $u_1 \in H_0$.
Then

(a) there exists a unique function $u(s) \in C^0([t, \infty[, H_1) \cup C^1([t, \infty[, H_0)$ satisfying the problem (7.2.1), (7.2.2) in H_{-1} almost everywhere on $[t, \infty[$;

(b) $\|u'(s)\|_0^2 + \|u(s)\|_1^2 = \|u_1\|_0^2 + \|u_0\|_1^2 - 2\int_t^s \langle A(s)u'(s), u'(s) \rangle \, ds + 2\int_t^s \langle f(s), u'(s) \rangle \, ds$

Assertion 7.2.2 Let $u_0 \in H_1$ and $u_1 \in H_1$, and let the function $f(s)$ be absolutely continuous, with $f(s) \in L_1([t, \infty[, H_0)$.
Then

(a) there exists a unique solution $u(s) \in C^0([t, \infty[, H_2) \cup C^1([t, \infty[, H_1) \cup C^2([t, \infty[, H_0)$ of the problem (7.2.1), (7.2.2),

(b) $\|u''(s)\|_0^2 + \|u'(s)\|_1^2 = \|u_2\|_0^2 + \|u_1\|_1^2 - 2\int_t^s \langle A(s)u''(s), u''(s) \rangle \, ds -$

$2\int_t^s \langle A'(s)u'(s), u''(s) \rangle \, ds + 2\int_t^s \langle f(s), u''(s) \rangle \, ds,$

where $u_2 = -A(t)u_1 - Bu_0 + f(t)$.

7.2.1 Generation of an asymptotically autonomous $\mathscr{D}^+$ process.

Let $M = H_1 \times H_0$ be provided with the norm $\|(u, v)\| = (\|u\|_1^2 + \|v\|_0^2)^{1/2}$. We define the mappings $S: R \times R_+ \times M \to M$ and $T: R_+ \times M \to M$ by

$$S(t, \tau)p = (u(t + \tau), u'(t + \tau)), \quad p = (u_0, u_1) \in M; \tag{7.2.5}$$

$$T(\tau)p = (v(\tau), v'(\tau)), \tag{7.2.6}$$

where $u(s)$ is the solution to the problem (7.2.1) (7.2.2) and $v(\tau)$ is the solution of the problem

$$v''(\tau) + Av'(\tau) + Bv(\tau) = 0, \quad v(0) = u_0, \quad v'(0) = u_1. \tag{7.2.7}$$

The problem (7.2.7) is a special case of (7.2.1), (7.2.2). The mappings (7.2.5) and (7.2.6) are completely defined. In terms of Assertion 7.2.1, it follows from (7.2.5) that $S(t, \tau)p \in M$ for all $t \in R$, $\tau \in R_+$ and $p \in M$.

Theorem 7.2.1 The mapping (7.2.5) is a $\mathscr{D}^+$ process on M and the mapping (7.2.6) is a $\mathscr{D}^+$ system on M.

Proof It is clear that

$$T^t(0)p = p \quad \forall p \in M \quad \forall t \in R,$$

$$T^t(\tau + \sigma)p = T^{t+\tau}(T^t(\tau)p) \quad \forall (t, \tau, p) \in R \times R_+ \times M.$$

In view of condition (1) of Assumption 7.2.1, (2.5), (2.6) and the positiveness of the operator $A(\xi)$, we have

$$\| S^t(\tau)p \|^2 \leqslant \| p \|^2 + 2 \int_0^t \| f(t + \xi) \|_0 \| S^t(\xi)p \| \, d\xi.$$

Applying the Gronwall–Bellman inequality, we arrive at

$$\| S^t(\tau)p \| \leqslant \| p \| + \int_0^\tau \| f(t + \xi) \|_0 \, d\xi. \tag{7.2.8}$$

Since the equation (7.2.1) is linear, the (7.2.6) implies that for any fixed $\tau \in R_+$ the one-parameter family of mappings $S^t(\cdot, \tau): M \to M$ $(t \in R)$ is equicontinuous. Therefore S is a $\mathscr{D}^+$ process on M.

The following condition is equivalent to condition (1) of Assumption 7.2.1.

$$\| v(\tau) \|_0^2 + \| v(\tau) \|_1^2 = \| u_1 \|_1^2 + \| u_0 \|_1^2 - 2 \int_0^t \langle Av'(s), u'(s) \rangle \, ds. \tag{7.2.9}$$

This relation implies

$$\| S(\tau)p \| \leqslant \| p \| \quad \forall t \in R_+. \tag{7.2.10}$$

Moreover,

$$T(0, p) = p \quad \forall p \in M,$$

$$T(\tau + \sigma, p) = T(\tau, T(\sigma, p)) \quad \forall p \in M, \quad \forall \tau, \sigma \in R_+.$$

Hence the mapping $T: R_+ \times M \to M$ is a $\mathscr{D}^+$ system. Thus the theorem is proved.

Theorem 7.2.2   The mapping $S(t, \tau): M \to M$ is an asymptotically autonomous $\mathscr{D}^+$ process with asymptotic hull $L[T]$.

Proof It is sufficient to show that

$$S(t, \tau)p \xrightarrow{M} T(\tau)p \quad \text{as } t \to \infty. \tag{7.2.11}$$

From the definitions of a $\mathscr{D}^+$ system and a $\mathscr{D}^+$ process, we have

$$S(t, \tau)p - T(\tau)p = (w(t + \tau), w(t + \tau)),$$

where

$$w''(s) + A(s)w'(s) + Bw(s) = f(s) - [A(s) - A]u'(s).$$

The following expression is equivalent of condition (1) of Assumption 7.2.1:

$$\| w'(s) \|_0^2 + \| w(s) \|_1^2 = -2 \int_t^s \langle A(\xi)w'(s), w'(s) \rangle \, d\xi + 2 \int_t^s \langle f(\xi), w'(\xi) \rangle \, d\xi$$

$$-2 \int_t^s \langle [A(\xi) - A]v'(\xi), w'(\xi) \rangle \, d\xi. \tag{7.2.12}$$

By virtue of (7.2.10), we get from (7.2.12)

$$\| (w(s), w'(s)) \|^2 \leqslant 2 \int_t^s [\| f(\xi) \|_0 + \| A(\xi) - A \|_{L(H_0)} \| x \|] \| (w(\xi), w'(\xi)) \| \, d\xi. \tag{7.2.13}$$

Using the Gronwall–Bellman inequality, we obtain

$$\| S(t, \tau)p - T(\tau)p \| \leqslant \int_0^\tau \{ \| f(t + \xi) \|_0 + \| A(t + \xi) - A \|_{L(H_0)} \| x \| \} \, d\xi.$$

The relation (7.2.9) follows from condition (1) of Assumption 7.2.1 and function integrability. Thus the theorem is proved.

7.2.2 Asymptotic stability

The following holds.

Theorem 7.2.3   Let the functional $V : R \times M \to R$ of the form

$$V(t, p) = \| p \| + \int_t^\infty \| f(s) \|_0 \, dt \tag{7.2.14}$$

be an L_4 functional of Dafermos type.

Then the corresponding limiting L functional has the form

$$W(T, t, p) = \| p \|. \tag{7.2.15}$$

Proof   It is clear that the one-parameter family $V(t, \cdot) : M \to M$ $(t \in R)$ is equicontinuous. Applying the estimate (7.2.8) and (7.2.14), we obtain

$$V'(t, p) = \lim_{\tau \to \infty^+} \sup \tau^{-1} [V(t + \tau, S(t, \tau)p) - V(t, x)]$$

$$= \lim_{\tau \to 0^+} \sup \left[\| S(t, \tau)p \| - \| x \| - \int_t^{t+\tau} \| f(s) \|_0 \, ds \right] \leqslant 0.$$

We get for every sequence $t_n \subset R_+$, $t_n \to \infty$, $n \to \infty$,

$$V(t + t_n, p) \to \|p\| \quad \text{as} \quad n \to \infty \quad \forall(t, p) \in R \times M. \qquad \blacksquare$$

The following result holds.

Theorem 7.2.4   Let

(1) $u(s)$ be a solution of the problem (7.2.1), (7.2.2), where $u_0 \in H_1$ and $u_1 \in H_0$;
(2) Ker $A \cap$ Ker $(B - \lambda I) = \{0\}$ $\forall \lambda \in R_+$.

Then

$$u(s) \xrightarrow{H_1} 0 \quad \text{as} \quad s \to \infty,$$

$$u'(s) \xrightarrow{H_0} 0 \quad \text{as} \quad s \to \infty. \qquad (7.2.16)$$

Proof   Consider the mapping $t \to S(t, \tau)p$ of the $\mathscr{D}^+$ process initiated from the point (t, p). We shall show that

$$S(t, \tau)p \xrightarrow{M} 0 \quad \text{as} \quad \tau \to \infty. \qquad (7.2.17)$$

The set of absolutely continuous functions in $L_1([t, \infty[, H_0)$ together with the first derivatives in $L_1([t, \infty[, H_0)$ is dense in $L_1([t, \infty[, H_0)$. In view of the estimate (7.2.8) and condition (1) of Assumption 7.2.1, we conclude that it is sufficient to show that (7.2.17) holds for $u_0 \in H_2$ and $u_1 \in H_1$ and that the function $f(s)$ is absolutely continuous, $f(s) \in L_1([t, \infty[, H_0)$.

From condition (2) of Assumption 7.2.1 we have

$$\|u''(s)\|_0^2 + \|u'(s)\|_1^2 \leqslant \|u_2\|_2^2 + \|u_1\|_1^2 + 2 \int_t^s [\|A(\xi)\|_{L(H_0)} \|u''(\xi)\|_0 + \|f(\xi)\|_0] \|u''(\xi)\|_0 \, d\xi. \qquad (7.2.18)$$

Applying the Gronwall–Bellman inequality, we get

$$\|u''(s)\|_0^2 + \|u'(s)\|_1^2]^{1/2} \leqslant (\|u_2\|_2^2 + \|u_1\|_1^2)^{1/2} + \int_t^s [\|A(\xi)\|_{L(H_0)} \|u'(\xi)\|_0 + \|f(\xi)\|_0] \, d\xi. \qquad (7.2.19)$$

Condition (1) of Assumption 7.2.1 and the estimates (7.2.8) and (7.2.19) show that the norm $\|(d/d\tau)S(t, \tau)p\|$ is uniformly bounded on R_+. From (7.2.1), we have

$$\|u(s)\|_2 = \|Bu(s)\|_0 \leqslant \|u''(s)\|_0 + \|A(s)\|_{L(H_0)} \|u'(s)\|_0 + \|f(s)\|_0. \qquad (7.2.20)$$

The estimates (7.2.19) and (7.2.20) imply that the norm $\|S(t, \tau)p\|_{H_2 \times H_1}$ is uniformly bounded on R_+. Therefore the motion of the $\mathscr{D}^+$ process uniformly continuous and its trajectories are precompact in M. The limiting set $\Omega(t, p)$ is nonempty and compact, and

$$S(t, \tau)p \to \Omega(t, p) \quad \text{as} \quad \tau \to \infty. \qquad (7.2.21)$$

Let $\tilde{q}(v_0, v_1) \in \Omega(t, p)$. Then

$$T(\tau)q \in \Omega(t, p) \quad \forall \tau \in R_+,$$

$$W_T(t + \tau, T(\tau)q) = \text{const},$$

where W_T is the limiting L functional (7.2.15). The relations (7.2.6) and (7.2.9) ensure that the solution of the problem (7.2.7) with the initial conditions $v(0) = v_0$ and $v'(0) = v_1$ satisfies

$$\langle Av'(\tau), v'(\tau) \rangle = 0 \quad \forall \tau \in R_+. \tag{7.2.22}$$

Since A is a positive selfconjugate operator, the condition (7.2.22) is equivalent to

$$Av'(\tau) = 0 \quad \forall \tau \in R_+, \tag{7.2.23}$$

and the problem (7.2.7) reduces to an equation of the type

$$v'(\tau) + Bv(\tau) = 0.$$

Every solution of this equation can be represented as

$$v(\tau) = \text{Re} \sum_{n=1}^{\infty} c_n \exp(i\sqrt{\lambda_n}\, \tau)v_n, \tag{7.2.24}$$

where $\{\lambda_n\}$ is a set of eigenvalues and $\{v_n\}$ is the corresponding set of eigenvectors. Substituting (7.2.24) into (7.2.23) and using the property of almost-periodic functions, we get

$$\text{Im} \sum_{n=1}^{\infty} \sqrt{\lambda_n}\, c_n \exp(i\sqrt{\lambda_n}\, \tau)Av_n = 0 \quad \forall \tau \in R_+, \quad \text{or} \quad c_n Av_n = 0. \tag{7.2.25}$$

Then, by condition (2) of Assumption 7.2.1, we find from (7.2.25) that $c_n = 0$ for all n, or $v(\tau) \equiv 0$. Therefore, $\Omega(t, \tau) = \{0\}$, and, furthermore, (7.2.21) implies (7.2.17). Thus the theorem is proved. ∎

7.3   The L Functional Semicontinuous from below for a Hyperbolic Equation

In this section we shall deal with the problem

$$v_t + f(v)_x = 0 \quad \forall (t, x) \in R_+ \times R, \tag{7.3.1}$$

$$v(0, x) = v_0(x) \quad \forall x \in R,$$

which generates a uniform $\mathcal{D}^+$ process. Stability-like properties of solutions to this problem are examined in terms of the generalized Lyapunov direct method and using an L functional of Dafermos type relative to a uniform $\mathcal{D}^+$ process and semicontinuous from below.

The problem (7.3.1) is referred to as a *hyperbolic conservation law*.

Let f be a continuously differentiable function and $v_0(\cdot)$ a measurable and essentially bounded function on R, i.e.

$$\exists c_i, \quad i = 1, 2, \quad \text{such that} \quad c_1 \leqslant v_0(x) \leqslant c_2 \tag{7.3.2}$$

almost everywhere.

Definition 7.3.1   A bounded measurable function $v(t, x)$ on $R_+ \times R$ is called an *admissible weak solution to the problem (7.3.1)*, if for any convex function $g(v)$ and any smooth nonnegative function $\varphi(t, x)$ with compact support on $R_+ \times R$ the inequality

$$\int_0^\infty \int_{-\infty}^\infty [g(v)\varphi_t + F(v)\varphi_x] \, dx \, dt + \int_{-\infty}^\infty g(v_0)\varphi(0, x) \, dx \geq 0, \qquad (7.3.3)$$

is satisfied, where

$$F(v) = \int_0^v f'(\tau) \, dg(\tau).$$

Assertion 7.3.1   The following hold:

(a) there exists a unique admissible weak solution $v(t, x)$ of the problem (3.1);
(b) $u(t, \cdot) \in C^0(R_+, L_1^{loc}(R))$;
(c) $v(0, x) = v_0(x)$ almost everywhere on R.

Assertion 7.3.2   Let $v(t, x)$ and $w(t, x)$ be admissible weak solutions of equation (7.3.1) with initial values $v_0(x)$ and $w_0(x)$, satisfying the condition (7.3.2).
Then

(a) for any $x_1, x_2 \in R$ and any $t \in R_+$

$$\int_{x_1}^{x_2} |v(t, x) - w(t, x)| \, dx \leq \int_{x_1 - kt}^{x_2 + kt} |v_0(x) - w_0(x)| \, dx; \qquad (7.3.4)$$

(b) $v_0(x) \leq w_0(x) \quad \forall x \in R \Rightarrow v(t, x) \leq w(t, x) \quad \forall (t, x) \in R_+ \times R$;

(c) ess $\sup\limits_{x \in R} v(t, x) \leq$ ess $\sup\limits_{x \in R} v_0(x) \quad \forall t \in R_+$;

(d) ess $\inf\limits_{x \in R} v(t, x) \geq$ ess $\inf\limits_{x \in R} v_0(x) \quad \forall t \in R_+$.

Assertion 7.3.3   If $v(t, x)$ is an admissible weak solution of the problem (7.3.1), where $v_0(\cdot)$ is a periodic function with period β, then

(a) for any convex function $g(v)$

$$\lim_{s \to \infty} (2s)^{-1} \int_{-s}^s g(v(t, x)) \, dx \leq \lim_{s \to \infty} (2s)^{-1} \int_{-s}^s g(v_0(x)) \, dx \quad \forall t \in R_+;$$

(b) for $g(v) = \pm v$

$$\lim_{s \to \infty} (2s)^{-1} \int_{-s}^s v(t, x) \, dx = \lim_{s \to \infty} (2s)^{-1} \int_{-s}^s v_0(x) \, dx \quad \forall t \in R_+.$$

7.3.1   *Generation of a uniform $\mathscr{D}^+$ process*

We shall consider the solution of the problem (7.3.1) with periodic initial conditions. Let $v_0(\cdot)$ be a periodic function with period ω. Then for every fixed $t \in R_+$ the function $v(t, \cdot)$ is also periodic, with period $\omega^* > 0$.

Let M be a set of functions $v_0(\cdot)$ measurable on R with period $\omega > 0$, satisfying (7.3.2) and having mean value

$$a = \lim_{r \to \infty} (2r)^{-1} \int_{-r}^{r} v_0(x)\,dx = \omega^{-1} \int_0^{\omega} v_0(x)\,dx \quad \forall a \in [c_1, c_2]. \qquad (7.3.5)$$

We introduce a metric on M by

$$d(v_0, w_0) = \omega^{-1} \int_0^{\omega} |v_0(x) - w_0(x)|\,dx \quad \forall v_0, w_0 \in M. \qquad (7.3.6)$$

It can easily be verified that (M, d) is a complete metric space.

Theorem 7.3.1   Let $v(t, p)$ be an admissible weak solution of the problem (7.3.1), and $v_0 \in M$. Let $T(t)v_0 = v(t, p)$.
Then the mapping $T: R_+ \times M \to M$ is a uniform $\mathscr{D}^+$ system on M.

Proof   It is clear that $T(v)v_0 = v_0$ for all $v_0 \in M$ and $T(\sigma + \tau)v_0 = T(\tau)T(\sigma)v_0$ for all $(t, \tau) \in R_+$. Suppose that $v_0, w_0 \in M$. In view of (7.3.5), we have the estimate

$$d(T(\tau)v_0, T(\tau)w_0) \leqslant d(v_0, w_0) \quad \forall \tau \in R_+. \qquad (7.3.7)$$

This inequality shows that the one-parameter family of mappings $T(\tau): M \to M$ is equicontinuous.

7.3.2   Convergence of solutions

The following result is known.

Theorem 7.3.2   Let the functional $V: M \to R$ be given by

$$V(v_0) = \operatorname{ess\,sup}_{x \in R} v_0(x) \quad \forall v_0 \in M. \qquad (7.3.8)$$

Then $V(v_0)$ is an L functional of Dafermos type relative to the uniform $\mathscr{D}^+$ system $T(t)$ and is semicontinuous from below.

Proof   If $v_n(x)$ is a sequence converging to $v_0(x) \in L_1^{\mathrm{loc}}(R)$ then the estimate

$$\operatorname{ess\,sup}_{x \in R} v_0(x) \leqslant \lim_{n \to \infty} \inf \{ \operatorname{ess\,sup}_{x \in R} v_n(x) \} \qquad (7.3.9)$$

is satisfied. Hence it follows that

$$V\left(\lim_{n \to \infty} \theta_n \right) \leqslant \lim_{n \to \infty} \inf V(\theta_n). \qquad (7.3.10)$$

Moreover, because of (7.3.8), we have for any $t \in R_+$

$$V(T(t, v_0)) = \operatorname{ess\,sup}_{x \in R} v(t, x) \leqslant \operatorname{ess\,sup}_{x \in R} v_0(x) = V(v_0). \qquad (7.3.11)$$

Therefore,

$$V(T(\tau, \theta)) \leqslant V(\theta) \quad \forall \theta \in M \quad \forall \tau \in R^+. \tag{7.3.12}$$

Thus V is an L functional for $T(t)$ semicontinuous from below.

Theorem 7.3.3. Let

(1) $v(t, x)$ be an admissible weak solution of the problem (7.3.1), where $v_0(\cdot)$ is a periodic function with period ω, satisfying (7.3.2) with locally bounded variation;

(2) the function f have a relative extremum on the set where there are no limiting points in R.
   Then

$$v(t, x) \xrightarrow{\quad L_1^{\text{loc}}(R) \quad} a \quad \text{as } t \to \infty, \tag{7.3.13}$$

where a is defined by (7.3.5).

Proof Since by the conditions the trajectory of a $\mathscr{D}^+$ system is precompact, $T(t)\theta \to \Omega(\theta)$ as $\tau \to +\infty$. We shall show that if $w_0 \in \Omega(v_0)$ then $w_0(x) = a$ on R. Since $w_0 \in M$, it is sufficient to show that $w_0(x) = \text{const}$ on R. Since $v_0(x)$ is of locally bounded variation on R, is continuous from the left and satisfies (7.3.4) we have

$$\underset{[x_1, x_2]}{\text{Var }} v(t, x) \leqslant \underset{[x_1 - kt, x_2 + kt]}{\text{Var }} v_0(x),$$

$$k = \max \{|f'(\cdot)| \forall x \in [c_1, c_2]\}. \tag{7.3.14}$$

Furthermore, the periodicity of this function $v_0(\cdot)$ gives

$$\omega^{-1} \underset{}{\text{Var }} v(t, x) = \lim_{r \to \infty} (2r)^{-1} \text{Var } v(t, x) \leqslant \lim_{r \to \infty} (2r)^{-1} \underset{[-r, r]}{\text{Var }} v_0(x) = \omega^{-1} \underset{[0^+, \omega]}{\text{Var }} v_0(x) \tag{7.3.15}$$

and by the Helly theorem, $v_0(x)$ is a function of bounded variation continuous from the left. The relation

$$V(T(\tau, \psi)) = V(\psi) \quad \forall \tau \in R_+$$

implies

$$\underset{x \in R}{\text{ess sup }} w(t, x) = \underset{x \in R}{\text{ess sup }} w_0(x) \equiv c \quad \forall t \in R_+, \tag{7.3.16}$$

provided that $w(t, x)$ is an admissible weak solution of equation (7.3.1) with initial condition $w_0(x)$. It follows from (7.3.16) that the function $w_0(x)$ is constant on R. Suppose the contrary. Then there exist $\varepsilon > 0$ and $b \in R$ such that $w_0(b) \leqslant c - 2\varepsilon$. Since $w_0(x)$ is continuous from the left, $w_0(x) \leqslant c - \varepsilon$ for $a < x \leqslant b$. On the other hand, there exists a $\delta > 0$ such that f' is strictly monotone on $]c - \delta, c]$. We set $\bar{c} = \max \{c - \delta, c\}$ and define

$$\bar{w}_0(x) = \begin{cases} \bar{c} & \text{for } x \in]a + np, b + np[, \\ c - b & \text{otherwise.} \end{cases}$$

We consider an admissible weak solution $\bar{w}(t, x)$ of the problem (7.3.1) with initial condition $\bar{w}_0(x)$. It is clear that $w_0(x) \leqslant \bar{w}_0(x)$ and

$$w(t, x) \leqslant \bar{w}(t, x) \quad \forall (t, x) \in R_+ \times R. \tag{7.3.17}$$

From (7.3.16) and (7.3.17), we have

$$\operatorname*{ess\,sup}_{x \in R} \bar{w}(t, x) \geqslant c \quad \forall t \in R,$$

and arrive at a contradiction. Thus the theorem is proved.   ∎

7.3.3   Construction of the L functional

By Theorem 7.3.2, the L functional (7.3.8) is an L functional of Dafermos type semicontinuous from below. In terms of (7.3.11), the functional

$$V(v_0) = - \operatorname*{ess\,sup}_{x \in R} v_0(x) \tag{7.3.18}$$

is also an L functional of Dafermos type semicontinuous from below.

Let M_1 be a set of functions from M of locally bounded variation with constant c. By the Helly theorem, the set M_1 is closed in M. The expression (7.3.15) implies that M_1 is positive invariant relative to T. Thus the narrowing T on $R_+ \times M$ is a uniform $\mathscr{D}^+$ system on M_1. In this case

$$V(v_0) = \lim_{r \to \infty} (2r)^{-1} \operatorname*{Var}_{[-r, r]} v_0(x) = \omega^{-1} \operatorname*{Var}_{[0^+, \omega]} v_0(x) \tag{7.3.19}$$

can be taken as the L functional. The L functional (7.3.19) is semicontinuous from below.

We now construct a continuous L functional for $T(t)$. It can easily be shown that for any convex function the functional

$$V(v_0) = \lim_{r \to \infty} (2r)^{-1} \int_{-r}^{r} g(v_0(x)) \, dx = \omega^{-1} \int_{0}^{\omega} g(v_0(x)) \, dx \tag{7.3.20}$$

is a continuous L functional for $T(t)$.

Functionals of the types (7.3.19) and (7.3.20) posses the property

$$V(u(t, w_0)) = V(w_0) \Rightarrow w_0(x) = \text{const} \quad \forall x \in R.$$

7.4   The Continuous L Functional for an Integro-Differential Equation

The generalized Lyapunov direct method will be used to study the asymptotic properties of a function $u(t)$ on R with delay,

$$u(\tau) = v(\tau) \quad \forall \tau \in R_- =]-\infty, 0], \tag{7.4.1}$$

satisfying on $R_+ = [0, +\infty[$ the integro-differential equation

$$\frac{d}{dt} [\rho u'(t)] + Cu(t) + \int_{-\infty}^{t} G(t - \tau) v(\tau) \, d\tau = 0. \tag{7.4.2}$$

The problem (7.4.1), (7.4.2) was examined by Dafermos [1].

Let H_0 be a separable space with a scalar product $\langle \cdot, \cdot \rangle$ and a norm $\| \cdot \|_0$, let ρ be a bounded selfconjugate operator on H_0, and let C and $G(t)$ be unbounded selfconjugate operators such that Dom C, Dom $G(t)$ (with Dom $C \subset$ Dom $G(t)$ for all $t \in R_+$) are dense in H_0. The domain Dom C of the operator C is provided with a norm $\|w\|_2 \equiv \|Cw\|_0$, which induces a Hilbert space H_2. Similarly, the Hilbert space H_1 is the domain of definition Dom $C^{1/2}$ of the operator $C^{1/2}$ provided with the norm $\|w\|_1 = \|C^{1/2}w\|_0$. Finally, H_{-1} is the space conjugate to H_1 by the product. Clearly, $H_2 \subset H_1 \subset H_0 \subset H_{-1}$. Suppose that the inclusion of H_2 into H_0 is a compact mapping. Then the inclusions of H_i into H_{i-1} are also compact mappings for $i = 0, 1, 2$.

Assumption 7.4.1   We assume the following conditions.

(1) There exists a $c_0 > 0$ such that $\langle \rho w, w \rangle \geq c_0 \|w\|_0^2$ for all $w \in H_0$.
(2) There exists a $c_1 > 0$ such that $\langle Cw, w \rangle \geq c_1 \|w\|_0^2$ for all $w \in$ Dom C.
(3) $\langle G(t)w, w \rangle \leq 0$ for all $(t, w) \in R_+ \times$ Dom $G(t)$.
(4) $G(t) \in C^0(R_+, L(H_2, H_0)) \cap L^1(R_+, L(H_2, H_0))$.
(5) $G(t) \in C^0(R_+, L(H_1, H_{-1})) \cap L^1(R_+, L(H_1, H_{-1}))$.
(6) $G(t) \in C^0(R_+, L(H_1, H_{-1})) \cap L^1(R_+, L(H_1, H_{-1}))$.

We now discuss the existence of solutions of the problem (7.4.1), (7.4.2). It is easy to see that there exists a decreasing function $h(t) \in C^0(R)$, with $h(0) = 1$, $h(t) \to 0$ as $t \to \infty$, such that

$$\int_0^\infty [\|G(t)\|_{L(H_1, H_{-1})} + \|G(t)\|_{L(H_1, H_{-1})}] h^{-2}(t) dt < \infty. \tag{7.4.3}$$

Let C^k, $k = 0, 1, 2$ be the Banach space of functions $w(\tau) \in C^k(R_-, H_1) \cap C^{k+1}(R_-, H_0)$ such that

$$\|w\|_{C^k} = \sum_{i=0}^k \sup_{\tau \in R_-} [h(-\tau) \|w^{(i)}(\tau)\|_1] + \sum_{i=0}^{k+1} \sup_{\tau \in R_-} [h(-\tau) \|w^{(i)}(\tau)\|_0] < \infty,$$

where $u^{(k)}(t)$ is the kth derivative of $u(t)$ (we shall write $u'(t)$ for $u^{(1)}(t)$ and $u''(t)$ for $u^{(2)}(t)$). Let B_k, $k = 0, 1, 2$, be the Banach space of functions $w(\tau) \in C^k(R_-, H_2) \cap C^{k+1}(R_-, H_1) \cap C^{k+2}(R_-, H_0)$ such that

$$\|w\|_{B_k} = \sum_{i=0}^k \sup_{\tau \in R_-} \|w^{(i)}(\tau)\|_2 + \sum_{i=1}^{k+1} \sup_{\tau \in R_-} \|w^{(i)}(\tau)\|_1 + \sum_{i=0}^{k+2} \sup_{\tau \in R_-} \|w^{(i)}(\tau)\|_0 < \infty$$

It is obvious that $B_k \subset C^{k+1} \subset C^k$, $k = 0, 1, 2$. We can show that the inclusion of B_k into C^k is compact.

We have the following assertions about the existence of solutions to the problem (7.4.1), (7.4.2).

Assertion 7.4.1   Let $v \in C^k$ and $t_0 > 0$.
Then

(a) there exists a unique function $u(t) \in C^k(]-\infty, t_0], H_1) \cap C^{k+1}(]-\infty, t_0], H_0)$, satisfying (7.4.1) on R_- and (7.4.2) on $]0, t_0]$;

(b) $\displaystyle\sum_{i=0}^{k} \sup_{[0,t_0]} \|w^{(i)}(t)\|_1 + \sum_{i=0}^{k+1} \sup_{J[0,t_0]} \|u^{(i)}(t)\|_0 \leqslant c\|\|v\|\|_{C^k}$,   (7.4.4a)

where c does not depend on v.

Assertion 7.4.2   Let $v \in B_k$ and $t_0 \geqslant 0$.
Then

(a) there exists a unique function $u(t) \in C^k(] - \infty, t_0], H_2) \cap C^{k+1}(] - \infty, t_0], H_1) \cap C^{k+2}(] - \infty, t_0], H_0)$ satisfying (7.4.1) on R_- and (7.4.2) on $[0, t_0]$;

(b) $\displaystyle\sum_{i=0}^{k} \sup_{[0,t_0]} \|u^{(i)}(t)\|_2 + \sum_{i=0}^{k+1} \sup_{[0,t_0]} \|u^{(i)}(t)\|_1 + \sum_{i=0}^{k+2} \sup_{[0,t_0]} \|u^{(i)}(t)\|_0 \leqslant c\|\|v\|\|_{B_k}$,   (7.4.4b)

where c does not depend on v.

7.4.1   Generation of a weak $\mathscr{D}^+$ system

Let us show that the problem (7.4.1), (7.4.2) generates a weak $\mathscr{D}^+$ system (in the sense of Definition 7.1.2) on the spaces B_k and C^k. We deal with the mapping $T: R_+ \times C^k \to C^k$ for a fixed k, defined as

$$T(s, v)\tau = u(s + \tau), \qquad \tau \in R_-$$ (7.4.5)

where the function $u(t)$ satisfies the condition (7.4.1) on R_- and (7.4.2) on the segment $[0, s]$. In terms of Assertion 7.4.1, such a mapping (7.4.5) can be defined. Similarly, in terms of Assertion 7.4.2 a mapping $T: R_+ \times B_k \to B_k$ can also be defined.

The following result hold.

Theorem 7.4.1   Let

$$\langle G'(t)w, w \rangle \geqslant 0 \quad \forall(t, w) \in R_+ \times H_1,$$ (7.4.6)

$$\int_0^\infty \|G(t)\|_{L(H_2, H_0)} \, dt = m_\infty < 1,$$ (7.4.7)

and let there exist $a_0 > 0$ such that

$$\langle Aw, w \rangle \geqslant a_0 \|w\|_1^2 \quad \forall w \in H_1,$$ (7.4.8)

where

$$A = c + \int_0^\infty G(t) \, dt.$$

Then the mapping $T: R_+ \times C_k$ defined by (7.4.5) is a weak $\mathscr{D}^+$ system $T(t): C^k \to C^k$ in the sense of Definition 7.1.2.

Proof   It is clear that

$$T(0, v) = v \quad \forall v \in C^k,$$
$$T(s_1 + s_2, v) = T(s_1, T(s_2, v)) \quad \forall v \in C^k \quad \forall s_1, s_2 \in R_+.$$ (7.4.9)

From Assertions 7.4.1 and 7.4.2, for a fixed t the mapping T is continuous on C^k and B_k. However, this mapping is not continuous on $R_+ \times C^k$ or on $R_+ \times B_k$. Therefore T is a weak $\mathcal{D}^+$ system in the sense of Definition 7.1.2. Thus the theorem is proved.

7.4.2 Asymptotic stability

Consider the functional $V: C^k \to R$ given by

$$V_k(t, v) = \tfrac{1}{2} \langle \rho u^{(k+1)}(t), u^{(k+1)}(t) \rangle + \tfrac{1}{2} \langle A u^{(k)}(t), u^{(k)}(t) \rangle$$

$$- \tfrac{1}{2} \int_{-\infty}^t \langle G(t-\tau)[u^{(k)}(t) - u^{(k)}(\tau)], u^{(k)}(t) - u^{(k)}(\tau) \rangle \, d\tau \geqslant 0, \quad (7.4.10)$$

where $u(t)$ is a solution of the problem (7.4.1), (7.4.2).

Theorem 7.4.2   The mapping $V: C^k \to R$ defined by (7.4.10) is an L functional.

Proof   In terms of (7.4.3), the functional (7.4.10) is continuous on $R_+ \times C^k$. We first assume that $V \in C^{k+1}$. Differentiating (7.4.2) i times ($i = 0, 1, \ldots, k$), we get

$$\frac{d}{dt}[\rho u^{(i+1)}(t)] + C u^{(i)}(t) + \int_{-\infty}^t G(t-\tau) u^{(i)}(\tau) \, d\tau = 0, \qquad (7.4.11)$$

where $u^{(k)}(t)$ denotes the kth derivative. Differentiating (7.4.10) with respect to t and using (7.4.11) for $i = k$, we obtain

$$V_k'(t, v) = -\tfrac{1}{2} \int_{-\infty}^t \langle G'(t-\tau)[u^{(k)}(t) - u^{(k)}(\tau), u^{(k)}(t) - u^{(k)}(\tau) \rangle d\tau \leqslant 0.$$

However, C^{k+1} is dense in C^k. Therefore, in terms of Assertion 7.4.1(b), the relation (7.4.12) is established for $v \in C^k$. The expressions (7.4.10) and (7.4.12) imply that (7.4.10) is an L functional for the $\mathcal{D}^+$ system $T(t)$. Thus the theorem is proved.

Consider the eigenvalue problem

$$Cw - \lambda \rho w = 0. \qquad (7.4.13)$$

Since $C^{-1} \in L(H_0, H_2)$ is a positive compact operator on H, the problem (7.4.13) has a sequence λ_n of eigenvalues, and the corresponding sequence w_n of eigenfunctions is complete in H_0.

Theorem 7.4.3   Let

(1) for every eigenfunction w_n of the problem (7.4.13) there exist at least one value $\xi_n \in R_+$ such that

$$G'(\xi_n)w_n \neq 0; \qquad (7.4.14)$$

(2) $u(t)$ be a solution of (7.4.1), (7.4.2) and, moreover, $v \in C^m$, $m = 0, 1, 2, \ldots$.

Then

$$u^{(i)}(t) \xrightarrow{H_1} 0 \quad \text{as } t \to \infty, \quad i = 0, \ldots, m, \tag{7.4.15}$$

$$u^{(i+1)}(t) \xrightarrow{H_0} 0 \quad \text{as } t \to \infty. \tag{7.4.16}$$

Proof We shall prove the theorem under the strict or condition $v \in B$. In terms of (7.4.5), (7.4.10), (7.4.12) and condition (1) of Assumption 7.4.1, we have

$$\sup_{R_+} \| u^{(k)}(t) \|_1 \leqslant [2a_0^{-1} V_k(0, v)]^{1/2}, \quad k = 0, \ldots, m+1; \tag{7.4.17}$$

$$\sup_{R_+} \| u^{(k+1)}(t) \|_0 \leqslant [2\rho_0^{-1} V_k(0, v)]^{1/2}, \quad k = 0, \ldots, m+1. \tag{7.4.18}$$

It follows from (7.4.4) and (7.4.9) that the norm $\| u^{(k)}(t) \|_2$ is uniformly bounded on R_+, $k = 0, \ldots, m$. The motion through v is uniformly bounded, and its trajectory is bounded in B_m and precompact on C^m. Then the limiting set is nonempty, compact and invariant, and by localization of the limiting set, (Theorem 7.1.3) the relation

$$T(t, v) \xrightarrow{C^m} \theta \quad \text{as } t \to \infty \tag{7.4.19}$$

is satisfied, where θ is the largest invariant set contained in the zero set of the derivative,

$$\{ V_m(v) = 0 \}. \tag{7.4.20}$$

We shall show that $\theta = \{0\}$. In fact, let $\bar{u}(t)$ be a solution of the problem (7.4.1), (7.4.2) and $v \in E$. From (7.4.6) and (7.4.12), we have

$$G'(t - \tau)[u^{(m)}(t) - u^{(m)}(\tau)] = 0 \quad \forall t \in R \quad \forall \tau] - \infty, t]. \tag{7.4.21}$$

Integrating (7.4.21) with respect to τ on $] - \infty, t]$ and integrating by parts, we arrive at

$$\int_{-\infty}^{t} G(t - \tau) u^{(m+1)} w(\tau) \, d\tau = 0 \quad \forall t \in R. \tag{7.4.22}$$

Using (7.4.11) for $i = m + 1$, we obtain

$$\frac{d}{dt} [\rho \bar{u}^{(m+2)}(t)] + C \bar{u}^{(m+1)}(t) = 0 \quad \forall \bar{u} \in H_{-1}. \tag{7.4.23}$$

Hence

$$\bar{u}^{(m+1)}(t) = \text{Re} \sum_{n=1}^{\infty} \alpha_n \exp(i\sqrt{\lambda_n} t) w_n, \tag{7.4.24}$$

where λ_n is the sequence of eigenvalues and w_n the sequence of eigenfunctions of the problem (7.4.13).

Substituting (7.4.24) into (7.4.21) and putting $t - \tau = \xi$, we find that

$$\text{Im} \sum_{n=1}^{\infty} \alpha_n (\sqrt{\lambda_n})^{-1} [1 - \exp(-i\sqrt{\lambda_n} \xi)] \exp(i\sqrt{\lambda_n} t) G'(\xi) w_n = 0 \quad \forall t \in R$$

$$\forall \xi \in R_+. \tag{7.4.25}$$

Using the properties of almost-periodic functions, we get

$$\alpha_n[1 - \exp(-i\sqrt{\lambda_n}\xi)]G'(\xi)w_n = 0 \quad \forall \xi \in R_+ \quad \forall n. \tag{7.4.26}$$

Condition (1) of the theorem implies that $\alpha_n = 0$ for all n and $\bar{u}^{(m+1)}(t) \equiv 0$. Similarly,

$$\bar{u}^{(m)}(t) \equiv 0, \quad \ldots, \quad \bar{u}'(t) \equiv 0, \quad \bar{u}(t) \equiv 0.$$

Therefore (7.4.16) and (7.4.17) are true for $v \in B_m$. In the same manner, these conclusions can be shown to be valid for $v \in C^m$ as well. Thus the theorem is proved. ∎

7.4.3   An Application

Consider the viscoelasticity equation

$$\frac{\partial}{\partial t}\left[\rho(x)\frac{\partial u_i}{\partial t}\right] = \frac{\partial}{\partial x_j}\left[C_{ijkl}(x)\frac{\partial u_k}{\partial x_l} - \int_{-\infty}^{t} G_{ijkl}(x, t-\tau)\frac{\partial u_k}{\partial x_l}d\tau\right] \tag{7.4.27}$$

$$u(t, x) = 0, \quad x \in \partial\mathscr{D}, \quad t \in R, \tag{7.4.28}$$

where there is an implied summation over repeated indices and $\mathscr{D}$ is a smooth domain in R^n.

Assumption 7.4.2   The continuously differentiable tensors $C_{ijkl}(x)$ and $G_{ijkl}(x, t)$ satisfy the following conditions:

(1) $C_{ijkl}(x) = C_{klij}(x), \quad x \in \mathscr{D}.$

(2) $G_{ijkl}(t, x) = G_{klij}(t, x), \quad (t, x) \in R_+ \times \mathscr{D}.$

(3) There exists a $c_2 > 0$ such that

$$\int_{\partial} C_{ijkl}\frac{\partial w_i}{\partial x_j}\frac{\partial w_k}{\partial x_l}ds \geq c_2 \int_{\mathscr{D}}\frac{\partial w_i}{\partial x_j}\frac{\partial w_i}{\partial x_j}ds \quad \forall w \in C_0^\infty(\mathscr{D}).$$

(4) $\displaystyle\int_{\mathscr{D}} G_{ijkl}(t, x)\frac{\partial w_i}{\partial x_j}\frac{\partial w_k}{\partial x}ds \geq 0, \quad t \in R_+ \quad \forall w \in C_0^\infty(\mathscr{D}).$

(5) $\displaystyle\int_{\mathscr{D}} G'_{ijkl}(t, x)\frac{\partial w_i}{\partial x_j}\frac{\partial w_k}{\partial x}ds \leq 0, \quad t \in R_+ \quad \forall w \in C_0^\infty(\mathscr{D}).$

We also assume that $\rho(x)$ is measurable, bounded on $\mathscr{D} \subset R^n$ and there exists a $c > 0$ such that $\rho(x) \geq c$ for all $x \in \mathscr{D}$.

Let $H_0 = L^2(\mathscr{D})$, wher $L^2(\mathscr{D})$ is the space of square-integrable n-dimensional functions on $\mathscr{D}_0$. From condition (3) of Assumption 7.4.2 and the properties of elliptic systems, the spaces H_1 and H_2 defined in Section 7.4.1 are equivalent to the ordinary Sobolev spaces $\overset{\circ}{W}_1$ and $\overset{\circ}{W}_2^2$, and the compactness of the inclusion of H_1 into H_0 follows from the Rellich theorem.

Under these conditions, equation (7.4.27) is an equation of the type (7.4.1), and conditions (1)–(3) of Assumption 7.4.1 and (7.4.5) are automatically satisfied. The other conditions (4)–(6) of Assumption 7.4.1 and (7.4.6) and (7.4.7) can be easily reformulated in terms of the functions $C_{ijkl}(x)$ and $G_{ijkl}(t, x)$. Then, by Theorem 7.4.3, the solutions of the problem (7.4.27), (7.4.28) are asymptotically stable.

Let us consider a special case of the problem (7.4.17), (7.4.18), namely a one-dimensional homogeneous problem of the type

$$\rho\frac{\partial^2 u}{\partial t^2} = C\frac{\partial^2 u}{\partial x^2} - \int_{-\infty}^{t} g(t - \tau)\frac{\partial^2 u}{\partial x^2}\,d\tau, \quad (t, x)\in R_+ \times [0, 1], \qquad (7.4.29)$$

$$u(t, 0) = u(t, 1) = 0, \quad t\in R, \qquad (7.4.30)$$

where ρ and C are positive constants, $g(\tau)$ does not depend on x, and the desired function $u(t, x)$ is defined on $R_- \times [0, 1]$. Assume that the function g satisfies the condition

$$c - \int_0^\infty g(\tau)\,d\tau > 0. \qquad (7.4.31)$$

Conditions (4) and (5) of Assumption 7.4.2 imply that the function $g(t)$ is nonnegative and monotonically nonincreasing on R_+. Here the condition (7.4.8) is transformed into the condition (7.4.31), and

$$\|G(t)\|_{L(H_2, H_0)} = c^{-1}g(t). \qquad (7.4.32)$$

Thus (7.4.7) is reduced to (7.4.31). Under the above assumptions, the conditions of Theorem 7.4.3 are satisfied if $g(t) \not\equiv 0$, and the solutions of the problem (7.4.29), (7.4.30) are asymptotically stable.

7.5 Comments and References

7.0 Nonautonomous dynamical systems have been studied by many authors (see Section 1.9 and, for example, Bronstein [1, 2]; Bronstein, Chernyi [1]; Bebutov [1, 2]; Benavides [1]; Hale [2]; Hopf [1]; Karacostas [1, 2]; Lax [1]; Marcus, Mizel [1]; Zheng [1]. For the development of the Bebutov–Miller–Sell concept of abstract compact and uniform dynamical processes on metric spaces see Dafermos [1–6].

Limit sets for a compact dynamical process are invariant with respect to its limiting process. This allows limiting Lyapunov functionals to be applied in the investigation of stability-like properties of compact processes.

The results presented in this chapter were obtained by Dafermos [1–6] and Shestakov [1–6].

7.1 The Basic results of this section are due to Dafermos [2, 4, 5] and Shestakov [1, 6]. These results constitute the generalized Lyapunov direct method for uniform and compact processes.

7.2. The problem on asymptotic stability for linear wave equation has been considered in the work of Dafermos [3].

7.3   The problem (7.3.1) (hyperbolic conservation law) is examined in terms of generalized Lyapunov direct method and using an *L* functional of Dafermos type (see Dafermos [6]).

7.4   The problem of studying of the asymptotic properties of solutions for an integro-differential equation with delay was examined by Dafermos [1, 3] using generalized Lyapunov direct method.

8 Stability in Abstract Dynamical Processes on Convergence Space

8.0 Introduction

This chapter addresses limiting processes for abstract dispersive dynamical processes on a Fréchet convergence space and the generalized Lyapunov direct method proposed by Ball. In contrast to ordinary Lyapunov functionals, Ball-type Lyapunov functionals do not necessarily increase along the motion of the dynamical process. For example, Lyapunov functionals of Ball type that increase along the motion of an asymptotically autonomous process can be applied to study stability properties. Functionals of this type need not take the same value of all trajectories of the limiting set.

8.1 Abstract Dynamical Processes on Convergence Space, and Reduction of a Dynamical Process to a Dynamical System

We introduce the definition of a nonsingular abstract dynamical process on a Fréchet convergence space, and demonstrate that under fairly general conditions to every such abstract dynamical process there corresponds an abstract autonomous dynamical process (dynamical system) on an appropriate phase space. The reduction of a nonautonomous dynamical process to a dynamical system is of a great theoretical importance. Various classes of nonautonomous equations can be considered from a unified point of view: ordinary differential and functional differential equations, integro-differential equations and, generally, nonautonomous evolutionary equations in a Banach space E. Moreover, the reduction of a dynamical process to a dynamical system allows application of methods of topological dynamics in studying stability-like properties of motion.

8.1.1   Semidynamical process on convergence space

Definition 8.1.1   An arbitrary set X is called a *Fréchet convergence space* if a class of sequences (called *convergent*) is distinguished, and to every sequence $\{x_n\}$ of the class there correspond an element $x = \lim_{n \to \infty} x_n$ such that the following conditions are satisfied:
(1) if $x_n = x$ for all n then $x_n \to x$;
(2) if $x_n \to x$ and $\{x_k\} \subset \{x_n\}$ then $x_k \to x$.

A convergence space X is called *Hausdorff* if any convergent sequence has a unique limit. If X is a convergence space and $A \subseteq X$, the set A is called *precompact* if any sequence from A has a subsequence convergent to a point of X. If $B \subseteq X$ the *closure* $\bar{B}$ is defined as

$$\bar{B} = \{p \in X : \exists p_n \subseteq B \text{ such that } p_n \xrightarrow{X} p\}.$$

A mapping $F: X \to Y$, where X and Y are convergence spaces, is called *continuous* if $F(p_n) \xrightarrow{Y} F(p) \Leftrightarrow p_n \xrightarrow{X} p$. A functional $v: X \xrightarrow{X} R$ is called *semicontinuous from below* if $p_n \to p \Rightarrow v(x) \leqslant \lim_{n \to \infty} v(p_n)$. We provide the set $\Phi = X^{R_+}$ of all X-valued mappings $X: R_+ \to X$ with the structure of pointwise convergence, i.e. $x_n \xrightarrow{\phi} x \Leftrightarrow x_n(t) \xrightarrow{X} x(t)$. We denote by ϕ^0 the set of all subsets of Φ.

Definition 8.1.2   The mapping $\Pi(\tau): R \to \Phi^0$ is called a (*nonunique*) *abstract semidynamical process* ($\mathscr{D}^+$ *process*) if

(1) for $\tau \in R$, $x \in \Pi(\tau)$ and $\sigma \in R_+$ the σ translation of $x(t)$, $x^\sigma = x(t + \sigma)$, satisfies $x^\sigma \in \Pi(\tau + \sigma)$;
(2) for $\tau \in R$ and $x_n \in \Pi(\tau)$ (the sequence $x_n(0)$ converges) there exist a $x \in \Pi(\tau)$ and a subsequence $x_k \subset x_n$ such that $x_k \xrightarrow{\phi} x$.

The set Φ^0 (or Φ) is called a *motion space*.
A $\mathscr{D}^+$ process $\Pi(\tau): R \to \Phi^0$ is given by a two-parameter family of operators of the type

$$S(\Pi, t, \tau, \mathscr{A}) = \bigcup_{x, x(0)} x(t) \quad \forall x \in \Pi(\tau) \quad \forall x(0) \in \mathscr{A} \tag{8.1.1}$$

where $\mathscr{A}$ is an arbitrary set in X. If $\mathscr{A} = \{p\}$, instead of (8.1.1) we write $S(t, \tau, p)$.
A function $x \in \Pi(\tau)$, $x(0) = p \in X$ is called a *motion through the point* $(\tau, p) \in R \times X$.

Definition 8.1.3   A unique abstract $\mathscr{D}^+$ process on X is an abstract $\mathscr{D}^+$ process on X in the sense of definition 8.1.2 such that only one motion passes through each point $(\tau, p) \in E \times X$.

It follows from definition 8.1.2 and (8.1.1) that

$$S(\Pi, 0, \tau, \mathscr{A}) \subseteq \mathscr{A} \quad \forall \tau \in R \quad \mathscr{A} \subseteq X, \tag{8.1.2}$$

$$S(\Pi, t + s, \tau, \mathscr{A}) \subseteq S(\Pi, t, \tau + s, S(\Pi, s, \tau, \mathscr{A})) \quad \forall (t, s) \in R_+, \quad \tau \in R, \quad \mathscr{A} \subseteq X. \tag{8.1.3}$$

If $\tau \in R, s \in R_+$, $x_1 \in \Pi(\tau), x_2 \in \Pi(\tau + s)$ and $x_1(s) = x_2(0)$, then $x(t) = x_1(t)$, $0 \leqslant t \leqslant s$. The function $x(t) = x_2(t - s)$ for $t \geqslant s$ does not necessary belong to $\Pi(\tau)$; Therefore in (8.1.2) and (8.1.3) the equalities may not hold. For a unique abstract $\mathscr{D}^+$ process and the equalities do hold in (8.1.2) and (8.1.3), i.e.

$$S(\Pi, 0, \tau, \mathscr{A}) = \mathscr{A} \quad \forall \tau \in R \quad \mathscr{A} \subseteq X, \tag{8.1.2'}$$

$$S(\Pi, t + s, \tau, \mathscr{A}) = S(\Pi, t, \tau + s, S(\Pi, s, \tau, \mathscr{A})) \quad \forall (t, s) \in R_+, \quad \tau \in R, \quad \mathscr{A} \subseteq X. \tag{8.1.3'}$$

Definition 8.1.2 of a $\mathscr{D}^+$ process is more convenient than the definition based on (8.1.2) and (8.1.3) for a nonsingular $\mathscr{D}^+$ process and that based on (8.1.2') and (8.1.3') for a singular $\mathscr{D}^+$ process. It is obvious that if $\mathscr{A}$ is precompact then so $S(\Pi, t, \tau, \mathscr{A})$ for every pair $(t, \tau) \in R_+ \times R$.

Example 8.1.1 Consider an ordinary differential equation

$$\frac{dx}{dt} = f(t, x),$$

where $f \in C(R \times R^n, R^n)$. For $\tau \in R$ and $X = R^n$, $x: [\tau, \infty[\to R^n$. Let $x^\tau(t) = x(t + \tau)$ for all $t \in R$. We define a mapping $\Pi(\tau) = \{x^\tau \in \Phi : x : [\tau, \infty[\to R^n\}$ and a solution $x(\sigma)$ of this equation such that $|x(\sigma)| \leqslant r$ for all $\sigma \in [\tau, \infty[$, with $r > 0$. Then Π is a $\mathscr{D}^+$ process on $X = R^n$.

Definition 8.1.4 An abstract $\mathscr{D}^+$ process $\Pi(\tau)$ on X is called *autonomous* (or a $\mathscr{D}^+$ *system on X*) if it does not depend on τ or if the corresponding family of mappings $S(\Pi, t, \tau, q)$ does not depend on τ.

Definition 8.1.5 A subset $\mathscr{A}$ of X is called *positive invariant* relative to the $\mathscr{D}^+$-system $\{T(t)x, t \geqslant 0\}$ if $T(t, \mathscr{A}) \subseteq \mathscr{A}$ for all $t \in R_+$, and *invariant* if $T(t, \mathscr{A}) = \mathscr{A}$ for all $t \in R_+$.

We denote by P the set of all abstract $\mathscr{D}^+$ processes on X. Let us introduce convergence on P as follows: $\Pi_k \xrightarrow{P} \Pi$ iff for every sequence $\{\Pi_m\} \subset \{\Pi_k\}$ for all $\tau \in R$, and for every sequence $x_m \subset \Pi_m(\tau)$ such that $x_m(0)$ converges, there exists a $x \in \Pi(\tau)$ and a subsequence $x_n \subset x_m$ such that $x_n \xrightarrow{\Phi} x$. This convergence transforms the set P into a Fréchet convergence space. Obviously, if $\Pi_k \xrightarrow{P} \Pi$ and $\bar{\Pi} \in P$ are such that $\bar{\Pi}_{(\tau)} \supseteq \Pi(\tau)$ for all $\tau \in R$ then $\Pi_k \xrightarrow{P} \bar{\Pi}$. Therefore P is not a Hausdorff convergence space.

Let $\Pi \in P$ be a fixed abstract $\mathscr{D}^+$ process, and $\Pi^s(\tau) = \Pi(\tau + s)$ for all $\tau, s \in R$.

Definition 8.1.6 A subset $\mathscr{H}[\Pi]$ of the space P is called a *positive hull* of the abstract $\mathscr{D}^+$ process Π if

(1) for any sequence $S_k \subset R_+$ there exists a $\bar{\Pi} \in \mathscr{H}[\Pi]$ and a subsequence $S_m \subset S_k$ such that $\Pi^{S_m} \xrightarrow{P} \Pi$;

(2) $\Pi \in \mathscr{H}[\Pi]$;

(3) the subset $\mathcal{H}[\Pi]$ is invariant under shifts, i.e. if $\bar{\Pi} \in \mathcal{H}[\Pi]$ then $\bar{\Pi}^s \in \mathcal{H}[\Pi]$ for all $s \in R$.

It is clear that if the subset $\mathcal{H}[\Pi] \subset P$ satisfies (1) and (2) then $\tilde{\mathcal{H}}[\Pi] = \{\bar{\Pi}^s : \bar{\Pi} \in \mathcal{H}[\Pi], s \in R\}$ is a hull of the $\mathcal{D}^+$ process Π.

An abstract $\mathcal{D}^+$ process may have an infinite number of hulls or may have none.

Definition 8.1.7   The mapping

$$\tilde{T}(t)\bar{\Pi} = \bar{\Pi}^t \quad \forall (t, \bar{\Pi}) \in R_+ \times \mathcal{H}[\Pi] \tag{8.1.4}$$

is called the $\mathcal{D}^+$ *system of shifts* on $\mathcal{H}[\Pi]$.

By definition, $\mathcal{H}[\Pi]$ is a positive trajectory through the point $\Pi \in P$, precompact in $\mathcal{H}[\Pi]$.

Definition 8.1.8.   The limiting set of a positive trajectory through $\Pi \in P$ is called the *asymptotic hull* of the element $\Pi \in P$ in $\mathcal{H}[\Pi]$ and is denoted by $\mathcal{L}[\Pi]$.

Assertion 8.1.1   The limiting set $\mathcal{H}[\Pi]$ is positive invariant relative to $\tilde{T}$. If, moreover, $\mathcal{H}[\Pi]$ is a Hausdorff space (relative to convergence in P) then $\mathcal{L}[\Pi]$ is invariant.

Definition 8.1.9   Every abstract $\mathcal{D}^+$ process belonging to the asymptotic hull $\mathcal{L}[\Pi]$ of a $\mathcal{D}^+$ process Π is referred to as a *limiting $\mathcal{D}^+$ process* for $\Pi \in P$.

8.1.2   *Reduction of an abstract $\mathcal{D}^+$ process*

We shall show that there are two generated $\mathcal{D}^+$ systems that may be compared with an abstract $\mathcal{D}^+$ process on X, namely the $\mathcal{D}^+$ system of shifts and the main $\mathcal{D}^+$ system. This will allow investigation of stability-like properties of the motion of a $\mathcal{D}^+$ process via the $\mathcal{D}^+$ systems generated by it using the methods of topological dynamics. Let $\Pi \in P$ be a fixed $\mathcal{D}^+$ process with hull $\mathcal{H}[\Pi]$. Consider the Cartesian product $\Phi \times \mathcal{H}[\Pi]$ and distinguish the set of all points $(x, \bar{\Pi})$ in it such that $x \in \bar{\Pi}(s)$ for some $s \in R$. This set is denoted by $\mathcal{P}[\mathcal{H}]$. introducing convergence on this set, we claim that $(x_n, \bar{\Pi}_n) \xrightarrow{P[\mathcal{H}]} (x, \bar{\Pi})$ if and only if there exists an $s \in R$ such that for $x_n \in \bar{\Pi}_\pi(s)$ and $x \in \bar{\Pi}(s)$, $\bar{\Pi}_n \xrightarrow{P} \bar{\Pi}$ and $x_n \xrightarrow{\phi} x$. O Frèchet bviously, $\mathcal{P}[\mathcal{H}]$ is a Frèchet convergence space.

Definition 8.1.10   Let the mapping $T : R_+ \times \mathcal{P}[\mathcal{H}[\Pi]] \to \mathcal{P}[\mathcal{H}[\Pi]]$, with

$$T(t)\{x, \bar{\Pi}\} = \{x^t, \bar{\Pi}^t\} \quad \forall (x, \bar{\Pi}) \in \mathcal{P}[\mathcal{H}[\Pi]], \tag{8.1.5}$$

be generated by a $\mathcal{D}^+$ system on $\mathcal{P}[\mathcal{H}[\Pi]]$. We say that the $\mathcal{D}^+$ process Π on X *generates* the $\mathcal{D}^+$ system (8.1.5).

The mapping (8.1.5) is sequentially continuous in $t \in R$ and posesses the semigroup property.

Definition 8.1.11   A subset $\mathscr{A} \subset X$ is said to be

(a) *positive unbiased* relative to $\Pi \in P$ if for any $p \in \mathscr{A}$ there exists a limiting $\mathscr{D}^+$ process $\bar{\Pi} \in \mathscr{L}[\Pi]$ and a motion $x \in \bar{\Pi}(0)$ such that $x(0) = p$ and $x(R_+) \subseteq \mathscr{A}$;
(b) *unbiased relative* to $\Pi \in P$ if for any $p \in \mathscr{A}$ there exists a limiting $\mathscr{D}^+$ process $\bar{\Pi} \in \mathscr{L}[\Pi]$ and a mapping $h: R \to \mathscr{A}$ such that $h(0) = p$ and $h^s \in \Pi(s)$ for all $s \in R$, where $h^s(t) = h(t + s)$.

8.2   The Generalized Lyapunov Direct Method for Abstract Autonomous $\mathscr{D}^+$ Processes

In this section we consider the generalized Lyapunov direct method for $\mathscr{D}^+$ systems on a convergence space X in terms of a Ball-type Lyapunov functional.

We consider the one-parameter family of continuous transformations $T(t): X \to X$ with properties

$$T(0)x = X \quad \forall x \in X, \quad T(s + t)x = T(s)T(t)x \quad \forall x \in X \quad \forall s, t \in R_+.$$

This family is a $\mathscr{D}^+$ system on X. The *positive trajectory* $\gamma^+(x)$ through the point $x \in X$ and the *positive limiting set* $\Omega(x)$ for the point $x \in X$ are defined respectively by

$$\gamma^+(x) = \bigcup_{t \in R_+} T(t)x,$$

$$\Omega(x) = \{y \in X : \exists t_n \to \infty, \; T(t_n)x \xrightarrow{X} y\}.$$

The following result holds.

Assertion 8.2.1   The positive limiting set $\Omega(x)$ is positive invariant for all $x \in X$. If $\gamma^+(x)$ is precompact then $\Omega(x)$ is nonempty. If X is a Hausdorff space, then $\Omega(x)$ is invariant.

Proof   This is standard.   ∎

Theorem 8.2.1   Let $x \in X$ and $V: X \to R$ satisfy the following conditions.

(1) For any sequence $t_n \to \infty$ such that $T(t_n)x \to y$ and any $t \in R_+$ the inequality

$$V(y) - V(T(t)y) \leqslant \lim_{n \to \infty} [V(T(t_n)x) - V(T(t_n + t)x)] \qquad (8.2.1)$$

is satisfied.
(2) For all $\tau \in R$

$$\lim_{t \to \infty} [V(T(t)x) - V(T(t + \tau)x)] \leqslant 0. \qquad (8.2.2)$$

Then for all $y \in \Omega(x)$ the mapping $t \to V(T(t)y)$ does not increase on R_+.

Proof   Let $t_n \to \infty$, $T(t_n)x \to y$, $t \in R_+$, Then $T(t_n + t)x \to T(t)y$. By condition (1),

$$V(y) - V(T(t)y) \leqslant \lim_{n \to \infty} [V(T(t_n)x) - V(T(t_n + t)x)] \leqslant 0.$$

This and the fact that $\Omega(x)$ is positive invariant prove the theorem.

To prove subsequent results, we need the following two propositions, whose proofs are omitted.

Assertion 8.2.2   Let

$$a = \lim_{t \to \infty} \inf f(t) \qquad b = \lim_{t \to \infty} \sup f(t),$$

and let

(1) the function $f : R_+ \to R$ be continuous;
(2) $\lim_{t \to \infty} [f(t) - f(t + s)] = 0$ for all $s \in R_+$;
(3) $b > -\infty$;
(4) $c \in [a, b]$, with $c = \mp \infty$ if $a = -\infty$ or $b = +\infty$.

Then these exists a sequence $t_n \to \infty$ such that $f(t_n + \tau) \to c$ uniformly in τ on compact sets from R_+.

Assertion 8.2.3   Let

(1) the function $f : R_+ \to R$ be continuous on R;
(2) $\lim_{t \to \infty} [f(t) - f(t + s)] = 0$ for all $s \in R$.

Then the difference $f(t) - f(t + s) \to 0$ as $t \to +\infty$ uniformly in s on compact sets from R_+.

Theorem 8.2.2   Let

$$a = \lim_{t \to +\infty} \inf V(T(t)x), \qquad b = \lim_{t \to +\infty} \sup V(T(t)x),$$

$$\Omega(V, x) = \{y \in X : \exists t_n \to \infty, \ T(t_n)x \to y, \ V(T(t_n)x) \to V(y)\},$$

$$\cdot \quad V^{-1}(c) = \{y \in X : V(T(t)y) = c \forall t \in R_+\}$$

and let the following condition be satisfied:

(1) the vector $x \in X$ and the functional $V : X \to R$ satisfy the condition that if $t_n \to \infty$, $T(t_n)x \to y$ and $V(T(t_n)x) - V(T(t_n + \tau)x) \to 0$ uniformly in τ on compact sets from R_+ then

$$V(T(t_n)x) \to V(y) \quad \text{as } t_n \to \infty; \tag{8.2.3}$$

(2) trajectory $\gamma^+(x)$ is precompact;
(3) the mapping $t \to V(T(t)x)$ is continuous on $]0, \infty[$;
(4) for all $\tau \in R_+$

$$\lim_{t \to +\infty} \inf [V(T(t)x) - V(T(t + \tau)x)] \geqslant 0 \tag{8.2.4}$$

and $b > -\infty$.

Then

(a) $\infty > b \geqslant a > -\infty$;
(b) $\Omega(V, x) \cap V(c) \neq \phi \quad \forall c \in [a, b].$ \tag{8.2.5}

Proof Let the conditions of Theorem 8.2.2 be satisfied and $f(t) = V(T(t)x)$ and $c \in [a, b]$, with $c \neq \mp \infty$. We consider a sequence t_n with the properties from Assertion 8.2.2. Since the positive trajectory $\gamma^+(x)$ is precompact, there exists a subsequence $s_n \subset t_n$ and an element $y \in \Omega(x)$ such that $T(s_n)x \to y$. For any $\tau \in R_+$ we have $T(s_n + \tau)x \to T(\tau)y$. From Assertion 8.2.2, we get

$$V(T(s_n + \tau)x) - V(T(s_n + \tau + \bar{\tau})x) \to 0$$

uniformly in $\bar{\tau}$ on compact sets from R_+. Therefore, by condition (1) of Theorem 8.2.2,

$$V(T(s_n + \tau)x) \to V(T(\tau)y) = c.$$

Thus $\Omega(V, x) \neq \phi$. Assume that $b = +\infty$ $(a = -\infty)$. Let $\{t_n\}$ be a sequence from Assertion 8.2.2 corresponding to $c = +\infty$ $(c = -\infty)$. Following the same arguments as above, we conclude that there exists an element $y \in \Omega(x)$ such that $V(T(t_n)x) \to V(y)$. This leads to a contradiction. Thus the theorem is proved. ∎

Theorem 8.2.3 Let conditions (1)–(3) of Theorem 8.2.2 be satisfied, and, moreover, for every $\tau \in R_+$

$$\lim_{t \to \infty} [V(T(t)x) - V(T(t + \tau)x] = 0. \tag{8.2.6}$$

Then

(a) $\infty > b \geqslant a > -\infty$;
(b) property (8.2.5) holds;
(c) $\Omega(x) = \Omega(V, x) \subseteq \bigcup_{c \in [a, b]} V^{-1}(c)$.

Proof Assume that the conditions of the Theorem are satisfied. Let

$$y \in \Omega(x), \quad t_n \to \infty, \quad T(t_n)x \to y.$$

Assertion 8.2.3 and condition (1) of Theorem 8.2.2 imply

$$V(T(t_n + \tau)x) \to V(y) \quad \forall \tau \in R_+.$$

Thus we have

$$\Omega(x) = \Omega(V, x) \subseteq \bigcup_{c \in [a, b]} V^{-1}(c), \quad c > -\infty.$$

Theorem 8.2.3 now follows from Theorem 8.2.2. ∎

Assume that X is a Hausdorff topological space and that only a finite number of sets

$$V^{-1}(c) \cap \bar{\gamma}^+(x), \quad c \in [a, b]$$

are nonempty. Under the conditions of Theorem 8.2.2, $a = b$, and we have by Theorem 8.2.3.

$$\Omega(x) = \Omega(V, x) \subseteq V^{-1}(c).$$

If it is known that the set $V^{-1}(c)$ consists only of a finite number of points and if the mapping $t \to T(t)x$ is continuous on R_+ then, by virtue of the connectedness of the set

$\Omega(x)$, it consists of the only point of the y state of equilibrium of the $\mathscr{D}^+$ system $T(t): X \to X$. Then we have

$$T(t)x \to y, \quad V(T(t)x) \to V(y) \quad \text{as } t \to +\infty.$$

8.3 The Generalized Lyapunov Direct Method for Abstract Asymptotically Autonomous $\mathscr{D}^+$ Processes

Assumption 8.3.1 We consider a family of two-parametric transformations $S(t, \tau): M \to M$, where (M, d) is a metric space with the following properties.

(1) $S(0, \tau)p = p$ for all $(\tau, p) \in R \times M$.
(2) $S(t + s, \tau)p = S(t, \tau + s)S(s, t)p$ for all $t, s \in R_+$ and all $(\tau, p) \in R \times M$.
(3) For a fixed $\tau_0 \in R$ and $t \in R$ the family of operators $S(t, \tau)$ with the parameter $\tau \in [\tau_0, \infty[$ is equicontinuous, i.e. for all $\varepsilon > 0$ and all $p \in M$ there exists a $\delta > 0$ for all $\tau \in [\tau_0, \infty[$ such that $d(p, q) < \delta$ implies

$$d(S(t, \tau)p, S(t, \tau)q) < \varepsilon. \tag{8.3.1}$$

Definition 8.3.1 A two-parameter family of transformations $S(t, \tau): M \to M$ with the properties (1)–(3) of Assumption 8.3.1 is called an *abstract evolutionary process* on M (a $\mathscr{D}^+$ *process* on M).

Obviously, $S(t, \tau)p$ is a solution value at time $t + \tau$, if it took the value p at time τ.

Definition 8.3.2 A $\mathscr{D}^+$ process $S(t, \tau)$ on M is called *asymptotically autonomous* if there exists a family of operators $\{T(t), t \geqslant 0\}$ such that

$$S(t, s_n) \to T(t)p, \quad s_n \to \infty \quad \forall t \in R_+ \quad \forall p \in M. \tag{8.3.2}$$

Definition 8.3.2 implies that the set $\{T(t), t \geqslant 0\}$ is a weak $\mathscr{D}^+$ system on M, i.e. the mapping $(t, p) \to T(t)p$ is continuous relative to $p \in M$ only.

Assertion 8.3.1. Let the trajectory $\gamma^+(\tau, p) = \bigcup_{t \in R_+} S(t, \tau)p$ be precompact. Then the limiting set

$$\Omega(\tau, p) = \{q \in M : \exists t_n \to \infty \text{ such that } S(t_n, \tau) \to q\}$$

is nonempty, invariant and precompact. Moreover,

$$d(S(t, \tau)p, \Omega(\tau, p)) \to 0 \quad \text{as } t \to +\infty. \tag{8.3.3}$$

Provided the motion $t \to S(t, \tau)p$ is continuous, the set $\Omega(\tau, p)$ is connected.

Let

$$\Omega(V, \tau, p) = \{q \in M : \exists t_n, t_n \to \infty \text{ such that } S(t_n, \tau)p \to q, V((t_n, \tau)p) \to V(q)\}.$$

Theorem 8.3.1 Let

(1) there exist a limiting continuous L functional $V: M \to R$ for an asymptotically autonomous $\mathscr{D}^+$ process $S(t, \tau)$ such that

$$V(T(p)) \leqslant V(p) \quad \forall t \in R_+ \quad \forall p \in M;$$

(2) the trajectory $\gamma^+(\tau, p)$ be precompact.

Then

$$\Omega(\tau, p) \cap V^{-1}(c) \neq 0 \quad \forall c \in [a, b], \tag{8.3.4}$$

$$V^{-1}(c) = \{ y : V(T(t)y) = c \forall t \in R_+ \}, \quad c \in R, \tag{8.3.5}$$

where

$$a = \lim_{t \to \infty} \inf V(S(t, \tau)p), \quad b = \lim_{t \to \infty} \sup V(S(t, \tau)p). \tag{8.3.6}$$

Proof Let $V(t) = V(S(t, \tau)p)$. We demonstrate that

$$\lim_{t \to \infty} \inf [V(t) - V(t + s)] \geqslant 0 \quad \forall s \in R_+. \tag{8.3.7}$$

In fact, if (8.3.7) does not hold, there exists a sequence $t_n \to \infty$ such that $V(t_n) - V(t_n + s) \leqslant \varepsilon < 0$ for all n.

Without loss of generality, we have

$$S(t_n, \tau)p \to q,$$

$$S(t_n + s, \tau)q \to T(s)q.$$

Then $V(q) - V(T(s)q) \leqslant \varepsilon < 0$, which leads to a contradiction. Therefore the inequality (8.3.7) is established. We apply Assertion 8.2.2 to the function $f(t) = V(t)$. Then for every $c \in [a, b]$ there is a sequence $t_n \to \infty$ such that

$$S(t_n, \tau)p \to q \in \Omega(\tau, p), \quad V(S(t_n + t, \tau)p) \to c \quad \forall t \in R_+.$$

Therefore

$$V(T(t)q) = c \quad \forall t \in R_+, \quad q \in \Omega(\tau, p) \cap V^{-1}(c), \quad c \in R.$$

Thus the Theorem is proved. ∎

Theorem 8.3.2 Let

(1) the conditions of Theorem 8.3.1 be satisfied;
(2) the motion $t \to s(t, \tau)p$ be continuous for all sufficiently large t;
(3) the set $\bigcup_{c \in [a, b]} V^{-1}(c)$ be finite.

Then

(a) $a = b$;
(b) there exists an equilibrium state $x_e \in V^{-1}(a)$ such that $S(t, \tau)p \to x_e$ as $t \to +\infty$.

Proof Since the union of sets $\bigcup V^{-1}(c)$ is finite, $a = b$. But $V(t)$ is continuous and $\Omega(\tau, p)$ is invariant, $\Omega(\tau, p) \in V^{-1}(a)$. Because the set $\Omega(\tau, p)$ is connected, $\Omega(\tau, p) = \{x_e\}$, where x_e is an equilibrium state, i.e. an invariant set consisting of one point. Thus the theorem is proved. ∎

The following result is a corollary of Theorem 8.3.2.

Theorem 8.3.3   Let

(1) $V: M \to R$ be a limiting L functional (not necessary continuous) for an asymptotically autonomous $\mathscr{D}^+$ process $S(t, \tau): M \to M$ such that

$$\lim_{t \to +\infty} [V(S(t, \tau)p) - V(S(t + s, \tau)p)] \geqslant 0 \quad \forall s \in R_+;$$

(2) the conditions

$$S(t_n, \tau)p \to q, \quad V(S(t_n, \tau)p) - V(S(t_n + t, \tau)p) \to 0$$

uniformly in t for compact sets from R_+ imply

$$V(S(t_n, \tau)p) \to V(q).$$

Then $\Omega(V, \tau, p) \cap V^{-1}(c) \neq \varnothing$ for all $c \in [a, b]$.

Example 8.3.1   Consider the boundary-value problem

$$u_t = u_{xx} + f(t, u, x), \quad 0 < x < 1, \quad t > s, \tag{8.3.9}$$

$$u_x(t, 0) = u_x(t, 1) = 0, \quad t > s, \quad u(s, x) = \psi(x),$$

where $\psi \in C([0, 1])$.

The corresponding autonomous boundary-value problem is

$$\bar{u}_t = \bar{u}_{xx} + \bar{f}(\bar{u}, x), \quad \bar{u}_x(t, 0) = \bar{u}_x(t, 1) = 0, \quad \bar{u}(x, 0) = \bar{\psi}(x) \tag{8.3.10}$$

Let f and $\bar{f}$ be regular and let there exist $a(s)$ and $c(\rho, s)$ such that

$$\sup_{x \in [0,1]} vf(t, 0, x) \leqslant 0, \quad |v| \geqslant v(v), \quad t \geqslant s;$$

$$\| f(t, v, \cdot) \|_{C^2([0,1])} \leqslant c(\rho, s), \quad |v| \leqslant \rho, \quad t \geqslant s$$

Also, let

$$\lim_{t \to \infty} \int_t^{t+1} \sup \| f(\tau, v, \cdot) - \bar{f}(v, \cdot) \|_{C([0,1])} d\tau = 0 \quad \forall p > 0$$

We consider for $u \in C^1([0, 1])$ the functional

$$V(u) = \int_0^1 [\tfrac{1}{2} u_x^2 - \bar{F}(u, x)] dx, \quad \bar{F}(u, x) = \int_0^u \bar{f}(s, x) ds.$$

Let $S(t, s)\psi = u(t, \cdot)$, where u is a solution of the boundary-value problem (8.3.9) and $\psi \in C^1([0, 1])$. Also, let $T(t)\psi = \bar{u}(t, \cdot)$, where $\bar{u}$ is a solution of the problem (8.3.10) and $\bar{\psi} \in C^1([0, 1])$. It can easily be seen that the mapping $S(t, s)\psi$ is an asymptotically autonomous $\mathscr{D}^+$ process on $M = C^1([0,1])$, corresponding in the $\mathscr{D}^+$ system $T(t)$. Moreover, if $\psi \in C([0,1])$ then $S(t,x)\psi \in C^1([0,1])$ for all $t > 0$. Therefore, to study the asymptotic properties of the solutions of the boundary-value problem (8.3.9), we assume $\psi \in C^1([0,1])$. We show that all solutions of (8.3.9) are bounded in $C([0,1])$ and precompact in $C^1([0,1])$. On the other hand, the L functional V is continuous, and the set $V^{-1}(c)$ consists of the equilibrium states only.

Theorem 8.3.2 yields the following result.

Theorem 8.3.4   Let

(1) u be a solution of the boundary-value problem (8.3.9);
(2) a and b be defined by (8.3.6).

Then

(a) $-\infty < a \leqslant b < \infty$;
(b) for all $c \in [a,b]$ there exists an equilibrium state $w(\cdot) \in \Omega(S,\psi)$ such that $V(w(\cdot)) = c$.

8.4   The Generalized Lyapunov Direct Method for an Abstract Nonautonomous $\mathscr{D}^+$ Process

In this section we present the generalized Lyapunov direct method for a nonautonomous $\mathscr{D}^+$ process on a Fréchet convergence space, i.e. the method of localization of the limiting set of motion of a $\mathscr{D}^+$ process using Ball-type L functionals.

8.4.1   Ball-type L functionals and limiting L functionals

Assumption 8.4.1 L functional $V: R \times X \to R$ may have the following properties.

(1) If

$$t_k \to \infty, \quad \Pi^{t_k} \xrightarrow{p} \Pi^* \in \mathscr{L}[\Pi] \quad \forall \tau \in R,$$

$$x_k \in \Pi^{t_k}(\tau), \quad x_k \xrightarrow{\phi} x^* \in \Pi^*(\tau),$$

and

$$V^{t_k + \tau}(0, x_k(0)) - V^{t_k + \tau}(t, x_k(t)) \to 0 \quad \forall t \in J, \quad J \subset R_+$$

uniformly in t on compact sets from R_+ then

$$V^{t_k + \tau}(0, x_k(0)) - V^\tau_{\Pi^*}(0, x^*(0)) \to 0 \quad \text{as } k \to \infty. \qquad (8.4.1)$$

(1') If

$$x_n \xrightarrow{X} x, \quad \Pi^* \in \mathscr{L}[\Pi]$$

and

$$V_{\Pi^*}(0, x_n) - V_{\Pi^*}(t, S(\Pi^*, t, 0, x_n)) \to 0 \quad \forall t \in R_+$$

then the limit

$$V_{\Pi^*}(0, x_n) \to V_{\Pi^*}(0, x) \tag{8.4.2}$$

holds.

(2) If

$$t_k \to +\infty, \quad \Pi^{t_k} \xrightarrow{p} \Pi^* \in \mathscr{L}[\Pi] \quad \forall \tau \in R_+$$

and

$$x_k \in \Pi^{t_k}(\tau), \quad x_k \xrightarrow{\Phi} \Pi^* \in \Pi^*(\tau), \tag{8.4.3}$$

then

$$V_{\Pi^*}^\tau(0, x^*(0)) - V_{\Pi^*}^\tau(t, x^*(t)) \leqslant \lim_{k \to \infty} [V^{t_k + \tau}(0, x_k^*(0)) - V^{t_k + \tau}(t, x^*(t))] \quad \forall t \in R_+.$$

(2′) For any $\Pi^* \in \mathscr{L}[\Pi]$ and any $t \in R_+$ the mapping

$$x \to V_{\Pi^*}(0, p) - V_{\Pi^*}(t, S(\Pi^*, t, 0, p)) \tag{8.4.4}$$

is semicontinuous from below on X.

(3) For $\tau \in R$ and $x \in \Pi(\tau)$

$$\lim_{t \to \infty} [V^\tau(t, x(t)) - V^\tau(t + s, x(t + s))] \leqslant 0 \quad \forall s \in R_+. \tag{8.4.5}$$

(4) For $\tau \in R$ and $x \in \Pi(\tau)$

$$\lim_{t \to \infty} [V^\tau(t, x(t)) - V^\tau(t + s, x(t + s))] \geqslant 0 \quad \forall s \in R_+. \tag{8.4.6}$$

(5) For $\tau \in R$ and $x \in \Pi(\tau)$

$$\lim_{t \to \infty} [V^\tau(t, x(t)) - V^\tau(t + s, x(t + s))] = 0 \quad \forall s \in R_+. \tag{8.4.7}$$

Definition 8.4.1   Let $\tau \in R$ and $x \in \Pi$. Then an *L* functional $V: R \times X \to X$ relative to an abstract $\mathscr{D}^+$ process Π on X is called a *Ball L functional of the first type* (respectively *second type*) if conditions (1) and (2) of Assumption 8.4.1 (respectively (1) and (5)) are satisfied.

Definition 8.4.2   A functional $V_{\Pi^*}(t, p): \mathscr{H}[\Pi] \times R \times X \to R$ such that

$$V_{\Pi^*}(t, p) = \lim_{n \to \infty} V(t + S_n, p) \quad \forall (t, p) \in R \times X \tag{8.4.8}$$

is called a *limiting L functional generated by the L functional V and the sequence* $s_n \subset R_+$, $\Pi^{s_n} \xrightarrow{p} \bar{\Pi}^*, n \to \infty$.

The following assertions hold.

Assertion 8.4.1   Let

(1) $s \in R$ and $x \in \Pi(s)$;
(2) $\Omega(x, \Pi)$ be a positive limiting set for $\{x, \Pi\}$ relative to the $\mathscr{D}^+$ system $T(t)$ generated by the $\mathscr{D}^+$ process Π according to (8.1.5).

Then

$$\Omega(x) = \{x(0): x \in E\},$$

where E is defined by

$$E = \{y \in \Phi : \exists \bar{\Pi} \in L(\Pi) \text{ such that } \{y, \bar{\Pi}\} \in \Omega(x, \Pi)\}.$$

Proof   This is obvious.   ∎

Assertion 8.4.2   Let

(1) $s \in R$ and $x \in \Pi(s)$;
(2) the positive trajectory $\gamma^+(x)$ be precompact in X.

Then the positive trajectory $\gamma^+(x, \Pi)$ of the $\mathscr{D}^+$ system $T(t)$ generated by the $\mathscr{D}^+$ process Π according to (8.1.5) is precompact in $\mathscr{P}[\mathscr{H}[\Pi]]$.

Proof   This follows immediately from Definitions 8.4.1 and 8.4.2.

Theorem 8.4.1.   Let

(1) $\Pi \in P$ be an abstract $\mathscr{D}^+$ process on X;
(2) $\tau \in R$ and $x \in \Pi(\tau)$;

Then the set $\Omega(x)$ is positive unbiased.
If the trajectory $\gamma^+(x)$ is precompact, the set $\Omega(x)$ is nonempty. If, moreover, $\mathscr{H}[\Pi]$ is Hausdorff, the set $(\Omega(x)$ is unbiased.

Proof   Let $p \in \Omega(x)$. According to Assertion 8.4.1, there exists $\{\bar{\Pi}, y\} \in \Omega(\Pi, x)$ such that $x(\cdot) = p$. By Assertion 8.2.1, the set $\Omega(\Pi, x)$ is positive invariant, so that, by Assertion 8.4.1, $\gamma^+(x) \subseteq \Omega(x)$. Since $y \in \bar{\Pi}^s(0)$, $\Omega(x)$ is positive unbiased. The rest of the theorem is proved in terms of Assertions 8.2.1, 8.4.1 and 8.4.2.   ∎

Theorem 8.4.2   Let

(1) conditions (2) and (3) of Assumption 8.4.1 be satisfied;
(2) $x \in \Pi(s)$ and $s \in R$.

Then for every $p \in \Omega(x)$ there exist $\bar{\Pi} \in \mathscr{L}(\Pi)$ and a trajectory $\bar{x} \in \Pi(0)$ such that

$$\bar{x}(0) = p, \quad \bar{x}(R_+) \subseteq \Omega(x),$$

and the mapping $t \to V_{\Pi}(t, x(t))$ does not increase on R_+.

Proof Let $y = \{\Pi, x\}$. We shall apply Theorem 8.2.1 to the $\mathscr{D}^+$ system $T(t)$ generated by the process Π. Consider the space $X^0 = \gamma^+(y)$ and the functional $\Lambda: X^0 \to R$ defined by $\Lambda(\bar{\Pi}, \bar{x}) = V_{\Pi}(s, \bar{x}(0))$.

Let $t_n \to \infty$. Then

$$\{\Pi^{t_n}, x^{t_n}\} \xrightarrow{\mathscr{P}[\mathscr{H}]} \{\Pi, \bar{x}\}, \quad \bar{x} \in \Pi(s).$$

By condition (2) of Assumption 8.4.1, we have

$$\Lambda(\bar{\Pi}, \bar{x}) - \Lambda(\bar{\Pi}^{\tau}, \bar{x}^{\tau})$$

$$= V_{\bar{\Pi}}(s, \bar{x}(0)) - V_{\bar{\Pi}}(s + \tau, \bar{x}(\tau)) \leqslant \lim_{n \to \infty} \inf [V(t_n + s, x^{t_n}(0)) - V(t_n + s + \tau, x^{t_n}(\tau))]$$

$$= \lim_{n \to \infty} \inf [\Lambda(\Pi^{t_n}, x^{t_n}) - \Lambda(\Pi^{t_n + \tau}, x^{t_n + \tau})].$$

Therefore condition (3) of Assumption 8.6.1 is satisfied. Let $p \in \Omega(x)$. By Theorem 8.4.1, there exist $\bar{\Pi} \in \mathscr{L}[\Pi]$ and a trajectory $\bar{x} \in \bar{\Pi}(0)$ such that $\bar{x}(0) = p$ and $\bar{x}(R_+) \subseteq \Omega(x)$. According to Assertion 8.4.1 and Theorem 8.2.1, the positive invariance of $\Omega(\Pi, x)$ implies that $t \to V_{\bar{\Pi}}(t, \bar{x}(t))$ does not increase on R_+. ∎

8.4.2 *Localization of the limiting set*

We define the sets

$$c \in R, \bar{\Pi} \in \mathscr{S}(\Pi): V_{\bar{\Pi}}^{-1}(c) = \{p \in X : \exists \bar{x} \in \bar{\Pi}(0), V_{\bar{\Pi}}(t, \bar{x}(t)) = c \quad \forall t \in R_+\};$$

$$\Omega(V, \bar{\Pi}, x) = \{p \in X : \exists t_k, t_k \to \infty, x(t_k) \xrightarrow{X} p; V_{\bar{\Pi}}(0, x(t_k)) \to V_{\bar{\Pi}}(0, p)\}$$

$$\tau \in R.$$

Theorem 8.4.3 Let

(1) condition (1) of Assumption 8.4.1 be satisfies for the $\mathscr{D}^+$ process Π on X;
(2) $\tau \in R$, $x(t) \in \Pi(\tau)$ and the trajectory $\gamma^+(x)$ be precompact in X;
(3) the mapping $t \to V(t + s, x(t))$ be continuous on $]0, \infty[$.

Then if condition (4) of Assumption 8.4.1 is satisfied and $b > -\infty$,

(a) $+\infty > b \geqslant a > -\infty$;
(b) for every $c \in [a, b]$ there exists a limiting $\mathscr{D}^+$ process $\bar{\Pi} \in \mathscr{L}[\Pi]$ such that

$$\Omega(V, \bar{\Pi}, x) \cap V_{\bar{\Pi}}^{-1}(c) \neq \varnothing, \tag{8.4.9}$$

where

$$a = \lim_{t \to \infty} V^{\tau}(t, x(t)), \quad b = \lim_{t \to +\infty} V^{\tau}(t, x(t)).$$

Theorem 8.4.4 Let conditions (1)–(3) of the theorem be satisfied.

Then if condition (5) of Assumption 8.4.1 is satisfied,

(a) $\infty > b \geqslant a > -\infty$;

(b) for every $c \in [a, b]$ there exists $\bar{\Pi} \in \mathscr{L}[\Pi]$ such that

$$\Omega(V, \bar{\Pi}, x) \cap V_{\bar{\Pi}}^{-1}(c) \neq \varnothing$$

$$\Omega(x) = \bigcup_{\bar{\Pi} \in \mathscr{L}[\Pi]} \Omega(V, \bar{\Pi}, x) \subseteq V_{\bar{\Pi}}^{-1}(c). \tag{8.4.10}$$

Proof of Theorem 8.4.3 and 8.4.4   Let Λ and y be as in the proof of Theorem 8.4.2. We consider the $\mathscr{D}^+$ system $T(t): X \to X$ generated by the $\mathscr{D}^+$ process Π. It can easily be seen that the condition (8.2.1) is satisfied. In fact, let $t_n \to \infty$, $\{\Pi^{t_n}, x^{t_n}\} \xrightarrow{\mathscr{P}[\mathscr{X}]} \{\Pi, \bar{x}\}$. By condition (5) of Assumption 8.4.1,

$$V(t_n + \tau, x^{t_n}(0)) - V(t_n + \tau + t, x^{t_n}(t)) \to 0$$

uniformly in t on compact sets from R_+. By virtue of condition (1) of Assumption 8.4.1, $\Lambda(\Pi^{t_n}, x^{t_n}) \to \Lambda(\bar{\Pi}, x)$, and Theorems 8.4.3 and 8.4.4 follow from Assertions 8.4.1 and 8.4.2 and Theorems 8.2.2, 8.2.3 and 8.4.1. ∎

Theorem 8.4.5   Let

(1) the passage to the limit in (8.4.8) be uniform in $p \in A$, where A is precompact in X;

(2) the functional $V_{\bar{\Pi}}(t, p)$ be continuous in p for every $\bar{\Pi} \in \mathscr{L}[\bar{\Pi}]$.

Then conditions (1) and (2) of Assumption 8.4.1 are satisfied.

Proof   For

$$s_n \to \infty, \quad \Pi^{s_n} \to \bar{\Pi} \in \mathscr{L}[\Pi], \quad p_n \xrightarrow{X} p, \quad t \in R,$$

we have

$$\lim_{n \to \infty} V(s_n + t, p_n) = \lim_{n \to \infty} V_{\bar{\Pi}}(t, p_n) = V_{\bar{\Pi}}(t, p). \quad ∎$$

Theorem 8.4.6   Suppose that

(1) condition (a) of Assumption 8.4.1 is satisfied;

(2) as $t \to \infty$, $V(t + s, x(t))$ does not increase in $t \in R_+$ for any $(s, x) \in R \times \bar{\Pi}(s)$.

Then conditions (1)–(3) of Theorem 8.4.3 imply conditions (3) of Assumption 8.4.1.

Proof   Let conditions (1)–(3) of Theorem 8.4.3 be satisfied. Let $\tau \in R_+$ and $t_n \to +\infty$. Without loss of generality, it may be assumed that

$$\Pi^{t_n} \xrightarrow{p} \bar{\Pi} \in \mathscr{L}(\Pi), \quad x^{t_n} \xrightarrow{\phi} \bar{x} \in \bar{\Pi}(s).$$

Then, by condition (2) of Assumption 8.4.1,

$$\lim_{n \to \infty} [V(t_n + s, x(t_n)) - V(t_n + \tau + s, x(t_n + \tau))] \geqslant V_{\bar{\Pi}}(s, \bar{x}(0)) - V_{\bar{\Pi}}(s + \tau, \bar{x}(\tau)).$$

According to the condition, the right-hand side of the last inequality is nonnegative. ∎

Theorem 8.4.7   Suppose that

(1) every $\bar{\Pi} \in \mathscr{L}(\Pi)$ is a $\mathscr{D}^+$ process on X;

(2) for $s_n \to \infty$, $\Pi^{s_n} \xrightarrow{P} \bar{\Pi} \in \mathscr{L}(\Pi)$ for all $t \in R_+$, $t \in R$, $x_n \in \Pi^{s_n}(\tau)$ and $x_n(0)$,

$$V^{\tau}(s_n + t, x_n(t)) - V^{\tau}(t, S(\bar{\Pi}, t, \tau), x_n(0)) \to 0. \tag{8.4.11}$$

Then conditions (1) and (2) of Assumption 8.41 imply conditions (1') and (2') respectively.

Proof   Let condition (1) of Assumption 8.4.1 be satisfied. For $t_n \to \infty$, we have

$$\Pi^{t_n} \xrightarrow{P} \bar{\Pi} \in \mathscr{L}(\Pi), \quad s \in R, \quad x_n \in \Pi^{t_n}(s), \quad x_n \xrightarrow{\phi} \bar{x} \in \bar{\Pi}(s),$$

and

$$V(t_n + s, x_n(0)) - V(t_n + s + t, x_n(t)) \to 0 \quad \forall t \in R_+.$$

Then

$$\lim_{n \to \infty} V(t_n + s, x_n(0)) = \lim_{n \to \infty} V_{\bar{\Pi}}(s, x_n(0) = V_{\bar{\Pi}}(s, \bar{x}(0)).$$

Thus condition (1') is satisfied.

Let condition (2) of Assumption 8.4.1 be satisfied. Assume that $t_n \to \infty$. Then

$$\Pi^{t_n} \xrightarrow{P} \bar{\Pi} \in \mathscr{L}(\Pi), \quad s \in R, \quad x_n \in \Pi^{t_n}(s), \quad x_n \xrightarrow{\Phi} x \in \bar{\Pi}(s), \quad t \in R_+$$

Applying (8.4.11) and the condition, we obtain

$$V_{\bar{\Pi}}(s, \bar{x}(0)) - V_{\bar{\Pi}}(s + t, \bar{x}(t)) = V_{\bar{\Pi}}(s, \bar{x}(0)) - V_{\bar{\Pi}}(s + t, s(\bar{\Pi}, t, s), \bar{x}(0))$$

$$\leqslant \lim_{n \to \infty} [V_{\bar{\Pi}}(s, x_n(0)) - V_{\bar{\Pi}}(s + t, s(\bar{\Pi}, t, s), x_n(0))]$$

$$= \lim_{n \to +\infty} [V(t_n + s, x_n(0)) - V(t_n + s + t, x_n(t))].$$

Thus condition (2') is satisfied.   ∎

Conditions (1') and (2') of Assumption 8.4.1 are advantageous, since they are expressed exclusively in terms of the limiting $\mathscr{D}^+$ process $\bar{\Pi} \in \mathscr{L}[\Pi]$ and can be verified easily.

8.5   Application to ODEs of Carathéodory Type

Consider the ordinary differential equation

$$\frac{dx}{dt} = f(t, x), \tag{8.5.1}$$

where the function $f : R \times R^n \to R^n$ is continuous in x for every fixed t and measurable in t for every fixed x. Moreover, we assume that for every compact $Q \subset R^n$ there exists a locally integrable function in $m(Q, t)$ such that

$$\|f(t, x)\| \leqslant m(Q, t) \quad \forall x \in Q. \tag{8.5.2}$$

A continuous function $x: [a, b] \to R^n$ satisfying the integral equation

$$x(t) = x(s) + \int_s^t f(\tau, x(\tau)) d\tau. \tag{8.5.3}$$

for all $s, t \in [a, b]$ is called a *solution of the equation (8.5.1) on the segment* $[a, b]$. We shall study the asymptotic properties of solutions of the nonautonomous equation (8.5.1) under the condition that $f(t, \cdot)$ tends to a continuous function $\bar{f}: R^n \to R^n$ as $t \to + \infty$.

Consider the following autonomous differential equation in R^n:

$$\frac{dx}{dt} = \bar{f}(x). \tag{8.5.4}$$

The solutions of this equation are defined similarly to those of the nonautonomous equation (8.5.1).

Assumption 8.5.1   The functions $\bar{f}$ and f satisfy the following conditions.

(1) For every compact $Q \subseteq R^n$ there exists a nondecreasing function $\mu(Q, t): R_+ \to R_+$ continuous for $t = 0$, with $\mu(Q, 0) = 0$, such that the integral $\int_a^b f(s, u(s)) ds$ is defined for all $a, b \in R$ and any continuous function $u: [a, b] \to Q$ and, moreover, the estimate

$$\left\| \int_a^b f(s, u(s)) ds \right\| \leqslant \mu(Q, b - a) \tag{8.5.5}$$

is satisfied.

(2) For any $a, b \in R$ from any sequence $u_k \to u_0$ in $C([a, b])$ and any sequence $t_k \to \infty$ the relation

$$\int_a^b f(s + t_k, u_k(s)) ds \to \int_a^b \bar{f}(u_0(s)) ds \tag{8.5.6}$$

is satisfied.

Let $X = R^n$, $r > 0$. We define for $s \in R$ and $u: [s, \infty[\to X$,

$$u^s(t) = u(s + t) \quad \forall t \in R_+,$$

$\Pi(s) = \{x^s \in X^{R_+} : x: [s, \infty[\to X, x \text{ is a solution of (8.5.1) such that } \|x(\sigma)\| \leqslant r \forall \sigma \in [s, \infty)\},$

$\bar{\Pi}(s) = \{x \in X^{R_+} : x \text{ is a solution of (8.5.4) such that } \|x(\sigma)\| \leqslant r \forall \sigma \in R_+\}.$

Definition 8.5.1   Let Π be a nonautonomous $\mathscr{D}^+$ process on X, and $\bar{\Pi}$ an autonomous $\mathscr{D}^+$ process on X. The process Π is called an *asymptotically autonomous $\mathscr{D}^+$ process on X* if $H = \{\bar{\Pi}\} \cup \Pi^\sigma$ is a hull of the process Π and $\mathscr{L}(\Pi) = \{\bar{\Pi}\}$.

It is evident that $\Pi^{s_n} \xrightarrow{p} \Pi$ for a subsequence $s_n \to \infty$. If $V_{\bar{\Pi}}$ does not depend on t then $V_{\bar{\Pi}}: X \to R$.

Assertion 8.5.1   Let Π and $\bar{\Pi}$ be the processes mentioned in Definition 8.5.1. Then Π is an asymptotically autonomous process.

Proof   This is standard, and is left for the reader.   ■

Theorem 8.5.1   Let

(1) continuously differentiable function $V: R^n \to R$ be such that

$$\left(\frac{\partial \bar{V}}{\partial x}\right)^T \bar{f}(x) \leqslant 0 \quad \forall x \in R^n; \tag{8.5.7}$$

(2) $s \in R$ and $x: [s, \infty[\to R^n$ be a bounded solution of the nonautonomous equation (8.5.1);
(3) $a = \liminf_{t \to \infty} \bar{V}(x(t))$ and $b = \limsup_{t \to \infty} \bar{V}(x(t))$;
(4) for $c \in R$,

$\bar{V}^{-1}(c) = \{y \in R^n : \exists$ a solution $x(t)$ of (8.5.4) such that $\bar{x}(0) = y, \bar{V}(\bar{x}(t)) = c \forall t \in R_+\}$.

Then

$$\Omega(x) \cap \bar{V}^{-1}(c) \neq \varnothing \quad \forall c \in [a, b]. \tag{8.5.8}$$

Proof   Let $r > 0$ be such that $x \in \Pi(s)$. We apply Theorem 8.4.3 in view of the fact that $V = V_{\bar{\Pi}} = \bar{V}$ and $x = p$. By Theorem 8.4.5, condition (1) and (2) of Assumption 8.4.1 are satisfied, and therefore, by Theorem 8.4.6, condition (4) of Assumption 8.4.1 is satisfied as well. Thus the theorem is proved.   ■

Theorem 8.5.2   Suppose that

(1) the function $\bar{V}$ is as in Theorem 8.5.1;
(2) $V: R \times R^n \to R$ satisfies the condition $\bar{V}(s_n + t, x) \to \bar{V}(x)$ for the sequence $s_n \to \infty$ and for $t \in R_+, n \to +\infty$, and, moreover, the passage to the limit is uniform in x for any bounded set from R^n;
(3) for every $\tau \in R_+$ the inequality

$$\liminf_{t \to \infty} [V(t, x(t)) - V(t + \tau, x(t + \tau))] \leqslant 0$$

holds.

Then the condition (8.5.8) is satisfied and

$$\Omega(x) \subseteq \bigcup_{c \in [a,b]} V^{-1}(c), \tag{8.5.9}$$

where

$$a = \liminf_{t \to \infty} V(t, x(t), \quad b = \limsup_{t \to \infty} V[t, x(t)).$$

Proof   This theorem is a corollary of Theorem 8.4.4.

Theorem 8.5.3   Suppose that

(1) the function $\bar{V}$ is as in Theorem 8.5.1;

(2) for every $c \in R$ the set $\bar{V}^{-1}(c)$ is either empty or contains a finite number of equilibrium states;

(3) every equilibrium state of equation (8.5.4) is isolated.

Then every bounded solution $x: [s, \infty[\to R^n$ of equation (8.5.1) tends to the equilibrium state of equation (8.5.4) as $t \to +\infty$.

Proof Let the conditions of the theorem be satisfied. Then only a finite number of the sets $\bar{V}^{-1}(c) \cap \{y \in R^n : \|y\| \le r,\ c \in [a, b]\}$ are nonempty. By Theorem 8.4.6, $\Omega(x)$ contains only a finite number of equilibrium states. Since $\Omega(x)$ is connected and every equilibrium state is isolated, the conclusion of the theorem is valid. ∎

The number of equilibrium states of the system described by equation (8.5.3) may be finite for every solution of equation (8.5.4) tending to some equilibrium state. However, a bounded solution of equation (8.5.1) may have nontrivial trajectories in its limiting set.

Example 8.5.1   Consider the system

$$\frac{dr}{dt} = -r(r-1)^2,$$

$$\frac{d\theta}{dt} = \cos^2 \theta + f(t), \tag{8.5.10}$$

where $f: R \to R$ is continuous, $f(t) \to 0$ as $t \to \infty$, and

$$\int_0^\infty f(s)\,ds = \infty.$$

(Here (r, θ) are polar coordinates.) Going to Cartesian coordinates $(x_1, x_2) = (r\cos\theta, r\sin\theta)$, it is easy to see that the system (8.5.10) satisfies conditions (1) and (2) of Assumption 8.5.1, and hence it generates an asymptotically autonomous $\mathcal{D}^+$ process. The limiting system for (8.5.10) has the form

$$\frac{dr}{dt} = -r(r-1)^2,$$

$$\frac{d\theta}{dt} = \cos^2 \theta. \tag{8.5.11}$$

Every solution of this limiting system tends to one of three equilibrium states: $(0,0)$, $(0, 1)$ or $(0, -1)$. However, (8.5.10) yields the estimate

$$\theta(t) \ge \theta(0) + \int_0^t f(s)\,ds,$$

so that every solution with initial values on the unit circle has this circle as its limiting set. The limiting function (8.5.11) does not have a Lyapunov function $\bar{V}$ such that the corresponding sets $\bar{V}^{-1}(c)$ contain only a finite number of equilibrium states. Therefore Theorem 8.5.3 is not applicable in this case. Every solution of equation (8.5.4) may tend to some equilibrium state, and a Lyapunov function $\bar{V}$ may exist that satisfies the conditions of Theorem 8.5.1, but a bounded solution of equation (8.5.1) may have nontrivial trajectories in its positive limiting set.

Example 8.5.2   Consider the system

$$\frac{dr}{dt} = -r(r-1)^2,$$

$$\frac{d\theta}{dt} = \begin{cases} f(t) & \text{for } \theta \in [-\tfrac{1}{2}\pi, \tfrac{1}{2}\pi], \\ \cos^2 \theta + f(t) & \text{for } \theta \in [\tfrac{1}{2}\pi, \tfrac{3}{2}\pi], \end{cases} \tag{8.5.12}$$

where the function f satisfies the same conditions as in Example 8.5.1. In this case the limiting system has the origin and the right half of the unit circle as its equilibrium states. Every solution tends to one of these equilibrium states. Let $h(\theta)$ be a smooth periodic function with period 2π such that $0 \leqslant h(\theta) < 1$ for all $\theta, h' > 0$ and all $\theta \in [\tfrac{1}{2}\pi, \tfrac{3}{2}\pi]$. We define

$$\bar{V}(r, \theta) = r^2[1 - h(\theta)]. \tag{8.5.13}$$

We can easily verify that $\bar{V}$ is a continuously differentiable Lyapunov function nonincreasing along the solutions of the autonomous system that are strictly decreasing (except for the equilibrium state). Similarly to Example 8.5.1, any solution of the nonautonomous system (8.5.12) with initial values on the unit circle has this circle as its limiting set. In this case the conditions of Theorem 8.5.1 are satisfied, and therefore

$$\Omega(x) \cap \bar{V}^{-1}(c) = \varnothing \quad \forall \subseteq \epsilon [a, b].$$

8.6   Comments and References

8.0   Limiting dynamical processes and stability in abstract dynamical processes have been considered in a number of papers (see e.g. Barbashin [1]; Budak [1]. Further developments of dynamical system theory together with original results can be found in the work of Bronstein [1, 2] Izman [1, 2], Nemytsky [1], Shestakov [3–5], Bhatia and Szego [1], Reed [1] and Roxin [1, 2]).

Ball [4] has developed the theory of limiting dynamical processes for abstract dynamical processes on Fréchet convergence spaces.

8.1   The definition of an abstract dynamical process on convergence space and reduction of a dynamical process to a dynamical system were given by Ball [4] and Shestakov [6].

8.2–8.3   The generalized Lyapunov direct method for abstract autonomous and asymptotically autonomous semidynamical processes has been developed by Ball [4].

8.4   The method of localization of the limiting set of a motion of a dynamical process has been developed by Ball [4] and Dafermos [2, 5].

8.5   The basic results of this section are due to Ball [4].

9 Limiting Lyapunov Functionals for Asymptotically Autonomous Evolutionary Equations of Parabolic and Hyperbolic Type in a Banach Space

9.0 Introduction

In the present chapter we utilize the results of Chapter 8 to study stability-like properties of solutions to asymptotically autonomous dynamical processess generated by nonautonomous evolutionary equations of the type

$$\frac{du}{dt} = Au + f(t, u)$$

in a Banach space E, where A is a generator of a strongly linear dynamical system $T(t)$ on E, and $f(t, u)$ is a nonlinear operator with a definite type of asymptotic autonomy for every bounded subset $Q \subset E$.

9.1 The Generalized Lyapunov Direct Method for Asymptotically Autonomous Evolutionary Problems of Parabolic Type

9.1.1 Evolutionary problems of parabolic type

Consider the evolutionary nonautonomous problem

$$\frac{du}{dt} = Au + f(t, u), \quad u(t_0) = u_0, \quad t \geqslant t_0 \tag{9.1.1}$$

in a space E, where A is a generator of a strictly continuous semigroup $T(t)$, $t > 0$, of bounded linear operators on E. If $f(t, u)$ does not depend on t, the evolutionary problem (9.1.1) can be expressed as

$$\frac{du}{dt} = A\bar{u} + g(\bar{u}), \quad \bar{u}(t_0) = u_0, \quad t \geq t_0. \tag{9.1.2}$$

Assumption 9.1.1   In this section A, f and g are restricted by the following conditions.

(1) The operator $T(t)$ is compact for every $t > 0$.
(2) The function $f(\cdot, u)$ is strictly measurable for every $u \in E$, the function $f(t, \cdot)$ is continuous for almost all $t \in R$, and for every bounded set $Q \subset E$ there exists a locally integrable function m_Q on R such that

$$\| f(t, u) \| \leq m_Q(t) \quad \text{for all } u \in Q \text{ and almost all } t \in R.$$

(3) For every $t_0 \in R$

$$\lim_{\delta \to 0} \sup_{t \geq t_0} \int_t^{t+S} m_Q(s)\, ds = 0.$$

(4) For every bounded subset $Q \subset E$

$$\lim_{t \to +\infty} \int_t^{t+1} \sup \| f(t, u) - g(u) \|\, ds = 0.$$

An evolutionary problem (9.1.1) for which conditions (1) and (2) are satisfied is referred to as an *evolutionary problem of parabolic type*.

We discuss the existence and precompactness of solutions to the problem (9.1.1).

Definition 9.1.1   A mapping $u: [t_0, t_1] \to E$ is called a *strict solution* of the evolutionary problem (9.1.1) if $u(t)$ is strictly differentiable on $]t_0, t_1[$ and satisfies equation (9.1.1) for $t_0 < t < t_1$ and $u(t) \to u(t_0)$ as $t \to t_0$.

If $u(t)$ is a strict solution of the problem (9.1.1), it is also a solution of the integral equation

$$u(t) = T(t - t_0)u_0 + \int_{t_0}^t T(t - s)f(s, u(s))\, ds, \ t > t_0. \tag{9.1.3}$$

It is clear that the solution of equation (9.1.3) is not necessarily a strict solution of the problem (9.1.1).

Definition 9.1.2   Functions satisfying equation (9.1.3) are referred to as *weak solutions* of the evolutionary problem (9.1.1).

Assertion 9.1.1   The mapping $u \in C([t_0, t_1], E)$ is a weak solution of the evolutionary problem (9.1.1) on $[t_0, t_1[$ iff

(1) the function $f(\cdot, u(\cdot))$ is locally integrable, i.e.

$$f(\cdot, u(\cdot)) \in \mathscr{L}_1([t_0, t_1), E);$$

(2) for every $v \in \mathcal{D}(A^*)$ the function $\langle u(t), v \rangle$ is absolutely continuous on $[t_0, t_1[$;

(3) the inequality

$$\frac{d}{dt}\langle u(t), v \rangle = \langle u(t), A^* v \rangle + \langle f(t, u(t)), v \rangle \qquad (9.1.4)$$

holds for almost all $t \in [t_0, t_1[$, where A^* is the operator conjugate with A.

Assertion 9.1.2 (existence of solutions) Let

(1) condition (1) and (2) of Assumption 9.1.1. be satisfied;

(2) $t_0 \in R$ and $u_0 \in E$.

Then there exists a weak solution of the evolutionary problem (9.1.1) on the maximal interval $[t_0, t_M[$, where $t_M > t_0$.

For every such solution the relation

$$\int_{t_0}^{t} \| f(s, u(s)) \| \, ds = \infty \qquad (9.1.5)$$

is valid when $t_M < +\infty$.

Assertion 9.1.3 (precompactness of solutions) For the evolutionary problem (9.1.1) let conditions (1)–(3) of Assumption 9.1.1 be satisfied.

Then every bounded weak solution $u: [t_0, \infty[\to E$ of the problem (9.1.1) is precompact on R_+.

Assertion 9.1.4 Suppose that

(1) condition (1)–(4) of Assumption 9.1.1 are satisfied;

(2) the function f is continuous;

(3) for $r > 0$, $s \in R$,

$$\Pi(s) = \{u^s \in \Phi : u : [s, \infty[\to E$$

is a weak solution of (9.1.1) such that $\| u(\sigma) \| \leqslant r \forall \sigma \in [s, \infty[\}$;

(4) $\bar{\Pi} = \{u \in \Phi : u$ is a weak solution of (9.1.2) such that $\| u(\sigma) \| \leqslant r \forall \sigma \in R_+ \}$;

(5) the space E is provided with the structure of a convergence space induced by the norm topology of E.

Then Π is an asymptotically autonomous $\mathcal{D}^+$ process on E, and $\bar{\Pi}$ is an autonomous $\mathcal{D}^+$ process on E of (9.1.1).

9.1.2 Localization of the limiting set

Consider the stability-like properties of the solutions of the evolutionary problem (9.1.1) under conditions (1) and (2) Assumption 9.1.1. Together with (9.1.1), we examine the autonomous evolutionary problem (9.1.2), where the mapping $g: E \to E$ satisfy condition (4) of Assumption 9.1.1. If this condition is satisfied then, according to

Assertion 9.1.4, equation (9.1.1) defines an asymptotically autonomous $\mathscr{D}^+$ process on E. In this case equation (9.1.1) is called *asymptotically autonomous*, and equation (9.1.2) is limiting.

Definition 9.1.3 A functional $V: E \rightarrow R$ continuously differentiable in the Fréchet sense is referred to as a *limiting L-functional* on the set Dom A for the problem (9.1.1) if

$$\langle Au + g(u), V'(y)\rangle \leqslant 0 \quad \forall u \in \text{Dom } A.$$

Theorem 9.1.1 A limiting L functional $V: E \rightarrow R$ on the set Dom A does not increase on R_+ along weak solutions $\bar{u}: R_+ \rightarrow E$ of the evolutionary problem (9.1.2).

Proof We shall demonstrate that $V(u(t)) \leqslant V(u(0))$ for all $t \in R_+$. Let $t_0 > 0$, $F(t) = f(t, u(t))$ and

$$F_n \rightarrow F \quad \text{in } C([0, t_0], E), \quad F_n \in C^1([0, t_0], E) \quad \forall n.$$

Let $v_{n0} \xrightarrow{E} u(t)$ and $v_{n0} \in \text{Dom } A$ for all n. Define

$$v_n(t) = T(t)v_{n0} + \int_0^t T(t - s)F_n(s)\, ds, \quad v_n \in C([0, t_0], E).$$

Then

$$v_n(t) \in \text{Dom } A, \quad v_n \in C^1([0, t_0], E),$$

$$\frac{du_n}{dt} = Au_n(t) + F_n(t) \quad \forall t \in [0, t_0].$$

Therefore

$$V(v_n(t)) - V(v_{n0}) = \int_0^t \langle Au_n(s) + F_n(s), V'(v_n(s))\rangle\, ds$$

$$\leqslant \int_0^t \langle F_n(s) - g(v_n(s)), V'(v_n(s))\rangle\, ds. \qquad (9.1.6)$$

For the difference $z_n(t) = v_n(t) - v(t)$ we have that there exists $d > 0$ such that

$$\| z_n(t)\| \leqslant de^{\alpha t} \| z_n(0)\| + \int_0^t de^{\alpha(t-s)} \| F_n(s) - F(s)\|\, ds.$$

Hence $v_n \rightarrow v$ in $C([0, t_0], E)$. In particular, the set $\{v_n(s): s \in [0, t_0] \forall n\}$ is precompact in E_0. Thus there exists $d_1 > 0$ such that

$$\| V'(U_n(s))\| \leqslant d_1 \quad \forall s \in [0, t_0] \quad \forall n,$$

Taking the limit in (9.1.6), we get

$$V(u(t)) \leqslant V(u(0)) \quad \forall t \in [0, t_0].$$

If $w \in x$ is an equilibrium state of the problem (9.1.2) then $w \in \mathscr{D}(A)$ and $Aw + g(w) = 0$.

■

We shall now present Theorems 9.1.2–9.1.4 similar to Theorems 8.5.1–8.5.3 for ordinary differential equations in R^n. The proofs of these results are similar to those of Theorems 8.5.1–8.5.3 and therefore are omitted.

Theorem 9.1.2 Let

(1) $s \in R$, $u: [s, \infty[\to E$ be a bounded weak solution of the problem (9.1.1);
(2) $V: E \to R$ be a limiting L functional for (9.1.1);
(3) $a = \lim_{t \to \infty} \inf V(u(t))$, $b = \lim_{t \to \infty} \sup V(u(t))$, $V^{-1}(c) = \{y \in E: \exists$ a weak solution $\bar{u}: R_+ \to E$ of (9.1.2) such that $\bar{u}(0) = 0$, $V(\bar{u}(t)) = c \forall t \in \varepsilon R_+\}$.

Then

$$\Omega(u) \cap V^{-1}(c) \neq \varnothing \quad \forall c \in [a, b].$$

The following result is a corollary of Theorem 9.1.2.

Theorem 9.1.3 Suppose that

(1) the conditions of Theorem 9.1.1 are satisfied;
(2) for every $c \in R$ the set $V^{-1}(c)$ is either empty or contains only the equilibrium states of equation (9.1.2);
(3) every equilibrium state of (9.1.2) is isolated in E.

Then every bounded weak solution $u: [s, \infty[\to E$ of the evolutionary problem (9.1.1) converges to the equilibrium state of the problem (9.1.2) as $t \to +\infty$.

Theorem 9.1.4 Suppose that

(1) $G: E \to R$ is a limiting L functional for the evolutionary problem (9.1.1), continuously differentiable is the Fréchet sense and satisfying the condition

$$[\nabla G(u)]^T [Au + g(u)) \leqslant 0 \quad \forall u \in E;$$

(2) $V: R \times E \to R$ is L functional such that

$$\lim_{n \to \infty} V(s_n + t, u) = G(u) \quad \forall t \in R_+$$

for every sequence s_n, and the passage to the limit is uniform for u belonging to a compact subset of E;

(3) for $s \in R$, $x: [s, \infty[\to E$ is a precompact weak solution of the problem (9.1.1);
(4) $a = \lim_{t \to +\infty} \inf V(t, u(t))$, $b = \lim_{t \to +\infty} \sup V(t, u(t))$;
(5) $\lim_{t \to +\infty} \inf [V(t, u(t)) - V(t + \tau, u(t + \tau))] \leqslant 0 \quad \forall \tau \in R_+$.

Then

$$\Omega(u) \cap V^{-1}(c) \neq \varnothing \quad \forall c \in [a, b],$$
$$\Omega(u) \subseteq \bigcup_{c \in [a, b]} V^{-1}(c),$$

where

$$V^{-1}(c) = \{y \in E: \exists \text{ a weak solution } \bar{u}(t): R^n \to E \text{ of (9.1.2) such}$$
$$\text{that } \bar{u}(0) = u_0, \quad G(\bar{u}(t)) = c \forall t \in R_+\}.$$

9.1.3   The limiting L functional for a boundary-value problem of heat conduction

We construct a limiting L functional for the boundary-value problem of parabolic type

$$u_t = \Delta u + g(t, u) \quad \forall x \in \mathscr{D} \subseteq R^n \quad \forall t > s, \qquad (9.1.7)$$

$$u|_{\partial\mathscr{D}} = 0, \quad u(s, x) = u_0(x), \qquad (9.1.8)$$

using an autonomous boundary-value problem of the same type,

$$u_t = \Delta u + q(u) \quad \forall x \in \mathscr{D} \quad \forall t > 0, \qquad (9.1.9)$$

$$u|_{\partial\mathscr{D}} = 0, \quad u(0, x) = \bar{u}_0(x). \qquad (9.1.10)$$

Assumption 9.1.2   The following conditions hold.

(1) The mappings $g(\cdot, u)$ and $g_u(\cdot, u)$ are measurable for all $u \in R$ and almost all $t \in R$, and the mapping $g(t, \cdot)$ is continuously differentiable.
(2) $g(t, 0) = 0$ for almost all $t \in R$ and $q(0) = 0$.
(3) There exists a nonnegative locally integrable function $m(t)$ such that

$$\lim_{s \to 0, \, t \geq t_0} \sup \int_t^{t+s} m(t)\, dt = 0 \quad \forall t \in R,$$

$$\frac{|g(t, u)|}{1 + |u|} + |g_u(t, u)| \leq m(t) \quad \forall u \in R. \qquad (9.1.11)$$

(4) The mapping q is continuously differentiable and there exists a nonnegative locally integrable function $m_1(t)$ such that

$$\lim_{t \to \infty} \int_t^{t+1} m_1(t)\, dt = 0,$$

$$\frac{|g(t, u) - q(u)|}{1 + |u|} + |g_u(t, u) - \bar{q}(u)| \leq m_1(t) \quad \forall u \in R \qquad (9.1.12)$$

for almost all $t \in R$.

If $n = 1$, the inequalities (9.1.11) and (9.1.12) can be replaced by the inequalities

$$|g(t, u)| + |g_u(t, u)| \leq m(t)\alpha(u), \qquad (9.1.13)$$

$$|g(t, u) - \bar{q}(u)| + |g_u(t, u) - \bar{q}_u(u)| \leq m_1(t)\alpha(u), \qquad (9.1.14)$$

respectively, where $\alpha(u)$ is a continuous function of u.

Let X be the Sobolev space $\overset{\circ}{W}{}_2^1(\mathscr{D})$. We define

$$\text{Dom}\, A = \{u \in X : \Delta u \in X\}, \quad A = \Delta.$$

Then A is a generator of the $\mathscr{D}^+$ system $T(t): X \to X$ such that the operator $T(t)$ is compact for $t > 0$. Define the functions f and $\bar{f}$ by $f(t, u)(x) = g(t, u(x))$,

$$\bar{f}(u)(x) = q(u(x)).$$

By conditions (1)–(3) of Assumption 9.1.2, $f\colon R \times X \to X$ and $\bar{f}\colon X \to X$. It can easily be shown that these conditions also imply conditions (1)–(5) of Assumption 9.1.1. Now we write (9.1.7) in the form

$$\frac{du}{dt} = Au + f(t, u) \tag{9.1.15}$$

and follow the previous procedure.

Obviously,

$$(\Delta v, \Phi) = (v, \Delta \Phi) \quad \forall v \in \text{Dom } A, \quad \Phi \in C_0^\infty(\mathscr{D}),$$

where $(\cdot, \cdot)$ is the scalar product in $\mathscr{L}^2(\mathscr{D})$. Therefore, a weak solution u of the evolutionary equation (9.1.15) satisfies (9.1.7) in the distributional sense.

We define an L functional $V\colon \mathring{W}_2^1(\mathscr{D}) \to R$ for the problem (9.1.9), (9.1.10) in the form

$$V(u) = \int_{\mathscr{D}} \left[\tfrac{1}{2}|\nabla u(x)|^2 - \beta(u(x))\right] dx, \tag{9.1.16}$$

$$\beta(u) = \int_0^u q(u)\, du.$$

Let $\Gamma = \{u \in \mathring{W}_2^1(\mathscr{D}), \Delta u \in \mathring{W}_2^1(\mathscr{D})\}$. We can easily show that

$$\frac{dV}{dt} = \langle \Delta u + q(u), \quad V'(u)\rangle = \int_{\mathscr{D}} |\Delta u(x) + q(u)(x)|^2\, dx \leqslant 0,$$

$$\bar{f}(u)(x) = q(u)(x) \quad \forall u \in \Gamma.$$

Using the L functional (9.1.16) and Theorem 9.1.2 and 9.1.3 on localization, we come to the following conclusion.

Theorem 9.1.5   Let $u\colon [s, \infty[\to \mathring{W}_2^1(\mathscr{D})$ be a weak solution of the boundary-value problem (9.1.7) (9.1.8), bounded in the norm, and let

$$a = \lim_{t\to\infty} \inf V(u(t)), \quad b = \lim_{t\to\infty} \sup V(u(t)).$$

Then for every $c \in [a, b]$ there exists an equilibrium state v of the boundary-value problem (9.1.9), (9.1.10) belonging to the limiting set $\Omega(u)$ and such that $V(v) = c$.

If the equilibrium states of the problem (9.1.9), (9.1.10) are isolated in $\mathring{W}_2^1(\mathscr{D})$ then the solution $u(t)$ converges to the unique equilibrium state as $t \to \infty$.

9.2   The Generalized Lyapunov Direct Method for Asymptotically Autonomous Evolutionary Equations of Hyperbolic Type

9.2.1   *Evolution problems of hyperbolic type*

Consider the evolutionary problem

$$\frac{du}{dt} = Au + f(t, u), \quad u(t_0) = u_0, \quad t \geqslant t_0, \tag{9.2.1}$$

in the space E and the autonomous evolutionary problem

$$\frac{du}{dt} = Au + g(u), \qquad u(t_0) = u_0, \quad t \geqslant t_0, \tag{9.2.2}$$

where A satisfies the conditions formulated in Section 9.1.

Assumption 9.2.1 In this section A, f, g and E are restricted by the following conditions.

(1) The space E is reflexive.
(2) The function $f(\cdot, u)$ is strictly measurable for every $u \in E$, the function $f(t, \cdot)$ is sequentially weakly continuous for almost all $t \in R$, and for every bounded set Q in E there exists a locally integrable function m_Q on Q such that the inequality from condition (2) of Assumption 9.1.1 is satisfied.
(3) The mapping $g: E \to E$ is sequentially weakly continuous, and for every $Q \subset E$ the relation from condition (4) of Assumption 9.1.1 is satisfied.
(4) Condition (3) of Assumption 9.1.1 is satisfied.

An evolutionary problem (9.2.1) for which these conditions are satisfied is called a problem of *hyperbolic type*.

Consider the existence and compactness of solutions of the problem (9.1.1). The following results hold.

Assertion 9.2.1 Let conditions (1) and (2) of Assumption 9.2.1 be satisfied.
Then there exists a weak solution of the problem (9.1.1) on the maximal interval $[t_0, t_M[, t_M > t_0$. For every such solution, the relation (9.1.5) holds for $t_M < \infty$.

Assertion 9.2.2 Let
(1) E_w be the Banach space E with the weak topology, and $Y = C([t_0, t_1], E_w)$ be the space of continuous functions from $[t_0, t_1]$ to E_w;
(2) the set $\mathscr{A} \subset Y_k$ such that for all $c \in R$,

$$\| u(t) \| \leqslant 0 \quad \forall u \in \mathscr{A} \quad \forall t \in [t_0, t_1];$$

(3) for every $x^* \in E^*$ the set of mappings $\{ \langle u(\cdot), x^* \rangle : u \in \mathscr{A} \}$ is equicontinuous in $C([t_0, t_1])$.

Then the set $\mathscr{A}$ is sequentially precompact.

Assertion 9.2.3 Suppose that

(1) conditions (1)–(4) of Assumption 9.2.1 are
(2) for $r > 0$ and $s \in R$ $\Pi(s) = \{ u^s \in E^{R+} : u : [s, \infty[+ E$ is a weak solution of (9.1.1) such that $\| u(\sigma) \| \leqslant r \forall \sigma \in R_+ \}$, $\Pi = \{ u \in E^{R+} : u$ is a weak solution of (9.1.2) such that $\| u(\sigma) \| \leqslant r \forall \sigma \in R_+ \}$;
(3) the space E has a convergence space structure induced by the weak convergence in E.

Then Π is an asymptotically autonomous $\mathscr{D}^+$ process on E.

9.2.2 Localization of the limiting set

Assumption 9.2.2 We shall need the following conditions. There exist a continuously differentiable L functional $V: E \to R$ and a continuous nonnegative functional $W: E \to R_+$ such that

(1) $\langle Au + g(u), V'(u) \rangle = -W(u) \quad \forall u \in \mathrm{Dom}\, A;$ (9.2.3)

(2) the functional W is sequentially weakly semicontinuous from below;
(3) the mapping $V': E \to E^*$ takes bounded sets into bounded sets.

Theorem 9.2.1 Let

(1) conditions (1) and (2) of Assumption 9.2.2 be satisfied;
(2) $u: [t_0, \infty[\to E$ be a bounded weak solution of the evolutionary problem (9.2.1) such that

$$\lim_{t \to \infty} \int_t^{t+1} W(u(s))\, ds = 0.$$

Then if $p \in \Omega(u)$, there exists a weak solution $\bar{u}$ of the evolutionary problem (9.2.2) such that

$$\bar{u}(0) = p, \quad W(\bar{u}(t)) = 0 \quad \forall t \in R_+,$$ (9.2.4)

where

$$\Omega(u) = \{p \in E : \exists t_n, \text{ such that as } t_n \to \infty, u(t_n) \xrightarrow{\text{weakly}} p\}.$$

Proof Let the conditions of the theorem be satisfied, $\|u(\tau)\| \leqslant r$ for all $\tau \in [t_0, \infty[$ and $V(t, \cdot) = V_{\bar{n}}(\cdot) = V(\cdot)$ for all $t \in R$. Let us show that condition (2) of Assumption 9.1.1 is satisfied. Let $u_n \in \Pi^{t_n}(s)$, $u_n \xrightarrow{\Phi} \bar{u}$, $t_n \to \infty$, for all $s \in R$, and $t \in R_+$. Applying the same technique as in the proof of Theorem 9.1.1, we easily obtain

$$V(u_n(t)) - V(u_n(0)) = -\int_0^t W(u_n(t))\, dt$$

$$+ \int_0^t \langle f(s + t_n + t, u_n(t))$$

$$- g(u_n(t), V'(u_n(t)) \rangle\, dt.$$

By virtue of condition (4) of Assumption 9.2.1 and condition (3) of Assumption 9.2.2, the second integral vanishes as $n \to \infty$. Therefore, by the factor lemma and condition (2) of Assumption 9.2.2,

$$\liminf_{n \to \infty} [V(u_n(0)) - V(u_n(t))] = \liminf_{n \to \infty} \int_0^t W(u_n(t))\, dt$$

$$\geqslant \int_0^t W(\bar{u}(t))\, dt = V(\bar{u}(0)) - V(\bar{u}(t)).$$

Hence condition (2) of Assumption 9.1.1 is satisfied.

Following the same arguments, we arrive at

$$\lim_{t \to +\infty} [V(u(t)) - V(u(t + \tau))] = 0 \quad \forall \tau \in R_+. \tag{9.2.5}$$

Thus the conditions of Theorem 8.4.2 are satisfied. Since the set $B_n(0)$ is sequentially weakly closed, there exists a weak solution $\bar{u}$ of the problem (9.2.2) such that

$$\bar{u}(0) = p \in \Omega(u), \quad \int_0^t W(\bar{u}(s)) \, ds = 0 \quad \forall t > 0.$$

Since the mapping $s \to W(\bar{u}(s))$ is continuous on R_+, $W(\bar{u}(t)) = 0$ for all $t \in R_+$.

Theorem 9.2.2   Let there exist an L functional $U_1 : E \to R$ continuously differentiable in the Fréchet sense and a continuous L functional $U_2 : E \to R$ bounded from below on a bounded set from E such that

(1) the operator $U_1 : E \to R$ is sequentially weakly continuous and $U_1' : E \to E^*$ maps bounded sets into bounded sets;

(2) $\langle Au + f(u), U_1'(u) \rangle = U_2(u) \quad \forall u \in \text{Dom } A$;

(3) if $p_n \xrightarrow{\text{weakly}} p$, $W(p_n) \to 0$ then $U_2(p) \leqslant \lim U_2(p_n)$; if, moreover, $U_2(p_n) \to U(p)$ then $U(p_n) \to U(p)$.

Let $t_0 \in R$ and let $u : [t_0, \infty[\to E$ be a bounded weak solution of the evolutionary problem (9.2.1). Define a and b as in Theorem 9.1.2.
Then

(a) $\infty > b \geqslant a > -\infty$;

(b) $\Omega(U, u) \cap V^{-1}(c) \neq \varnothing \quad \forall c \in [a, b]$,

where

$$\Omega(U, u) = \{p \in E : \exists \{t_n\{t_n\} \subset R_+, t_n \to \infty, u_n(t) \xrightarrow{\text{weakly}} p, U(u(t_n)) \to U(p)\}$$

$$V^{-1}(c) = \{p \in R : \exists \text{ a weak solution } \bar{u}(t), \bar{u}(0) = p, \text{ of (9.2.2) such that } V(\bar{u}(t)) = c,$$
$$W(\bar{u}(t)) = 0 \forall t \in R_+\}.$$

If, moreover, for every $c \in [a, b]$ the set $V^{-1}(c)$ is finite in the weak closure of the set $\text{Range } u$ then

(a') $a = b$;

(b') there exists $p_0 \in V^{-1}(a)$ such that

$$u(t) \xrightarrow{\text{weakly}} p_0, \quad V(u(t)) \to V(p_0) \quad \text{as } t \to +\infty.$$

Proof   Let the conditions of the theorem be satisfied. Let $\| u(\sigma) \| \leqslant r$ for all $t \in [t_0, \infty[$ and $U(t, \cdot) = U_{\bar{n}}(\cdot) = U(\cdot)$ for all $t \in R$. We shall demonstrate that condition (i) of Assumption 9.1.1 is satisfied. Let $u_n \in \Pi^{t_n}(s)$, $u_n \xrightarrow{\Phi} \bar{u} \in \Pi$, $t_n \to \infty$, $s \in R$. We have $V(u_n(0)) - V(u_n(t)) \to 0$ uniformly in t on compact sets from R_+. Assume that $V(u_n(0)) \nrightarrow V(\bar{u}(0))$. Without loss of generality, we have

$$|V(u_n(0)) - V(\bar{u}(0))| \geqslant \varepsilon > 0 \quad \forall n.$$

Let $t_0 > 0$. It follows from the proof of Theorem 9.2.1 that

$$\lim_{n \to \infty} \int_0^{t_0} W(u_n(\tau)) \, d\tau = 0.$$

Since $W \geq 0$, there exists $u_k \subset u_n$ such that

$$W(u_k(\tau)) \to W(\bar{u}(\tau)) \quad \text{for almost all } \tau \in [0, t_0].$$

Similarly, t_0

$$\lim_{k \to \infty} \int_0^{t_0} U_2(u_k(\tau)) \, d\tau = \lim_{k \to \infty} [U_1(u_k(t_0)) - U_1(u_k(0))]$$

$$= U_1(\bar{u}(t_0)) - U_1(\bar{u}(0))$$

$$= \int_0^{t_0} U_2(\bar{u}(\tau)) \, d\tau.$$

In this case the second equality follows from the sequential weak continuity of U_1.
  Let

$$\mathscr{F} = \{\tau \in [0, t_0] : U_2(\bar{u}(\tau)) \leq \lim_{k \to \infty} U_2(u_k(\tau))\}.$$

Since U_2 is bounded from below on bounded sets, condition (3) of the theorem and the Fatou lemma imply that $\mathscr{F}$ has zero measure. Then, without loss of generality, we can conclude that there exists $t_1 \in [0, t_0]$ such that

$$W(u_k(t_1)) \to 0, \qquad U_2(u_k(t_1)) \to U_2(\bar{u}(t_1)).$$

Thus, by virtue of condition (3), $V(u_k(t_1)) \to V(\bar{u}(t_1))$. But $V(u_k(t_1)) - V(u_k(0)) \to 0$ and $V(\bar{u}(t_1)) = V(u(0))$. Then $V(u_k(0)) \to V(\bar{u}(0))$, which leads to a contradiction. Therefore condition (1) of Assumption 9.1.1 is satisfied. The mapping $t \to V(u)(t)$ is continuous, because V and u are continuous. Clearly, the set $u([t_1, \infty))$ is sequentially weakly precompact. Besides, the proofs of Theorem 8.4.6 and 9.2.1 imply that condition (4) of Assumption 9.1.1 is satisfied. In view of the relation

$$V(y) = V(0) + \int_0^1 \langle y, V'(s, y) \rangle \, ds$$

and condition (3) of Assumption 9.2.2, we find that V takes bounded sets from E into bounded sets. Therefore $b > -\infty$.
  Thus, by Theorem 8.4.3, we have

$$-\infty < a \leq b < \infty, \qquad \Omega(V, u) \cap V^{-1}(c) \neq \varnothing \quad \forall c \in [a, b].$$

The last assertion of the theorem follows from the weak connectivity of $\Omega(V, u)$.

9.2.3   The limiting L functional for an evolutionary equation for damped oscillations

  Let B be a positive selfconjugate operator densely defined on a Hilbert space H with scalar product $(\cdot, \cdot)$ and norm $\| \cdot \|$. We assume that B^{-1} is defined on H and is compact. Let $H_B = \text{Dom}(B^{1/2})$, H_B be the Hilbert space with the scalar product $(\cdot, \cdot)$ and norm

$\|\cdot\|_B$, where $\|v\|_B = \|B_v^{1/2}\|$ and $v \in H_B$, and $E = E_B \times H$. Clearly, the set H_B is dense in H, and H is dense in H_B^*, with both inclusions being compact. We introduce the norm

$$\|\{w, v\}\|_E = \|w\|_B^2 + \|v\|^2)^{1/2}.$$

In this case E is a Hilbert space.

We study the stability-like properties of solutions of the nonautonomous equation of damped oscillations

$$w'' + Bw + F(t, w, w') = 0 \qquad (9.2.6)$$

Assumption 9.2.3   For equation (9.2.6) the following conditions hold.

(1) $F: R \times H_B \times H \to H$ for any $\{w, v\} \in E$, the mapping $F(\cdot, w, v): R \to H$ is strongly measurable, and for almost all $t \in R$ the mapping $F(t, \cdot, \cdot): E \to H$ is sequentially weakly continuous.

(2) For any bounded subset $\mathscr{A} \in E$ there exists a locally integrable function $m_{\mathscr{A}}$ on R such that for all $\{w, v\} \in E$

$$\|F(t, w, v)\| \leqslant m_{\mathscr{A}}(t) \quad \text{for almost all } t \in R_+.$$

Also,

$$\limsup_{s \to 0, t \geqslant t_0} \int_t^{t+s} m_{\mathscr{A}}(\tau)\, d\tau = 0 \quad \forall t_0 \in R.$$

We also consider the autonomous equation

$$w'' + Bw + C(w, w') + \bar{F}(w) = 0 \qquad (9.2.7)$$

Assumption 9.2.4   For equation (9.2.7) the following conditions hold.

(1) The mappings $C: E \to H$ and $\bar{F}: H_B \to H$ are sequentially weakly continuous, and for any bounded subset $\mathscr{A} \subset E$

$$\lim_{t \to 0} \int_t^{t+1} \sup_{\{w, v\} \in \mathscr{A}} \|F(s, w, v) - C(w, v) - \bar{F}(w)\|\, ds = 0.$$

(2) There exists a functional $W: H_B \to R$ continuously differentiable is the Fréchet sense such that $W'(w) = \bar{F}(w)$ for all $w \in H_B$.

(3) The mapping $\{w, v\} \to (C(w, v), v)$ is sequentially weakly continuous from below on E. Moreover, for any bounded subset $Q \subset H_B$ there exists a strictly increasing continuous function $K_Q: R_+ \to R_+$, $K_Q(0) = 0$, such that

$$\inf_{w \in Q} (C(w, v), v) \geqslant K_Q(\|v\|) \quad \forall v \in H.$$

The following result holds

Theorem 9.2.3   Let

(1) Assumptions 9.2.3 and 9.2.4 be valid;
(2) $\{w_1, w_t\}: [t_0, \infty[\to E$ be a weakly bounded solution of the nonautonomous equation (9.2.6).

Then

(a) $-\infty < a \leqslant b < \infty$;
(b) $\Omega(s,w) \cap \mathscr{L}_c(V) \neq \varnothing \quad \forall c \in [a,b]$.

If, moreover, for every $c \in [a,b]$ the set of $\mathscr{L}_c(V)$ contains only a finite number of elements in the weak closure of the domain $\{w, w_t\}$ then

(c) $a = b$ and there exists $\{y, 0\} \in \mathscr{L}_c(V)$

such that

$$\{w(t), w_t(t)\} \xrightarrow{E} \{y, 0\} \quad \text{as } t \to +\infty. \tag{9.2.8}$$

Here we have defined the functional $V(w,v)$, numbers a and b and sets $\Omega(s,w)$ and $\mathscr{L}_c(V)$ by

$$V(w,v) = \tfrac{1}{2} \| \{w,v\} \|_E^2 + \Phi(w),$$

$$a = \liminf_{t \to +\infty} V(w(t), w_t(t)), \qquad b = \limsup_{t \to +\infty} V(w(t), w_t(t)),$$

$$\Omega(s,w) = \{ y \in H_B : \exists t_n, t_n \to \infty, \{w, w_t\}(t_n) \xrightarrow{E} \{y, 0\}, \tag{9.2.9}$$

$$\mathscr{L}_c(V) = \{ \{y, 0\} \subset E : y \in \mathrm{Dom}\, B, B_y + \bar{F}(y) = 0, V(y, 0) = c \} \quad \forall c \in R.$$

Proof Let the conditions of the theorem be satisfied, and, moreover,

$$\mathrm{Dom}\, A = \mathrm{Dom}\, B \times H_B \subseteq E, \qquad A(\{w, v\}) = \{v - Bw\}.$$

It is clear that A is a generator of the strongly continuous $\mathscr{D}^+$ system (group) on E, and $f: R \times E \to E$, $f(t, \{w, v\}) = \{-F(t, w, v), 0\}$. If $u = \{w, w_t\}$ then (9.2.6) becomes $u' = Au + f(t, u)$ and $\bar{f}(\{w, u\}) = \{0, -c(w, v) - \bar{F}(w)\}$. Then (9.2.7) becomes $u' = Au + \bar{f}(t, u)$. Conditions (1) and (2) of Assumption 9.2.3 and condition (1) of Assumption 9.2.4 imply that Assumption 9.2.1 is satisfied.

Consider the functional $V: E \to R$ defined by (9.2.8) and the functional

$$W_2(w, u) = (c(w, v), v), \qquad \{w, v\} \in E.$$

It can easily be verified that V is a continuously differentiable functional and condition (1) of Assumption 9.2.2 is satisfied. Since $c(\cdot)$ is sequentially weakly continuous, the functional W_2 is continuous. The property that V' takes bounded sets into bounded sets results from the sequential weak continuity of $\bar{F}$. Then, Assumption 9.2.2 is satisfied.

Define $U_1: E \to R$ and $U_2: E \to R$ by

$$U_1(w, v) = -(w, v),$$
$$U_2(w, v) = \| w \|_B^2 + (C(w, v), w) + (\bar{F}(w), w) - \|v\|^2,$$

where $\{w, u\} \in E$. It can easily be verified that U_1 is a continuously differentiable mapping, U_1' and U_2 take bounded sets into bounded sets, and

$$\langle Au + \bar{f}(u), U_1'(u)) \rangle = U_2(u) \quad \forall u \in \mathrm{Dom}\, A.$$

Since the inclusion of H_B into H is compact, U_1 is sequentially weakly continuous. Let

$$w_n \xrightarrow{H_B} w, \quad u_n \xrightarrow{H} 0.$$

Then

$$w_n \xrightarrow{H} w, \quad C(w_n, u_n) \xrightarrow{H} C(w, 0), \quad \bar{F}(w_n) \xrightarrow{H} F(w), \quad \|w\|_B^2 \le \liminf_{n \to \infty} \|w_n\|_B^2.$$

Therefore,

$$U_2(w, 0) \le \liminf_{n \to \infty} U_2(w_n, u_n).$$

We assume that $U_2(w_n, u_n) \to U_2(w, 0)$. Then $\|w_n\|_B \to \|w\|_B$ as $n \to \infty$. Therefore $w_n \xrightarrow{H_B} w$, and then $V(w_n, u_n) \to V(w, 0)$. By the same token, conditions (1)–(3) of the theorem are satisfied.   ∎

It is easy to show that the solutions of the autonomous equation

$$u' = Au + \bar{f}(u)$$

are strictly precompact and bounded if $\bar{f} : E \to E$ is compact and there exists a μ such that

$$\| T(t) \| \le \mu e^{\alpha t} \quad \forall t \in R_+, \quad \alpha < 0.$$

By means of this method, some results can be obtained for equations (9.2.6) and (9.2.7).

9.3   Applications

Example 9.3.1   Let $\mathscr{D}$ be a nonempty bounded open set from R^n with boundary ∂D. Let $H = \mathscr{L}_2(\mathscr{D})$. We consider the boundary-value problem of hyperbolic type

$$w_{tt} + a(t, w)w_t - \Delta w + \varphi(t, w) = 0 \quad \forall x \in \mathscr{D}, \quad t > s, \quad w|_{\partial \mathscr{D}} = 0,$$
$$w|_{t=s} = \bar{w}_1, \quad w_t|_{t=s} = \bar{w}_2$$

and the corresponding autonomous boundary-value problem

$$w_{tt} + \bar{a}(w)w_t - \Delta w + \bar{\varphi}(w) = 0 \quad \forall x \in \mathscr{D}, \quad t > 0,$$
$$w|_{\partial \mathscr{D}} = 0, \quad w|_{t=0} = \bar{w}_1, \quad w_t|_{t=0} = \bar{w}_2,$$

where $\bar{w}_i$ are given functions.

Assumption 9.3.1   The following conditions are satisfied.

(1) The function $a : R \times R \to R$, $a(\cdot, w)$ is measurable for all $w \in R$ and $a(t, \cdot)$ is continuous for almost all $t \in R$; $\bar{a} : R \to R$ is continuous and $\bar{a}(w) \ge \delta > 0$ for all $w \in R$, where δ is a constant.

(2) The function $\varphi: R \times R \to R$ is measurable in t for any fixed w and continuous in w for almost all $t \in R$; $\bar{\varphi}: R \to R$ is continuous.

(3) There exists a nonnegative locally integrable function such that

$$\limsup_{s \to 0, t \geq t_0} \int_t^{t+s} m(\tau)\, d\tau = 0 \quad \forall t_0 \in R$$

and such that if $n = 1$ then for almost all $t \in R$

$$|a(t, w)| + |\varphi(t, w)| \leq m(t)\theta_1(w) \quad \forall w \in R,$$

where the function $\theta_1: R \to R$ is continuous. If $n > 1$ then for almost all $t \in R$

$$|a(t, w)| + \frac{|\varphi(t, w)|}{1 + |w|^\gamma} \leq m(t) \quad \forall w \in R,$$

where $1 \leq \gamma + \infty$ for $n = 2$, and $1 \leq \gamma \leq n(n-2)$ for $n > 2$.

(4) There exists a nonnegative locally integrable function $\eta(t)$ such that

$$\lim_{t \to \infty} \int_t^{t+1} \eta(\tau)\, d\tau = 0$$

and such that if $n = 1$ then for almost all $t \in R$

$$|a(t, w) - \bar{a}(w)| + |\varphi(t, w) - \bar{\varphi}(w)| \leq \eta(t)\theta_2(w) \quad \forall w \in R$$

where $\theta_2: R \to R$ is continuous. If $n > 1$ then for almost all $t \in R$

$$|a(t, w) - \bar{a}(w)| + \frac{|\varphi(t, w) - \bar{\varphi}(w)|}{1 + |w|^\gamma} \leq \eta(t) \quad \forall w \in R,$$

where $1 \leq \gamma < \infty$ for $n = 2$, and $1 \leq \gamma \leq n(n-2)$ for $n > 2$.

It is easy see that Assumptions 9.2.3 and 9.2.4 are satisfied for equations (9.3.1) and (9.3.2) provided that

$$B = -\Delta, \quad \mathrm{Dom}\, B = \{w \in \overset{\circ}{W}_2^1(\mathscr{D}): \Delta w \in \mathscr{L}_2(\mathscr{D})\}, \quad H_B = \overset{\circ}{W}_2^1(\mathscr{D});$$

$$F(t, w, v) = a(t, w(\cdot))v(\cdot) + \varphi(t, w(\cdot));$$

$$C(w, u) = \bar{a}(w(\cdot))v(\cdot), \quad F(w) = \bar{\varphi}(w(\cdot));$$

$$\Phi(w) = \int_{\mathscr{D}} \int_0^{w(x)} \bar{\varphi}(r)\, dr\, dx.$$

If in the problem (9.3.1) the function $a(t, w)$ depends on $x \in \Omega$ as well and is zero outside some compact set Ω then condition (3) of Assumption 9.2.4 does not hold. This is an example of 'weak' damping.

Example 9.3.2 Consider the boundary-value problem

$$w_{tt} + c_1 w_{xxx} - \left[c_2 + a(t) + c_3 \int_0^B w_\xi^2(t, \xi)\, d\xi \right] w_{xx} + [c_4 + b(t)]w_t = 0,$$

$$w = w_x = 0 \quad \text{for } x = 0, l,$$

$$w|_{t=s} = w_0, \quad w_t|_{t=s} = w_t^0, \tag{9.3.3}$$

where $a(\cdot)$ and $b(\cdot)$ are measurable functions such that

$$\lim_{t \to \infty} \int_t^{t+1} [|a(s)| + b(s)|] \, ds = 0 \qquad (9.3.4)$$

and $c_i, i = 1, \ldots, 4$, are constants, with $c_1 > 0$, $c_3 > 0$ and $c_4 > 0$. We introduce the notation

$$\mathscr{D} = \,]0, l[, \qquad H = \mathscr{L}_2(\mathscr{D}),$$

$$B = c_1 \frac{d^4}{dx^4}, \qquad \text{Dom } B = \{w \in \mathring{W}_2^2(\mathscr{S}) : w_{xxxx} \in L_2(\mathscr{D})\}.$$

Then $H_B = \mathring{W}_2^2(D)$. We set

$$C(w, v) = c_4 v(\cdot),$$

$$F(t, w, v) = [c_4 + b(t)]v(\cdot) - [c_2 + a(t) + c_3 \| w_x \|^2] w_{xx}(\cdot),$$

$$\bar{F}(w) = -(c_2 + c_3 \| w_x \|^2) w_{xx}(\cdot) \qquad \Phi(w) = -\tfrac{1}{2} c_2 \| w_x \|^2 + \tfrac{1}{4} c^3 \| w_x \|^4.$$

It is easy to verify that Assumptions 9.2.3 and 9.2.4 are satisfied and that for the problem (9.3.3) there are only a finite number of equilibrium states. Therefore the last assertion of Theorem 9.2.3 holds.

Example 9.3.3   We consider the autonomous boundary-value problem

$$w_{tt} + (1 + \| \nabla x \|^2) w_t - \Delta w + \bar{\varphi}(w) = 0,$$

$$w \|_{\partial \mathscr{D}} = 0, \qquad \{w, w_t\}(0) \in E, \qquad (9.3.5)$$

where $\mathscr{D}$ and $\bar{\varphi}$ are as in Example 9.3.1.

Let $H = L_2(\mathscr{D})$ and $E = \mathring{W}_2^1(\mathscr{D}) \times L_2(\mathscr{D})$, where $\mathring{W}_2^1(\mathscr{D})$ is the Sobolev space. We suppose for the sake of simplicity that there are only a finite number of solutions $y \in \mathring{W}_2^1(\mathscr{D})$ of the equation

$$\Delta y = \bar{\varphi}(y). \qquad (9.3.6)$$

Similarly to the above, we write equation (9.3.5) in the form

$$\frac{du}{dt} = Au + \bar{f}(u),$$

where

$$A(\{w, v\}) = \{u, \Delta w\},$$

$$\bar{f}(\{w, v\}) = \{0, -(1 + \| \nabla w \|^2)u - \bar{\varphi}(w)\}.$$

However, $\bar{f}$ is not sequentially weakly continuous; therefore the above theory cannot be applied. Nevertheless, we shall demonstrate that for a solution $y \in \mathring{W}_2^1(\mathscr{D})$ of equation (9.3.6) any weak solution $\{w, w_t\}(t)$ of the problem (9.3.5) converges strongly in E to $\{y, 0\}$ as $t \to \infty$. Indeed, let E be provided with a convergence structure generated by weak convergence. Let $X = \{u \in E^{R^+} : u = \{w, w_t\}$ is a weak solution of (9.3.5)$\}$ and let $\gamma(\tau)(u)$ be the τ shift u^τ of the solution $u \in X$. It is easy to show that γ has a the following weakened property of continuity:

$$t_n \to \infty \text{ and } \gamma(t_n)u \xrightarrow{E^{R_+}} v \text{ then for } u = \{w, 0\} \text{ and } \gamma(t)$$

$$\gamma(t_n)u \xrightarrow{E^{R_+}} \gamma(t)v \quad \forall t \in R_+. \tag{9.3.7}$$

To prove this, we write the energy equation

$$\tfrac{1}{2}\|\{w, w_t\}\|^2 + \Phi(w) + \int_s^t (1 + \|\nabla w\|^2)\|w_t\|^2 \, d\tau = 0,$$

which is valid for weak solutions of the problem (9.3.5). Hence $\|\nabla w\|^2(t)$ is bounded in $t \in R_+$, and

$$\int_0^\infty \|w_t(\tau)\|^2 \, d\tau < \infty.$$

Using the established facts and the variation-of-constants formula (9.1.3), we get the desired property (9.3.7). The strict convergence

$$\{w, w_t\}(t) \to \{y, 0\} \quad \text{as } t \to \infty$$

is established in the same manner as in Example 9.3.1.

9.4 Comments and References

9.0 There have been many investigations of parabolic and hyperbolic equations (Haraux [1, 2]; Belonosov, Vishnevsky [1]; Belonosov, Zelenyak [1]; Vishnevsky [1, 2]; Gutowski [1]; Pazy [1]; Plaut [1]; Ralstone [1]; Russell [1]; Rauch [1]; Chafee [1, 2]; Chafee, Infante [1]).

The theory of limiting equations for the above classes of equations has been developed by Ball [1–3], Ball and Peletier [1] and Ball and Knowles [1]. Many results obtained in this direction are due to Shestakov [1]. Some stability problems have been presented by Martynyuk and Gutowski [1].

9.1 The presentation of the generalized Lyapunov direct method for asymptotically autonomous parabolic evolutionary equations is according with Ball [4].

9.2 The generalized Lyapunov direct method for asymptotically autonomous hyperbolic evolutionary equations has been considered in the work of Ball [4].

9.3 The basic results of this section are due to Ball [4], Ball and Peletier [1], Ball and Knowles [1].

References

Andreev A. S.
[1] Asymptotic stability and instability of nonautonomous systems. *Appl. Math. Mech.* **43** (1979) 796–805 (Russian).
[2] Asymptotic stability and instability of zero solutions of nonautonomous systems. *Appl. Math. Mech.* **48** (1984) 225–232 (Russian).
[3] Optimal stabilization of motions of controlled mechanical systems. *Theses of Reports of 5th All-Union Conference on Control in Mechanical Systems, Kazan, 21–22 June 1985.* KAI, Kazan (1985), p. 60 (Russian).
Artstein Z.
[1] Continuous dependence of solutions of Volterra integral equations. *SIAM J. Math. Anal.* **6** (1975) 446–456.
[2] Limiting equations and stability of nonautonomous ordinary differential equations. Appendix to LaSalle J. P., the stability of dynamical systems. *CBMS Regional Conference Series in Applied Mathematics.* SIAM, Philadelphia (1976), pp. 57–76.
[3] The limiting equations of nonautonomous ordinary differential equations. *J. Diff. Eqns* **25** (1977) 184–202.
[4] Topological dynamics of an ordinary differential equation. *J. Diff. Eqns* **23** (1977) 216–223.
[5] Topological dynamics of ordinary differential equations and Kurzwell equations. *J. Diff. Eqns* **23** (1977) 224–243.
[6] Uniform asymptotic stability via the limiting equations. *J. Diff. Eqns* **27** (1978) 172–189.
[7] Stability, observability and invariance. *J. Diff. Eqns* **44** (1982) 224–248.
Azbelev N. V., Tzaluk Z. B.
[1] Integral inequalities. *Math. Coll.* **56** (1962) 325–342 (Russian).
Ball J. M.
[1] Initial boundary value problems for an extensible beam. *J. Math. Anal. Applies* **42** (1973) 61–90

[2]   Saddle point analysis for ordinary differential equations in a Banach space, and an application to dynamic buckling of a beam. *Nonlinear Elasticity* (ed. R. W. Dickey). New York (1973), pp. 28–34.

[3]   Stability theory for an extensive beam. *J. Diff. Eqns* **14** (1973) 399–418.

[4]   On the asymptotic behaviour of generalized processes, with applications to nonlinear evolution equations. *J. Diff. Eqns* **27** (1978) 224–265.

Ball J. M., Knowles V.

[1]   Lyapunov functions for thermomechanics with spatially varying boundary temperatures. *Arch. Rat. Mech. Anal.* **92** (1986) 116–131.

Ball J. M., Peletier L. A.

[1]   Global attraction for the one-dimensional heat equation with nonlinear time-dependent boundary conditions. *Arch. Rat. Mech. Anal.* **65** (1977) 193–201.

Barbashin E. A.

[1]   The theory of generalized dynamical systems. *Sci. Notes Moscow State Univ. Maths* **2** (1949) 110–133 (Russian).

[2]   *Introduction to Stability Theory.* Nauka, Moscow (1967) (Russian).

[3]   *Lyapunov Functions.* Nauka, Moscow (1970) (Russian).

Barnea B. I.

[1]   A method and new results for stability and instability of autonomous functional differential equations. *SIAM J. Appl. Math.,* **17** (1969), 681–697.

Bebutov M. V.

[1]   Dynamical systems in the space of continuous functions. *Dokl. Akad. Nauk SSSR* **27** (1940) 904–906 (Russian).

[2]   Dynamical systems in the space of continuous functions. *Bull. Moscow State Univ. Maths* **2** (1941) 1–52 (Russian).

Bellman R.

[1]   *Stability Theory of Differential Equations.* Academic Press, New York (1953).

Belonosov V. S., Vishnevsky M. P.

[1]   Some problems in the qualitative theory of boundary value problems for nonlinear parabolic systems. *Mathematical Problems in Chemistry,* Part 1, Novosibirsk (1975), pp. 132–138 (Russian).

Belonosov V. S., Zelenyak T. I.

[1]   Lyapunov stability of solutions of mixed problem for nonlinear parabolic equations. *Proceedings of All-Union Conference on Equations with Partial Derivatives, Dedicated to I. G. Petrovskij on this 75th Birthday.* Moscow (1978), pp. 42–45 (Russian).

Benavides I. D.

[1]   System for nonautonomous differential equations with Lipschiz operator. *J. Diff. Eqns* **34** (1979) 230–238.

Bhatia N. P., Szego G. P.

[1]   *Dynamical Systems: Stability Theory and Applications.* Springer, Berlin (1967).

Bogolyubov N. N., Mitropolsky Yu. A.

[1]   *Asymptotic Methods in the Theory of Nonlinear Oscillations.* Nauka, Moscow (1974) (Russian).

Bondi P., Moauro V., Visentin F.
[1]   Limiting equations in the stability problem. *J. Nonlin. Anal.* **1** (1977) 123–128.
[2]   Addendum to the paper "Limiting equations in the stability problem". *J. Nonlin. Anal.* **1** (1977) 701.

Bronstein I. U.
[1]   Dynamical systems without uniqueness as semigroups of nonunivalent mappings of topological spaces. *Izv. Akad. Nauk SSSR, Ser. Mat.* **1** (1963) 1–18 (Russian).
[2]   *Nonautonomous Dynamical Systems.* Shtiinza, Kishinev (1984) (Russian).

Bronstein I. U., Chernyi V. F.
[1]   On extensions of dynamical systems with uniformly asymptotically stable points. *Diff. Eqns* **10** (1974) 1225–1230 (Russian).

Bronstein I. U., Glavan V. A. Chernyi V. F.
[1]   Relation between some types of stability by means of extensions of dynamical systems. *Math. Res.* **10** (1975) 58–67 (Russian).

Budak B. M.
[1]   Dispersive dynamical systems. *Vestn. Moscow State Univ.,* **8** (1947) 135–137 (Russian).

Burton T. A.
[1]   *Volterra Integral and Differential Equations.* Academic Press, New York (1983).
[2]   Periodic solutions of linear Volterra equations. *Funkc. Ekvacioj* **27** (1984) 229–253.
[3]   *Stability and Periodic Solutions of Ordinary and Functional Differential Equations.* Academic Press, New York (1985).

Burton T. A., Grimmer R. C.
[1]   Oscillations, continuation, and uniqueness of retarded differential equations. *Trans. Am. Math. Soc.* **179** (1973) 193–209.

Chafee N. A.
[1]   Stability analysis for semilinear parabolic partial differential equations. *J. Diff. Eqns* **15** (1974) 522–540.
[2]   Asymptotic behaviour for solutions of a one-dimensional parabolic equation with Neumann boundary conditions. *J. Diff. Eqns* **18** (1975) 111–134.

Chafee N. A., Infante E. T.
[1]   A bifurcation problem for a nonlinear differential equation of parabolic type. *Appl. Anal.* **4** (1974) 17–37.

Chernetskaya L. N.
[1]   Integrality estimations of stability on finite intervals. *Process Stability and Applications.* VINITI 31.10.86, No. 7501-B86, Moscow (1986), pp. 40–52 (Russian).

Chetayev N. G.
[1]   *Stability of Motion.* Nauka, Moscow (1965) (Russian).

Chow S. N., Hale J. K.
[1]   Strongly limit-compact maps. *Funkc. Ekvacioj* **17** (1974) 31–38.

Coleman B. D., Mizel V. J.
[1]   On the stability of solutions of differential equations. *Arch. Rat. Mech. Anal.* **30** (1968) 173–196.

Conley C. C., Miller R. K.
[1]   Asymptotic stability without uniform stability: almost periodic coefficients. *J. Diff. Eqns* **1** (1965) 333–336.
Corduneanu C.
[1]   The application of differential inequalities in the theory of stability. *Annalele Stiintifice ale Univ. "Al. i Cusa" din Iasi, Sec. 1* **6** (1960) 47–58.
Corduneanu C., Lakshmikantham V.
[1]   Equations with unbounded delay: a survey. *Nonlin. Anal.* **4** (1980) 831–877.
Dafermos C. M.
[1]   Asymptotic stability in viscoelasticity. *Arch. Rat. Mech. Anal.* **37** (1970) 297–308.
[2]   An invariance principle for compact processes. *J. Diff. Eqns* **9** (1971) 239–252.
[3]   Applications of the invariance principle for compact processes. 1. Asymptotically dynamical systems. *J. Diff. Eqns* **9** (1971) 291–299.
[4]   Semiflows associated with compact and uniform processes. *Math. Syst. Theory* **8** (1975) 142–149.
[5]   Uniform processes and semicontinuous Lyapunov functional. *J. Diff. Eqns* **11** (1972) 401–415.
[6]   Applications of the invariance principle for compact processes. 2. Asymptotic behaviour of solutions of a hyperbolic conservation law. *J. Diff. Eqns* **11** (1972) 416–424.
Dannan F., Elaydi S.
[1]   Lipschitz stability of nonlinear systems of differential equations. *J. Math. Anal. Applics* **113** (1986) 562–577.
D'Anna A.
[1]   Limiting equations in the first approximation stability problem for non-autonomous differential equations. *Boll. UMI* **6** (1982) 39–46.
[2]   Total stability properties for an almost periodic equation by means of limiting equations. *Funkc. Ekvacioj* **27** (1984) 201–209.
D'Anna A., Maio A., Moauro V.
[1]   Global stability properties by means of limiting equations. *Nonlin. Anal. Theory, Methods Applics* **4** (1980) 407–410.
D'Anna A., Maio A., Monte A.
[1]   Limiting equations in asymptotic stability problems for nonautonomous differential equations. *Ann. Univ. Ferrara, Sez. 7, Sci. Mat.* **25** (1979) 75–97.
D'Anna A., Monte A.
[1]   Limiting equations as a method to prove sets stability. *Ann. Univ. Ferrara, Sez. 7, Sci. Mat.* **28** (1982) 67–79.
Dishliev A. B., Bainov D. D.
[1]   Investigation of the Lipschitz stability via limiting equations. *Dyn. Stab. Syst.* **5** (1990) 59–64.
Driver R. D.
[1]   Existence and stability of solutions of a delay differential system. *Arch. Rat. Mech. Anal.* **10** (1962) 401–426.
Duboshin G. N.
[1]   Stability with respect to persistent perturbations. *Proc. GAIS Moscow State Univ.* **14** (1940) 153–164 (Russian).

Filatov A. N.

[1] Methods of Averaging in Differential and Integro-differential Equations. *Tashkent, Fan,* 1971.

Fink A. M.

[1] *Almost Periodic Differential Equations.* Springer, Berlin (1974).

Fitzgibbon W. E.

[1] Weakly continuous accretive operators. *Bull. Am. Math. Soc.* **78** (1973) 473–474.

Freedman H. I.

[1] *Deterministic Mathematical Models in Population Ecology,* Hifr Consulting, Edmonton (1987).

Fréchet M.

[1] Les fonctions asymptotiquement presque-périodiques. *Rev. Scientifique,* **79** (1941) 341–354.

Gopalsamy K.

[1] Global asymptotic stability in a periodoc integro-differential system. *Tohoku Math. J.* **37** (1985) 323–332.

Gorahin S. I.

[1] On stability of motion under persistent perturbation. *Izv. AN Kaz. SSR.,* No. 56 (1948), 46–73.

Grebenikov E. A., Riabov Yu. A.

[1] *New Qualitative Methods in Celestial Mechanics.* Nauka, Moscow (1971) (Russian).

[2] *Constructive Methods of Analysis of Nonlinear Systems.* Nauka, Moscow (1979) (Russian).

Grimmer R., Seifert G.

[1] Stability properties of Volterra integro-differential equations. *Diff. Eqns* **19** (1975) 142–166.

Grujič Lj.T., Martynyuk A. A., Ribbens-Pavella M.

[1] *Large Scale Systems Stability under Structural and Singular Perturbations.* Springer, Berlin (1987).

Gutowski R.

[1] *Introduction sur la stabilité du mouvement des systèmes continus.* L'université de Poitiers (1988).

Haddock J. R.

[1] Stability theory for nonautonomous systems. *Dynamical Systems: An International Symposium.* Academic Press, New York (1976), pp. 271–274.

[2] On Liapunov functions for nonautonomous systems. *J. Math. Anal. Applics* **47** (1974) 599–603.

[3] Friendly spaces for functional differential equations with infinite delay. *Proceedings of 6th International Conference on Trends in the Theory and Practice of Nonlinear Analysis* (ed. V. Lakshmikantham). North-Holland, Amsterdam (1985), pp. 173–182.

Hahn W.

[1] *Stability of Motion.* Springer, Berlin (1967).

Halanay A.

[1] *Differential Equations: Stability, Oscillations, Time Lags.* Academic Press, New York (1966).

Hale J.

[1] *Oscillations in Nonlinear Systems.* McGraw-Hill, New York (1963)

[2] *Theory of Functional Differential Equations.* Springer, New York (1977).

[3] Some recent results on dissipative processes. *Functional Differential Equations and Bifurcations* (ed. A. F. Ize). Lecture Notes is Mathematics, Vol. 799. Springer, Berlin (1980), pp. 152–172.

Hale J. K., Kato J.

[1] Phase space for retarded equations with infinite delay. *Funkc. Ekvacioj* **21** (1978) 11–41.

Hamaya Y.

[1] Total stability property in limiting equations of integro-differential equations. *Funkc. Ekvacioj* (to appear)

Hamaya Y., Yoshizawa T.

[1] Almost periodic solutions of an integro-differential equation. *Proc. R. Soc. Edin.* **114A** (1990) 151–159.

Hapaev M. M.

[1] A Lyapunov-like theorem. *Dokl Akad. Nauk SSSR* **176** (1967) 1262–1265 (Russian).

Hapaev M. M. and Falin A. I.

[1] On investigation of stability of integro-differential equations by method of averaging. *Dokl. AN USSR*, **250** (1980), 295–299.

Haraux A.

[1] *Nonlinear Evolution Equations: global Behaviour of Solutions.* Lecture Notes in Mathematics, Vol. 841. Springer, Berlin (1981).

[2] On a uniqueness theorem of L. Amerio and G. Prouse. *Proc. R. Soc. Edin.* **96** (1984) 221–230.

Hatvani L.

[1] On partial asymptotic stability and instability. II. The method of limiting equations. *Acta Sci. Math.* **46** (1983) 143–156.

[2] On location of positive limit sets of solutions of nonautonomous systems. *Colloq. Math. Soc. Janos Bolyai*, Szeged (1984) 413–428.

[3] On partial asymptotic stability by the method of limiting equations. *Ann. Math. Pura Appl.* **139**(4) (1985) 65–82.

Hino Y.

[1] Stability and existence of almost periodic solutions of some functional differential equations. *Tohoko Math. J.* **28** (1976) 289–409.

[2] Stability properties for functional differential equations with infinite delay. *Tohoku Math. J.* **35** (1983) 597–605.

Hino Y., Murakami S.

[1] Favard's property for linear retarded equations with infinite delay. *Funkc. Ekvacioj* **29** (1986) 11–17.

Hino Y., Yoshizawa T.

[1] Total stability property in limiting equations for a functional differential equation with infinite delay. *Časopis pro pestovani matematiky* **111** (1986) 62–69.

Hopf E.

[1] On the right weak solutions of the Cauchy problem for a quasilinear equation of first order. *J. Math. Mech.* **19** (1969) 483–487.

Izman M. S.

[1] Stability of sets and attractors in dispersive dynamical systems. *Math. Res. Akad. Nauk Mold. SSR* 3(4) (1968) 51–77 (Russian).

[2] Application of the second Lyapunov method for the investigation of stability and asymptotic stability of sets in dispersive dynamical systems. *Diff. Eqns* 5 (1969) 1207–1217 (Russian).

Kappel F., Schappacher N.

[1] Some considerations to the fundamental theory of infinite delay equations. *Diff. Eqns* 37 (1980) 141–183.

Karakostas G.

[1] Uniform asymptotic stability of causal operator equations. *J. Integr. Eqns* 5 (1983) 59–71.

Karimzhanov A.

[1] Stability of integro-differential systems via limiting equations. *Dokl. Akad. Nauk SSSR. Ser. A* 12 (1985) 11–14 (Russian).

[2] Limiting equations in the problem of stability nonautonomous systems. *Appl. Mech.* 21 (1985) 110–117 (Russian).

[3] Development of the direct Lyapunov method via auxiliary systems in the investigation of nonautonomous systems. Candidate thesis, Institute of Mechanics, Kiev (1985) (Russian).

Karimzhanov A., Kosolapov V. I.

[1] Theory of stability of systems involving a small parameter. *Stability of Motion*. Nauka, Novosibirsk (1985), pp. 40–43 (Russian).

Kato J.

[1] Remarks on linear functional differential equations. *Funkc. Ekvacioj* 12 (1969) 89–98.

[2] Uniform asymptotic stability and total stability. *Tohoku Math. J.* 22 (1970) 254–269.

[3] On Liapunov-Razumikhin type theorems. Lecture Notes in Mathematics, Vol. 243. Springer, Berlin (1971), pp. 54–65.

[4] On Liapunov-Razumikhin type theorems for functional differential equations. *Funkc. Ekvacioj* 16 (1973) 225–239.

[5] Stability problem in functional differential equations with infinite delay. *Funkc. Ekvacioj* 21 (1978) 63–80.

[6] Stability in functional differential equations. Lecture Notes in Mathematics, Vol. 799. Springer, Berlin (1980), pp. 252–262.

[7] Liapunov's second method in functional differential equations. *Tohoku Math. J.* 32 (1980) 487–497.

[8] Phase space for functional differential equations. *Qualitative Theory of Differential Equations*. Bolyai Institute, Szeged (1988).

Kato J., Yoshizawa T.

[1] A relationship between uniformly asymptotic stability and total stability. *Funkc. Ekvacioj* 12 (1969) 233–238.

[2] Stability under perturbations by a class of functions. *Dynamical Systems: An International Symposium*, Vol. 2. Academic Press, New York (1976), pp. 217–222.

[3]   Remarks on global properties in limiting equations. *Funkc. Ekvacioj* **24** (1981) 363–371.

Kosolapov V. I.

[1]   Theory of stability of motion in the neutral case. *Dokl. Akad. Nauk SSSR, Ser A.* **1** (1979) 27–31 (Russian).

Krasovsky N. N.

[1]   *Stability of Motion.* Stanford University Press (1963). [Russian original published by Fizmatgiz, Moscow (1959).]

[2]   Problems of stabilization of controlled motion. *Theory of Stability of Motion* (ed. I. G. Malkin). Nauka, Moscow (1968), pp. 475–514 (Russian).

Ladde G. S., Lakshmikantham V., Zhang B. G.

[1]   *Theory of Differential Equations with Deviating Arguments.* Marcel Dekker, New York (1987).

Lakshmikantham V., Leela S.

[1]   *Differential and Integral Inequalities: Theory and Applications*, Vol. 1. Academic Press, New York (1966).

Lakshmikantham V., Leela S., Martynyuk A. A.

[1]   *Stability Analysis of Nonlinear Systems.* Marcel Dekker, New York (1989).

Lakshmikantham V., Salvadori L.

[1]   On Massera type converse theorem in terms of two different measures. *Boll. UMI*, **13A** (1976) 293–301.

LaSalle J. P.

[1]   Stability theory for ordinary differential equations. *J. Diff. Eqns* **4** (1968) 57–65.

[2]   *The Stability of Dynamical Systems.* SIAM, Philadelphia (1976).

[3]   Stability of nonautonomous systems. *J. Nonlin. Anal.* **1** (1976) 83–90.

[4]   Stability theory and invariance principles. *Dyn. Syst.* **1** (1976) 211–222.

Lax P. D.

[1]   Hyperbolic systems of conservation laws II. *Commun. Pure Appl. Maths* **10** (1957) 537–566.

Lions J. L.

[1]   Equations Differentielles-Operationnelles et Problems aux Limites. Berlin, Springer, 1961.

Lyapunov A. M.

[1]   *The General Problem of the Stability of Motion.* Taylor & Francis, London (1992). [Russian original published by the Kharkov Mathematical Society (1892).]

Malkin I. G.

[1]   On stability under persistent perturbations. *Appl. Math. Mech.* **8** (1944) 241–245 (Russian).

Marchetti F., Negrini P., Salvadori L., Scalia M.

[1]   Liapunov direct method in approaching bifurcation problems, *Ann. Mat. Pura Appl.* **108**(4) (1976) 93–103.

Marcus L.

[1]   Asymptotically autonomous differential systems. *Contributions to the Theory of Nonlinear Oscillations*, Vol. 3 (ed. S. Lefschetz), Princeton University Press (1956), pp. 17–29.

Marcus L., Mizel V.
[1] Limiting equations for problems involving long-range memory, *Mem. Am. Math. Soc.* **43** (1983) 1–60.

Martynyuk A. A.
[1] Method of averaging and principle of comparison in the theory of stability of motion. *Appl. Mech.* **7**(9) (1971) 64–69 (Russian).
[2] Unstable equilibria of multidimensional systems, including "neutral" unstable subsystems. *Appl. Mech.* **8**(6) (1972) 77–82 (Russian).
[3] Method of averaging in the theory of stability of motion. *Nonlin. Vibr. Probl.* **14** (1973) 71–79 (Russian).
[4] Qualitative investigation of behaviour of weakly connected oscillators in the neighbourhoods of equilibria. *Appl. Mech.* **9**(7) (1973) 122–126 (Russian).
[5] *Stability of Motion of Composite Systems.* Naukova Dumka, Kiev (1975) (Russian).
[6] Technical stability of nonlinear and control systems. *Nonlin. Vibr. Probl.* **19** (1979) 21–84.
[7] Method of averaging and optimal stabilization of motion of large scale systems. *Real Time Control of Large-Scale Systems, Greece* (1984), pp. 228–237.
[8] Practical stability and optimal stabilization of controlled motion. *Mathematical Control Theory.* Banach Centre Publications, Warsaw (1985), pp. 383–400.

Martynyuk A. A., Chernetzkaya L. N.
[1] Estimates of nonstationary motions on time-variable sets. *Appl. Mech.* **25**(3) (1989) 97–102 (Russian).

Martynyuk A. A., Gutowski R.
[1] *Integral Inequality and Stability of Motion.* Kiev, Naukova Dumka (1979) (Russian).

Martynyuk A. A., Karimzhanov A.
[1] Limiting equations and stability of nonautonomous motions, *Appl. Mech.* **23**(9) (1987) 101–106 (Russian).
[2] Limiting equations and stability of non-stationary motions. *J. Math. Anal. Applics* **132** (1988) 101–108.
[3] Stability of weakly connected large-scale systems. *Appl. Mech.* **25**(3) (1989) 129–133 (Russian).

Massera J. L.
[1] On Lyapunov's condition of stability. *Ann. Maths* **50** (1949) 705–721.

Miller R. K.
[1] Asymptotic stability properties of linear Volterra integro-differential equations. *J. Diff. Eqns* **10** (1971) 485–506.
[2] *Nonlinear Volterra Integral Equations.* Benjamin, Menlo Park (1971).

Miller R. K., Sell G R.
[1] Existence, uniqueness and continuity of solutions of integral equations. *Ann. Math. Pura Appl.* **80** (1968) 135–152.
[2] Topological dynamics and its relation to integral equations and nonautonomous systems. *Dynamical Systems: An International Symposium*, Vol. 1. Academic Press, New York (1976), pp. 223–249.

Morgan A. P., Narenda K. S.
[1]   On the uniform asymptotic stability of certain linear nonautonomous differential equations. *SIAM J. Control* **15** (1977) 5–24.
Murakami S.
[1]   Perturbation theorems for functional differential equations with infinite delay via limiting equations. *J. Diff. Eqns* **59** (1985) 314–335.
[2]   Almost periodic solutions of a system of integrodifferential equations. *Tohoku Math. J.* **39**(1) (1987) 71–79; **39**(2) (1987) 65–73.
Naito T.
[1]   On autonomous linear functional differential equations with infinite retardations. *J. Diff. Eqns* **21** (1971) 297–315.
Nemytsky V. V., Stepanov V. V.
[1]   *Qualitative Theory of Differential Equations.* Princeton University Press (1960).
Parks P. C.
[1]   A stability criterion for a panel flutter problem via the second method of Liapunov. *Proceedings of International Symposium, Puerto Rico.* New York (1967), pp. 287–298.
Pazy A.
[1]   A class of semilinear equations of evolution. *Israel J. Maths* **20**(3) (1975) 23–36.
Peterson L. D., Marle C. G.
[1]   Stability of solutions of nonlinear diffusion problems. *J. Math. Anal. Applics* **14** (1966) 221–241.
Plaut R. H.
[1]   Asymptotic stability and instability criteria for some elastic systems by Liapunov's direct method. *Q. Appl. Maths* **17** (1972) 535–540.
Ralstone J. V.
[1]   Solutions of the wave equation with localized energy. *Commun. Pure Appl. Maths* **22** (1969) 307–323.
Rauch J.
[1]   Qualitative behaviour of dissipative wave equations on bounded domains. *Arch. Rat. Mech. Anal.* **62** (1976) 77–85.
Razumikhin B. S.
[1]   Stability of systems with delay. *Appl. Math. Mech.* **20** (1956) 500–512 (Russian).
[2]   Application of Liapunov's method to problems of stability of delay systems. *Avtom. Telemekh.* **21** (1960) 740–748 (Russian).
Reed M.
[1]   *Abstract Nonlinear Wave Equations.* Lecture Notes in Mathematics. Springer, New York (1976).
Rezvan V.
[1]   *Absolute Stability of Automatic Delay Systems.* Nauka, Moscow (1983) (Russian).
Rim D. S.
[1]   Torsion differentials and deformations. *Trans. Am. Math. Soc.* **169** (1972) 257–278.
Rouche N., Mawhin J.
[1]   *Equations différentielles ordinaires,* Vol. 1. Masson, Paris (1973).
Roxin E.
[1]   Stability in general control systems. *J. Diff. Eqns* **1** (1965) 115–150.

[2] On generalized dynamical systems defined by conditional equations. *J. Diff. Eqns* **1** (1965) 188–205.

Rumyantsev V. V., Oziraner A. S.

[1] *Stability and Stabilization of Motion with Respect to Some Variables*. Nauka, Moscow (1987) (Russian).

Russell D. L.

[1] Decay rates for weakly damped systems in Hilbert space obtained with control-theoretic methods. *J. Diff. Eqns* **19** (1975) 344–370.

Sage A. P.

[1] *Methodology for Large Scale Systems*. McGraw-Hill, New York (1977)

Saperstone S. H.

[1] *Semidynamical Systems in Infinite Dimensional Spaces*. Springer, Berlin (1981).

Sawano K.

[1] Exponentially asymptotic stability for functional differential equations with infinite retardations. *Tohoku Math. J.* **31** (1979) 363–382.

Schumacher K.

[1] Existence and continuous dependence for functional-differential equations with unbounded delay. *Arch. Rat. Mech. Anal.* **67** (1978) 315–335.

Seifert G.

[1] Almost periodic solutions and asymptotic stability. *J. Math. Anal. Applics* **21** (1968) 136–149.

[2] Lyapunov–Razumikhin conditions for asymptotic stability in functional differential equations of Volterra type. *J. Diff. Eqns* **16** (1974) 289–297.

Sell G. R.

[1] Nouautonomous differential equations and topological dynamics I. The basic theory. *Trans. Am. Math. Soc.* **127** (1967) 241–262.

[2] Nonautonomous differential equations and topological dynamics II. Limiting equations. *Trans. Am. Math. Soc.* **127** (1967) 263–283.

[3] *Topological Dynamics and Ordinary Differential Equations*. Von Nostrand Reinhold, London (1971).

Shestakov A. A.

[1] Localization of limit sets of solutions to nonautonomous systems. *Proceedings of 5th All-Union Conference on the Qualitative Theory of Differential Equations*. Kishinev (1979), pp. 191–192 (Russian).

[2] Theory and applications of the generalized direct Lyapunov method for abstract dynamical systems. A survey of the present state of the geometrical approach to the direct Lyapunov method. *Diff. Eqns* **18** (1982) 2069–2097 (Russian).

[3] Generalized direct Lyapunov method for abstract semidynamical processes I. *Diff. Eqns* **22** (1986) 1475–1490 (Russian).

[4] Generalized direct Lyapunov method for abstract semidynamical processes II. *Diff. Eqns* **23** (1987) 371–387 (Russian).

[5] Generalized direct Lyapunov method for abstract semidynamical processes III. *Diff. Eqns* **23** (1987) 923–926 (Russian).

[6] The direct Lyapunov method as a method for localizing by Lyapunov functions the limit sets of nonautonomous dynamical processes on the basis of limiting equations and dynamical systems. *Lyapunov Functions and Their Applications*. Novosibirsk (1987), pp. 14–48 (Russian).

Sibirsky K. S.
[1]   *Introduction to Topological Dynamics.* Akad. Nauk Mold. SSR, Kishinev (1970) (Russian).

Sibuya Y.
[1]   A study on generation of nonuniqueness. Dynamical Systems: An International Symposium, II, Academic Press, 1976, 243–248.

Šiliak D. D.
[1]   *Large-Scale Dynamical Systems. Stability and Structure.* North-Holland, Amsterdam (1978).

Starzhinskij V. M.
[1]   Einige Probleme nichtlineare Schwingungen I. *Z. Angew. Math. Mech.* **51** (1971) 455–469.
[2]   Einige Probleme nichtlineare Schwingungen II. *Z. Angew. Math. Mech.* **53** (1971) 453–462.

Strauss A., Yorke J. A.
[1]   Perturbation theorems for ordinary differential equations. *J. Diff. Eqns* **3** (1967) 15–30.
[2]   Perturbing uniform asymptotically stable nonlinear systems. *J. Diff. Eqns.* **6** (1969) 452–483.

Visentin F.
[1]   Limiting equations in the problem of total stability. *Ric. Mat.* **28** (1979) 323–331.
[2]   Limiting equations and total stability. *Appl. Nonlin. Anal.* **5** (1979) 721–725.
[3]   Limiting processes in the stability problem. *Appl. Math. Comp.* **7** (1980) 81–91.

Vishnevsky M. P.
[1]   Conditional asymptotic stability of stationary solutions of parabolic systems. *Dyn. Solid Med.* **14** (1973) 95–99 (Russian).

Volosov V. M.
[1]   Averaging in ordinary differential equations. *Prog. Math. Sci.* **17**(6) (1962) 3–126 (Russian).

Wu J., Li Z., Wang Z.
[1]   Remarks on "Periodic solutions of linear Volterra equations". *Funkc. Ekvacioj* **30** (1987) 105–109.

Yoshizawa T.
[1]   Stability of sets and perturbed systems. *Funkc. Ekvacioj* **5** (1962) 18–24.
[2]   *Stability Theory by Liapunov's Second Method.* Mathematical Society of Japan, Tokyo (1966).
[3]   *Stability Theory and the Existence of Periodic Solutions and Almost Periodic Solutions.* Springer, New York (1975).
[4]   Asymptotically almost periodic solutions of an almost periodic system. *Funkc. Ekvacioj* **12** (1969) 23–40.

Zelenyak T. I.
[1]   Stationary solutions of mixed problems appearing in the investigation of some chemical processes. *Diff. Eqns* **2** (1966) 205–213 (Russian).
[2]   The problem of stability of solutions of mixed problems for one quasilinear equation. *Diff. Eqns* **3** (1967) 19–29 (Russian).

[3]   Stabilization of solutions of boundary-value problems for parabolic equations of second order with a single space variable. *Diff. Eqns* **4** (1968) 34–45 (Russian).

Zheng G.

[1]   A necessary condition for L^2-stability of quasilinear conservation laws. *Proc. Am. Math. Soc.* **96** (1986) 1–18.

Zhikov V. V.

[1]   Almost-periodic solutions of linear and nonlinear equations in Banach space. *Doklady Akad. Nauk SSSR* **112** (1970) 278–281 (Russian).

[2]   Monotonicity in the theory of almost-periodic solutions of nonlinear operator equations. *Math. Collect.* **90** (1973) 214–228 (Russian).

Zubov V. I.

[1]   Some sufficient indicators of stability of nonlinear systems of equations. *Appl. Math. Mech.* **17** (1953) 506–508 (Russian).

[2]   *The Lyapunov Method and its Applications.* Nordhoff, Dordrecht (1964). [Russian original published in Leningrad (1957).]

[3]   *Mathematical Methods for the Investigation of Automatic Control Systems.* Pergamon, Oxford (1963). [Russian original published in Leningrad (1959).]

Index

Printed and bound by CPI Group (UK) Ltd, Croydon, CR0 4YY
24/10/2024
01778291-0003